Alternative Lithography

Nanostructure Science and Technology

Series Editor: David J. Lockwood, FRSC
National Research Council of Canada
Ottawa, Ontario, Canada

Current volumes in this series:

Alternative Lithography: Unleashing the Potentials of Nanotechnology
Edited by Clivia M. Sotomayor Torres

Nanostructured Catalysts
Edited by Susannah L. Scott, Cathleen M. Crudden, and Christopher W. Jones

Polyoxometalate Chemistry for Nano-Composite Design
Edited by Toshihiro Yamase and Michael T. Pope

Self-Assembled Nanostructures
Jin Z. Zhang, Zhong-lin Wang, Jun Liu, Shaowei Chen, and Gang-yu Liu

---

A Continuation Order Plan is available for this series. A continuation order will bring delivery of each new volume immediately upon publication. Volumes are billed only upon actual shipment. For further information please contact the publisher.

Alternative Lithography

Unleashing the Potentials of Nanotechnology

Edited by

Clivia M. Sotomayor Torres

University of Wuppertal
Wuppertal, Germany

Kluwer Academic / Plenum Publishers
New York Boston Dordrecht London Moscow

Library of Congress Cataloging-in-Publication Data

Alternative lithography: unleashing the potentials of nanotechnology/edited by Clivia M. Sotomayor Torres.
p. cm. — (Nanoscience science and technology)
Includes bibliographical references and index.
ISBN 0-306-47858-7
1. Molecular electronics. 2. Microlithography. 3. Nanotechnology. I. Sotomayor Torres, C. M. II. Series.

TK7874.8.A48 2003
621.3815′31—dc22

2003060193

ISBN 0-306-47858-7

233 Spring Street, New York, New York 10013

http://www.wkap.com

10 9 8 7 6 5 4 3 2 1

A C.I.P. record for this book is available from the Library of Congress

Printed in the United States of America

To Marfa, Oreste, Drina,
Wolfgang and Samira

Contributors

Jouni Ahopelto
VTT Information Technology
Microelectronics
Tekniikant. 17
P.O.Box 1208
Fin-02044 VTT
Finland

Todd C. Bailey
University of Texas at Austin
Department of Chemical Engineering
CPE 4.452, Mail Code C0400
Austin, TX 78712
USA

Tatjana Borzenko
Physikalisches Institut
Universität Würzburg
Am Hubland
D-97074 Würzburg
Germany

Yong Chen
Laboratoire de Photonique et
Nanostructures
CNRS
Route de Nozay
F-91460 Marcoussis
France

B. J. Choi
University of Texas at Austin
Department of Chemical Engineering
CPE 4.452, Mail Code C0400
Austin, TX 78712
USA

S. Y. Chou
Department of Electrical Engineering
Princeton University
Princeton, NJ
USA

M.Colburn
University of Texas at Austin
Department of Chemical Engineering
CPE 4.452, Mail Code C0400
Austin, TX 78712
USA

J. G. Ekerdt
University of Texas at Austin
Department of Chemical Engineering
CPE 4.452, Mail Code C0400
Austin, TX 78712
USA

Alfred Forchel
Lehrstuhl für Technische Physik
Bayerische Julius-Maximilians
Universität Würzburg
Würzburg
Germany

Ricardo Garcia
Instituto de Microelectrónica
de Madrid, CSIC
SP-28760 Madrid
Spain

A. Grot
University of Texas at Austin
Department of Chemical Engineering
CPE 4.452, Mail Code C0400
Austin, TX 78712
USA

Tomi Haatainen
VTT Information Technology
Microelectornics
Tekniikant. 17
P.O.Box 1208
Fin-02044 VTT
Finland

Babak Heidari
Lund University
Div of Solid state Physics & the
Nanometer Consortium
S-221 00 Lund
Sweden

Laura J. Heyderman
Laboratory for Micro- and
Nanotechnology
Paul Scherrer Institut
CH-5232 Villingen PSI
Switzerland

Volkmar Hock
Physikalisches Institut
Universität Würzburg
Am Hubland
D-97074 Würzburg
Germany

Thomas Hoffmann
IMEC
Kapeldreef 75
B-3001 Leuven
Belgium

Juriaan Huskens
University of Twente
CT/SMCT
Po Box 217
7500 AE, Enschede
The Netherlands

Martin Kamp
Lehrstuhl für Technische Physik
Bayerische Julius-Maximilians
Universität Würzburg
Germany

A. Lebib
Laboratoire de Photonique et
Nanostructures
CNRS
Route de Nozay
F-91460 Marcoussis
France

S. P. Li
Laboratoire de Photonique et
Nanostructures
CNRS
Route de Nozay
F-91460 Marcoussis
France

Maik Liebau
University of Twente
CT/SMCT
PO Box 217
7500 AE, Enschede
The Netherlands

Ingo Martini
Lehrstuhl für Technische Physik
Bayerische Julius-Maximilians Universität
Würzburg
Germany

Laurens W. Molenkamp
Physikalisches Institut
Universität Würzburg
Würzburg
Germany

Lars Montelius
Lund University
Div of Solid state Physics & the Nanometer Consortium
Lund
Sweden

M. Natali
Laboratoire de Photonique et Nanostructures
CNRS
Route de Nozay
F-91460 Marcoussis
France

Anne Pépin
Laboratoire de Photonique
et Nanostructures
CNRS
Route de Nozay
F-91460 Marcoussis
France

David N. Reinhoudt
University of Twente
CT/SMCT
PO Box 217
7500 AE, Enschede
The Netherlands

Georg Schmidt
Physikalisches Institut
Universität Würzburg
Würzburg
Germany

Jörg Seekamp
Institut of Materials Science and Dept. of Electrical and Information
Engineering
University of Wuppertal
D-42097 Wuppertal
Germany

S.V. Screenivasan
University of Texas at Austin
Department of Chemical Engineering
CPE 4.452, Mail Code C0400
Austin TX 78712
USA

Helmut Schift
Laboratory for micro- and
Nanotechnology
Paul Scherrer Institut
Ch-5232 Villingen PSI
Switzerland

Massimo Tormen
National Nanotechnology
Laboratory (NNL), TASC-INFM at
Elettra Synchrotron Light Source
I-34012 Basovizza
Trieste, Italy

Clivia M. Sotomayor Torres
Institut of Materials Science and Dept. of
Electrical and Information
Engineering
University of Wuppertal
D-42097 Wuppertal
Germany

Christophe Vieu
LAAS-CNRS 7
7, avenue du colonel Roche
F-31077 Toulouse Cedex 4
France

C.G. Willson
Texas Materials Institute
The University of Texas at Austin
Austin, Texas
USA

Foreword

Good old Gutenberg could not have imagined that his revolutionary printing concept which so greatly contributed to dissemination of knowledge and thus today's wealth, would have been a source of inspiration five hundred years later. Now, it seems intuitive that a simple way to produce a large number of replicates is using a mold to emboss pattern you need, but at the nanoscale nothing is simple: the devil is in the detail. And this book is about the "devil". In the following 17 chapters, the authors—all of them well recognized and active actors in this emerging field— describe the state-of-the-art, today's technological bottlenecks and the prospects for micro-contact printing and nanoimprint lithography.

Many results of this book originate from projects funded by the European Commission through its "Nanotechnology Information Devices" (NID) initiative. NID was launched with the objective to develop nanoscale devices for the time when the red-brick scenario of the ITRS roadmap would be reached. It became soon clear however, that there was no point to investigate only alternative devices to CMOS, but what was really needed was an integrated approach that took into account more facets of this difficult undertaking. Technologically speaking, this meant to have a coherent strategy to develop novel devices, nanofabrication tools and circuit & system architectures at the same time. There is no point having an interesting device; we have to know how to produce them massively by millions and once we have achieved it we must find a way to interconnect them in a useful way. Thus, projects were therefore selected taking also into account a balance between these different technological areas. Two additional important issues had to be taken into the NID strategy. First, the NID budget was insufficient to fund all promising European research work. Second, excellent research was performed also outside Europe. To integrate these issues an international nanoelectronics network was established, called *Phantoms*. Phantom's mission is to twofold: to offer information on nanotechnology and electronics to scientists and, to support networking activities. An excellent example of the latter is the establishment of working groups on printing and self-assembly. Discussion within these working groups have contributed to gather technical facts on printing, to exchange experiences, to start novel collaborations and possibly motivated to gather their collective knowledge into this book. As the working groups are open clubs, it is not astonishing that authors of this book come from Europe and the USA.

Nanotechnology is supposed to be an emerging technology with high potential industrial impact and the underlying Nanoscience is meant to have some particular characteristics that differ from the classical disciplines and. If this is true, these two printing techniques are nanotechnologies at its best. A couple of comparisons between the common perception of nanotechnology and the facts for printing:

1. In the literature there is a debate on the definition of nanotechnology.

"Purists" consider that nanotechnology comprises the bottom-up creation of larger ensembles only, while a more relaxed view includes any technology below 100 nm. Here micro-contact priting fits in any definition. It is bottom-up and top-down at the same time, as the stamps structures are driven by miniturization and at the same time the ink relies upon the self-assembly properties of its molecules.

2. Nanotechnology relies upon a number of traditional sciences.

This is particularly true for printing technology. Physicists, engineers, chemists and computer scientist work hand in hand to develop suitable machines. Each chapter in this book reveals that interdisciplinary is a necessity. One of the many examples, the sticking of material cannot be solved without complementary knowledge about chemical composition and characteristics of the resist, the way the mould has been manufactured and simulations of its rheological behavior.

3. Nanotechnology has a large economical potential.

The beauty of printing technologies is that it allows patterning at different scales (from microns down to tens of nanometers) changing only marginally the set-up. Mixed large and small patterns are possible, increasing so the number of possible applications. The first applications are likely to be those where the technological restrictions are less demanding. Microfluidic systems and biosensors, where the critical sizes are mostly in the sub-micron range, might be the first killer products for larger market penetration. Equipment for biotechnological applications may then generate the revenues for further tool developments towards other markets such as optical applications, magnetic storage or electrical devices. The market prospects are still diffuse, but the commercial dynamics gives the impression race just begun: EV Group (Schärding, Austria) extends the range of products towards nanoimpring, Jenoptik (Jena, Germany) pooled its hot-embossing knowledge into its novel subsidiary company Mildendo or Obducat, a small Swedish company, which has doubled its number of employees in a couple of years.

Printing technology has delivered interesting science in the past and will possibly have a significant impact for nanofabrication. To speculate is not strong point of mine but I guess that if Gutenberg would have lived in our days he may have become a nanotechnologist.

Brussels, December 2002

Ramón Compañó

Preface

First serious encounters leave a mark, be new people or new technologies. I came across nanofabrication in my first visit to the Nanoelectronics Research Centre of the University of Glasgow in 1984. I was simply marvelled by the engineering skills applied to realize then semiconductors quantum wires and quantum dots. The enthusiasm of the people working there was overwhelming and contagious. The feeling was one of embarking on a new engineering enterprise, where the limits were fuzzy and the science exhilarating.

Later on, as I became a bit more familiar with the science and engineering of semiconductor nanostructures, I appreciated better the levels of complexity and cost involved in nanofabrication by electron beam lithography and reactive ion etching.

When I set up my laboratory in Wuppertal, it was clear that one had to look for alternative ways to make solid-state nanostructures. The work of Stephen Y. Chou was illuminating: reproducible, affordable and scalable feature sizes in the nanometer scale. What started out of necessity, became an increasingly open ended journey. As nanotechnology became recognised in disciplines beyond physics and electrical engineering, nanofabrication needs in biology, optics, electronics, to name a few, started to emerge and, with them, a range of chemical and physical approaches to nanofabrication.

My research group and I have the privilege of participating in international projects on nanoimprint lithography and its applications in optics and organic electronics, working in an interdisciplinary and remarkably open manner. In this context, the European Commission ESPRIT Programme (Microelectronic Research Initiative) and later on the Information Society Technology Programme (Future and Emerging Technologies) have been instrumental in the formation of working groups on, eg. Printing and Self-assembly, which acted as nucleation centres of discussion and activities and thus served to give an undeniable impulse to research in alternative nanofabrication techniques in Europe. It has been through these interactions that the idea of this book came about. The wealth of knowledge and breadth of experience are worth gathering and making available to other scientists and engineers who are working in nanotechnology. As the reader will see, the authors of the chapters offer an insight into a range of techniques and applications, as well as up-scaling and some of the fundamental issues remaining to be addressed.

The dynamism of the research field is demonstrated in several new techniques which emerged while this book was in preparation. I am sure that they will be duly compiled and made available to the community in the near future.

Clivia M. Sotomayor Torres
Velbert Neviges
December 2002

Acknowledgments

My thanks go to Steve Beaumont and Chris Wilkinson for introducing me to nanofabrication and to Hella-Cristin Scheer for accompanying me in the first steps of our joint work on nanoimprint lithography. I have learnt much on many aspects of nanofabrication with partners of our European IST FET projects NANOTECH and later on CHANIL, Thomas Hoffmann, Lars Montelius, Babak, Heidari, Gabi Gruetzner, Karl Pfeiffer, Freimut Reuther, Gerhard Bleidiessel, Christophe Cardinaud, Christian Mayer and, especially, Jouni Ahopelto. Ramon Compaño, our project officer, deserves special thanks for his support and his initiative to create an environment conducive to gelling the European experience in alternative nanolithography.

It is clear that without the collaboration of the authors of the following chapters, this book would not have materialised. I am grateful to them and hope they will be happy with the final product.

I am indebted to members of my group, Swetlana Kosinez, Patrick Ferrand, Sergiy Zankovych, Alicia Kam, Lars Eisenberg, and Furkan Bulut for their help in preparing this volume. Thanks to my editor, Ingrid Gnerlich, who showed unfaltering support even after deadlines elapsed.

Last but not least, I have a debt of gratitude with my family for their patience with me while working on this book, for respecting my closed door and accepting the interference with family holidays.

Contents

Alternative Lithography

1

Alternative Lithography

An introduction

CLIVIA M. Sotomayor Torres
Institute of Materials Science and Dept. Electrical & Information Engineering, University of Wuppertal, Germany

1.1. INTRODUCTION

Progress in nanotechnology depends upon the capability to fabricate nanostructures in a variety of materials with accuracy in the nanometre scale and sometimes in the atomic scale. Depending on application, there are varying degrees of strong specifications, which have to be met in industrially relevant processes due to manufacturability and costs considerations as, for example, in the electronics industry. However, it appears that less demanding conditions are needed for developments in optics, sensors and biological applications. In order for progress to be made, enabling nanofabrication techniques as tools for experiments to understand the underlying science and engineering in the nanometre scale, easily accessible and flexible nanofabrication approaches are required. These have to be suitable for investigations in, e.g., materials science, organic optoelectronics, nano-optics and the life sciences. Alternative techniques to cost-intensive or limited-access fabrication methods with nanometre resolution have been under development for nearly two decades. One clear example is the evolving set of scanning probes techniques, which has become ubiquitous in many research areas. If one considers planar structures, i.e., where nanostructuring is carried out on a surface, as distinct from a three-dimensional nanofabrication or multi-layer self-assembly, then several emerging nanofabrication techniques can be discussed. Their classification depends on whether the nature of the patterning is chemical or physical, or its modality in time is parallel or sequential, or a hard or a soft mould or stamp is used, etc. The literature on the subject is increasing very rapidly and recent reviews on, for example, progress in micro-contact printing,[1,2] scanning probe-based techniques[3] and nanoimprint-based lithography (NIL) techniques,[4] have been published.

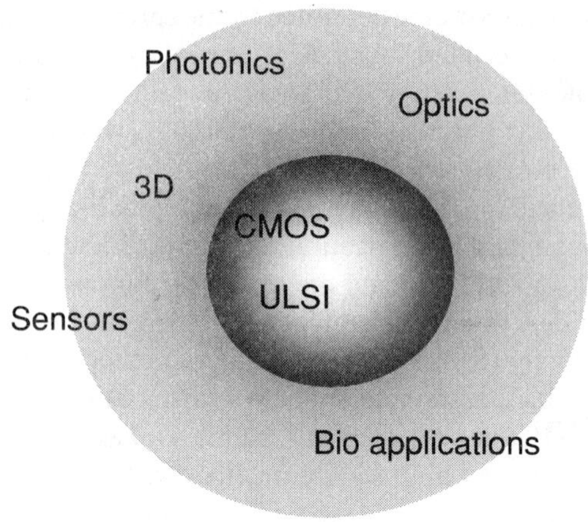

FIGURE 1.1. Connection between applications which have led developments in lithography, such as CMOS and ultra large scale integration, and a range of applications with probably less demands on the fabrication process such as those outside the front end of the CMOS process, the fabrication needs of which can be successfully covered by emerging lithographic methods. (Courtesy of J. Ahopelto).

Here we attempt to give an overview of the issues, with special reference to the contents of the following chapters.

Richard Feynmann's famous sentence "There is plenty of room at the bottom" was visionary. His line of discourse led to the suggestion "We would press the metal (features to be replicated) into a plastic material, make a mold of it, then peel the plastic off very carefully, evaporate silica onto the plastic to get a very thin film, then shadow it by evaporating gold at an angle against the silica... we would just need to press the same metal plate again into plastic and we would have another copy".[5] These comments were based on his knowledge of existing technologies in the late 1950s. LIGA, hot embossing and, later on, nanoimprinting, followed. The underlying philosophy of alternative lithography is illustrated in Figure 1.

Thus, we start with nanoimprint lithography (NIL) since perhaps it is the most developed of the alternative nanofabrication techniques. We then address micro-contact printing, followed by scanning probe methods and mention recent developments of alternative nanopatterning. We conclude with a brief look at the current status of the most advanced techniques and their perspectives.

1.2. MOULDING POLYMERS IN THE NANOMETER SCALE

The main technique is nanoimprint lithography (NIL) which is described in Chapter 2. To nanoimprint a surface, three basic components are required: a stamp with suitable feature sizes, a material to be printed and equipment for printing with adequate control of temperature, pressure and control of parallelism of the stamp and substrate. In essence, the process consists of pressing the stamp using a pressure in the range of about 50 to 100 bar, against a thin polymer film. This takes place when the polymer is held some 90 to 100 °C above its T_g, in a time scale of few minutes, during which the polymer can flow to fill in the volume delimited by the surface topology of the stamp. The stamp is detached from the printed substrate after cooling both the stamp and substrate. The cycle involves time, temperature and pressure. And here we have the main issues of nanoimprint lithography: polymer flow and rheology, which are addressed from the materials point of view in Chapters 3 and 4.

NIL has the advantage over conventional nanofabrication methods, of being a flexible, low-cost and biocompatible fabrication technique. There are several variations of NIL including and roll-to-roll NIL[6] and the most popular parallel process using wafer size stamps,[7] which is described in Chapter 5. The sequential process called Step-and-Stamp Imprint Lithography (SSIL)[8] is addressed in Chapter 6, while a variation which instead of temperature uses ultraviolet light polymer curing, namely, step and flash imprint lithography (SFIL) is presented in Chapter 7. The use of elastomeric PDMS stamps for imprinting, made by high resolution electron beam lithogtraphy is discussed in Chapter 8.

1.2.1. Highlights of Nanoimprint lithography

There are several key achievements in the development of NIL as a nanofabrication technique and its potential applications reported over the last seven years. In a chronological sequence these are:

(a) The first report of what is known as NIL as a potential nanofabrication technique appeared in 1995.[9]
(b) Feature sizes down to 6 nm by NIL are achieved in 1997 in PMMA.[10]
(c) The use of NIL to fabricate polymer-based optical devices is demonstrated in 1998.[131]
(d) Alignment accuracy of 1 μm is demonstrated using commercially available equipment.[12]
(e) In 1999 Yu et al.[13] reported on NIL-made metal-semiconductor-metal MSM photodetectors with no mobility decrease up to NIL pressures of 600 psi.
(f) Broadband waveguide metal polarisers with 190 nm period were reported by Wang et al.[16]
(g) NIL of 150 mm diameter wafers is achieved.[7]
(h) A sequential variation of NIL, Step-and stamp imprint lithography, is demonstrated using a commercial flip-chip bonder.[8]
(i) Polymers developed specifically for NIL[15] become commercially available in 2000 (mr-I-8000 and mr-I-9000).
(j) Low-cost stamp replication using NIL is demonstrated.[16]
(k) Mäkela et al. showed that the electrical conductivity of nanoimprinted conducting polymers is not impaired by imprint lithography.[17]
(l) A new resist suitable for NIL, which is also sensitive to electron beam and UV lithography, has been reported.[18]
(m) The first microfluidic device made by NIL was reported by Studer et al. in 2001.[19]
(n) An anti-adhesion treatment for stamps containing sub 100 nm features is reported.[20]

(o) A quality control method with in situ potentials for NIL based on optical microscopy is successfully tested in the laboratory.[21]

1.2.2. Stamps and moulds

In first approximation the ultimate resolution of NIL depends on the minimum feature size in the stamp. For high resolution stamps are usually made by electron beam lithography and dry etching, and for shallow stamps, by metal lift off. The material of choice is Si or SiO_2, which is sometimes electroplated. Depending on application, stamps in wafer-size scale have been studied by several groups. As expected the stamp design is intrinsically correlated to the polymer to be printed. However, it remains a non-trivial problem to model polymer flow for a given stamp design, in order to facilitate designs with complex mixtures of features and patterns. These are some of the aspects discussed in Chapters 3 and 4. Metal stamps with features smaller than 10 nm have been fabricated by electron beam lithography and lift-off. However, the metal roughness was found to be a problem for sizes below 10 nm due to the granularity of the evaporated metal. A promising development, with potential impact in a range of applications, is the syntheses of a new polymer, such as mr-L6000, developed by *mrt* GmbH. This particular polymer is suitable for NIL and at the same time UV and electron beam sensitive.[18] mr-L6000 has high electron sensitivity, $\sim 5\mu C/cm^2$ compared to $80\text{-}200\mu C/cm^2$ for PMMA, which results in shorter electron beam writing times. The limits of its resolution are still not found and currently these are around 20 nm. An example is shown in Figure 2, where several patterns were written by electron beam.

After development the features are very smooth. This polymer can be nanoimprinted. In Figure 3 AFM images of a stamp and a print using mr-L 6000 are shown with lateral dimensions of 40 nm and 30 nm high. More details, including a comparison of cross-linkable and further examples can be found in Ref. 22. Moreover, the appeal of

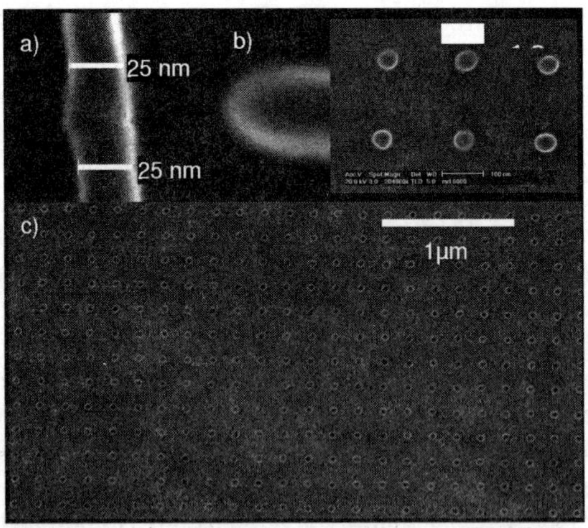

FIGURE 1.2. SEM micrograph of patterns written by electron beam lithography on mr-L 6000 (see text) spun on Si, after development: a wire of 25 nm width (a), a dash of 22 nm (b) and an array of dots of 50 nm diameter (c). (Courtesy of S. Zankovych)

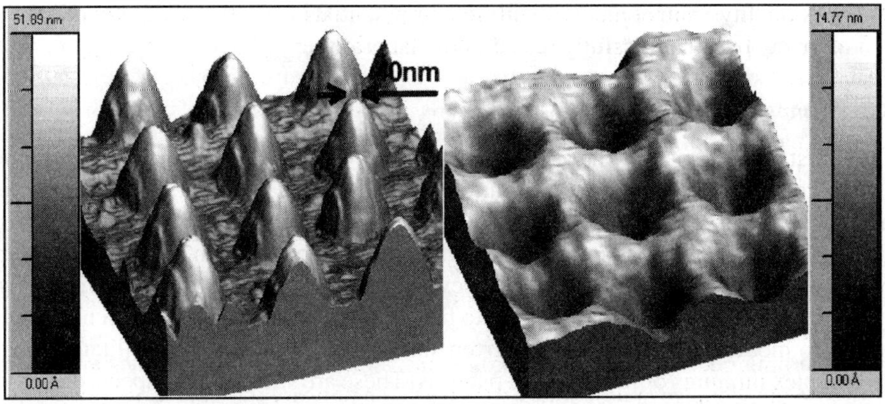

FIGURE 1.3. AFM images of a stamp made out of mr-L 6000 (see text) by electron beam lithography with top diameter of 40 nm, base diameter of 100 nm and 30 nm high; (left) and a print into mr-L 6000 (right). (Courtesy of Dr. J. Seekamp).

this material is its suitability for mix-and-match lithography with applications in MEMs, NEMs and microfluidics, to name but a few.

The need to keep the stamp parallel to the substrate during the process is an essential condition to obtain uniform thickness of the residual layer and the patterned features across the sample area. Likewise, thermal gradients must be avoided using suitable heating and cooling elements. These thermo-mechanical aspects are incorporated in the design of nanoimprint lithography equipment and are discussed in detail in Chapter 6.

Developments in the fabrication of elastomeric stamps by high resolution electron beam lithography for NIL are discussed in Chapter 8.

1.2.3. Road Blocks

One of the criticisms levied at NIL and other alternative lithographic techniques is the *multilevel issue*. For this to be overcome the first step is to find engineering solutions for x-y in-plane alignment to control overlay to, ideally, better than 10 nm. In general the stamp is considered the functional equivalent of the photo-mask in conventional projection lithography. At present the limit is $\sim$1 μm with a standard deviation of 0.4 μm over a 100 mm wafer using a commercially available aligner.[23] In first approximation alignment depends on stamp size, the thermal and mechanical stability of the equipment and polymer during alignment as well as the choice of stamp and substrate material. A compromise may be found in applications with few lithography and mask steps for specific applications. Other factors and possible solutions have been discussed in references 4 and 24. Alignment issues are discussed in Chapter 5 on wafer scale NIL, in Chapter 6 on SSIL and in Chapter 7 on SFIL.

The other key aspect is *throughput*. This is heavily dominated by cost and comparisons are usually made to the microelectronic industry with its 80 wafers per hour yardstick. Whereas a working definition of throughput in NIL as understood in conventional lithography is still lacking, a rough approximation would consider the actual printing time, loading, alignment and separation times, which currently add up to 10-15 minutes. Factors contributing to throughput include the stamp size, high density of features, absence of an anti-adhesive layer, polymer curing time, printing temperature and pressure and stamp lifetime, among others. One figure of merit used when discussing nanofabrication techniques is the "exposure rate", which for nanoimprinting is about 0.152 cm^2/s to print for 10 nm features over a 150 mm wafer in 20 minutes including heating and cooling cycles.

As any other technology coming under what is generically known as printing, NIL must face up to validation and standards. *Validation*, or the quality control issue, needs agreement as to what counts as tolerances for a good print, whereas *standards* depend strongly on design rules and it is too early to expect definitive statements.

Concerning quality control, one pioneering approach for potential on-line quality control of stamp damage and/or adhesion is based on optical lithography of fluorescence loaded printable polymers.[21] Preliminary tests show that the presence of particles down to 100 nm can be detected while tracking the status of the stamp. By recording the complementary fluorescence image the status of the printed substrate can be obtained. Optical microscope images are loaded digitally for fast image processing.

The problem of critical dimension (CD) is one that, depending the fabrication demands below 50 nm, can become a serious bottleneck for NIL. Studies using electron beam sensitive resist for NIL have shown that good CD control can be obtained in highly dense patterns with feature size down to about 75 nm.[25] Factors affecting CD control include the glass transition temperature of the polymer as well as its long-term viscoelastic properties. The design has to incorporate polymer shrinkage in temporal and thermal process steps, e.g., when the residual layer is removed, which affects the final size of the printed features. Below the 20 nm range, the polymer properties themselves come into play with relaxation taking place over weeks and months affecting not only the morphology but also the depth and/or aspect ratio of the printed features. It is here where combinatory fabrication approaches involving NIL and polymer development have a role to play. In particular the use of polymers, which lend themselves to be modified by electron beams, UV exposure and NIL.

1.2.4. Examples

Applications of NIL are the subject of several chapters in this book. Here we wish to mention two not discussed in these chapters, namely the use of nanoimprinted polymers as an etch mask and applications in the organic opto-electronics field.

1.2.4.1. NIL-made mask for dry etching. The development of printable polymers in order to use them as masks in order to replace masks made by sequential processes drove much of the initial work in NIL. Novel linear polymers and cross-linked polymers, which are printable and exhibit high etch resistance, were reported by Pfeiffer et al.[26] In particular, the selectivity of the printed polymer compared to SiO_2 was tested in a SiO_2 dry etching process and demonstrated that the printed polymers acted satisfactorily as masks.[27]

The possibility to pattern a mask by NIL, as it is done with electron beam lithography, and subsequently etching the substrate material to transfer the pattern into an optically active layer, is also a very attractive avenue. In particular, as optical devices increasingly require submicrometer features, the need for cost-efficient lithography

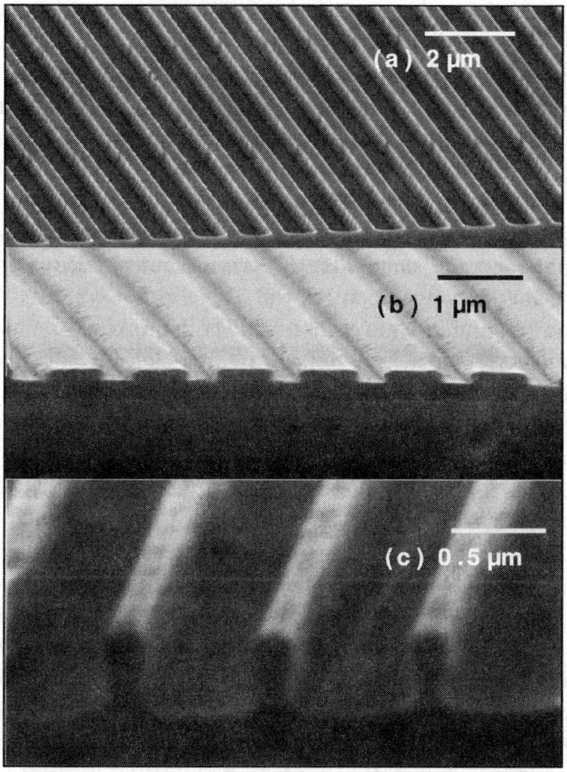

FIGURE 1.4. SEM micrographs of: (a) the Si stamp with 800 nm period used to print the structure in (b); (b) printed PMMA on GaAs and (c) GaAs wires after dry etching using PMMA NIL patterning and lift off, the wire width is approximately 200 nm. (Courtesy of M. V. Maximov)

increases. In this respect one issue to be addressed is whether the NIL cycle, in particular the pressure parameter, induces non-radiative recombination centres, thereby potentially limiting the radiative lifetime of a structured light emitting diode or semiconductor laser. To this effect, a silicon stamp containing parallel wires with 800 nm period was used to print PMMA spun on GaAs. After lift off the wire pattern was transferred to a GaAs substrate resulting in 200 nm ribs[28] as shown in Figure 4. The emission from GaAs in these wires is currently under study.

While this proves that printed PMMA using NIL can be used as a mask this is a preliminary result and several aspect need to be revisited, especially the wire width changes from stamp, through lift off and the dry etching step, potentially affecting the critical dimensions. An example of this is given in Chapter 7. While other polymers with higher resistance to dry etching can be developed and later on used, the issue of critical dimensions is one that begs attention. The polymer used in NIL so far, have been tested in the typical dry etching gas mixture for silicon and not for GaAs. This also remains to be done. Once the quality of the printed etch mask is improved, the change in lateral size during oxygen plasma etching of the nanoimprinted window to remove the residual layer, also needs to be better understood and controlled, and eventually be considered in the stamp design for a given process.

In a parallel experiment to determine if NIL results in optically-relevant damage of the underlying semiconductors, photoluminescence and photoluminescence excitation (PLE) spectroscopy were carried out at 20 K on multiple quantum well samples before and after NIL. The samples had quantum wells of different thickness as different depths from the surface to be imprinted. Two aspects were investigated: the integrated photoluminescence intensity and the emission energy and, in PLE, the changes of strain in the quantum wells. The full experimental details are given elsewhere.[29] The findings are very encouraging in that, using Si stamps, printing temperatures up 190 °C and applying pressures up to 200 bar for 5 minutes, resulted in no detectable change in the emission energy nor in the integrated emission intensity. The PLE data from the GaAs-GaAlAs and the GaInAs-InP quantum wells showed no evidence of strain as no changes were detected in the splitting of the heavy hole and light hole excitons after printing at a pressure of 200 bar.

Thus, active multilayer semiconductor samples subjected to NIL, show that their low temperature integrated emission, emission energy and strain state are not affected. This means that, once the critical dimension issue is overcome, NIL has the potential to replace electron beam lithography to write masks for semiconductor opto-electronic devices.

1.2.4.2. Organic opto-electronics NIL is a natural fabrication approach for organic opto-electronics and sensors. In addition to useful optical device structures printed in a single step in layers containing the small molecule Alq_3 as an example of a patterned light emitting organic layer,[11] PMMA and in polystyrene for gratings and 1-D photonic crystal structures[30] and the study of conductivity in printed polyaniline blend,[17] other reports on NIL for organic layers have appeared. In particular, printing patterns of nanoelectrode arrays down to 100 nm into oligomers, such α-sexithyophene,[31] and of UV curable oligomers in low temperature NIL, such as the modified triphenylamine analogues PDAS and PBAS, into structures which, after patterning still retain their conducting character.[32] Organic thin film transistor (OTFT) structures based in pentacene have been made with features down to 100 nm and their electrical characteristics measured in a NIL process combined with cold welding[33] as well as using the standard NIL process.[34] The cold welding approach has the advantage of reducing surface contact contamination, which is a main problem in the fabrication of organic electronic devices. In Figure 5 SEM micrographs of part of an electron beam written stamp after Au and Ti lift off containing an interdigitated nanoelectrode array overlapping over 4 μm in the central part and having channel length of 90 nm.

A section of the nanoimprinted 950K molecular weight PMMA is also shown. This process is now being applied to oligomers for thin film transistors.

The attractive feature of organic optoelectronics is the prospects of having both optical and electronic devices on a single platform. However, much engineering, science, testing and design work lay ahead for structures with feature size close to 50 nm.

1.3. MICROCONTACT PRINTING

Microcontact printing (MCP) is a front-runner alternative lithography technique, which is reviewed in Chapters 9 and 10. It competes with NIL in a number of aspects.

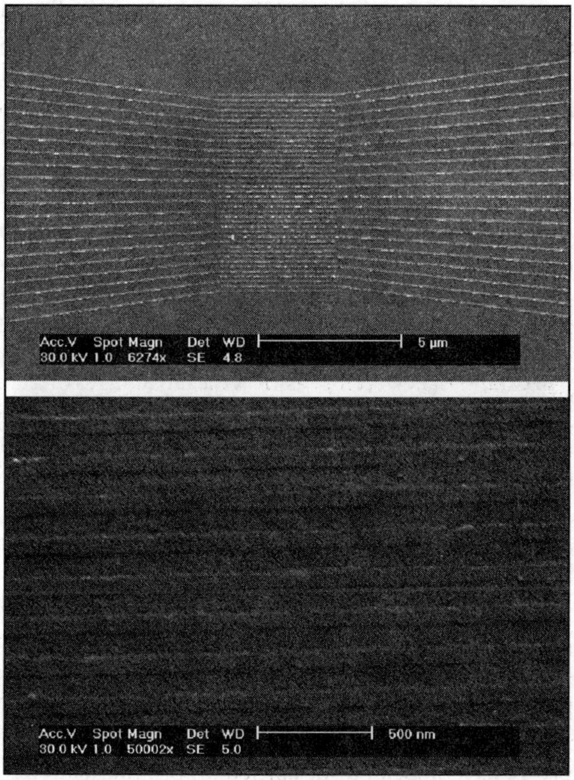

FIGURE 1.5. SEM micrographs of (top) part of an electron beam written stamp after Au and Ti lift off containing an interdigitated nanoelectrode array overlapping over 4 m and having channel length of 90 nm and (bottom) a section of the printed 950K molecular weight PMMA. (Courtesy of C. Clavijo-Cedeño).

MCP is basically an inking approach, using self-assembled monolayers (SAMs) as the ink and an elastomeric stamp to define the features. The SAMs are transferred to a substrate for spatial definition of chemical surface characteristics, to enable further processing and, when combined with other techniques in a mix-and-match approach, for molecular engineering. There are several issues concerning MCP and it is its flexibility for nanopatterning down to 100 nm, which makes it particularly attractive for applications in the liquid state. Chemistry aspects of MCP are dealt with in Chapter 9, while Chapter 10 reviews its potentials, the MCP stamp fabrication and selected applications. The use of SAMs in the liquid state, makes it particularly suitable for biology- and chemistry-related applications.

1.4. SCANNING PROBE APROACHES

The attraction of scanning probe approaches resides in its capacity to move single atoms or very small clusters of atoms. With its unbeatable resolution, the drawback is the throughput factor, which due to its sequential nature makes this approach rather limited for industrial uptake at present. While the "Millipede" of the IBM group gained attention as one multiple-tip scanning approach for high density magnetic memories,[3] it remains to be seen if scanning probe techniques will scale-up. Nevertheless, at the laboratory level, and for much of the basic science studies, nanofabrication by means of local oxidation, offers a high-resolution and accessible nanopatterning technique. Chapter 11 illustrates the basic science of this approach and explores the possibilities.

1.5. APPLICATIONS

In this book applications of alternative lithography techniques are explored in a non-exhaustive manner. The applications of NIL discussed here are in the area of nano-electronic devices in Chapter 12, which considers the advantages of a combined NIL+UV lithography approach. The high density and small feature size attainable with NIL are shown to be particularly suitable to study nanomagnetism[35] and data storage.[36] These topics are discussed in Chapter 13. MCP applications for electronic devices are discussed in Chapter 14. The prospects of NIL for fabrication of optical components and subsystems is discussed in Chapter 15, with special emphasis on photonic crystals which also require high density of small features.

The use of hot embossing in bioelectronics has been pioneered by the group at Glasgow University[37] and is revised in the context of biotechnology in Chapter 16. Microfluidics, nanofluidics and Lab-on-a-chip developments are thriving areas which need low-cost, high throughput nanofabrication techniques due to their intrinsic need to use microfluidics devices only once in medical testing. These and many other considerations are revised, as well as the technological development in the field in Chapter 17.

1.6. RECENT NANOFABRICATION EXPERIMENTS

The search for alternative nanofabrication techniques in general and for specific applications is an ongoing task. As more accurate tools, materials and engineering concepts addressing the nanometer scale become available, novel techniques emerge. Using ion beams, developments on a desk-top NanoFIB (focused ion beam) machine are in progress, which has shown resolution close to 10 nm.[38]

As well as ion beam, photon beams are also the subject of further developments with view to increased resolution. Among these two, the potential use of entangled (correlated) photon-number states to increase resolution to $\lambda/2N$, where N is the photon number in an N-photon absorbing substrate, takes the search to the realms of quantum interferometry.[39] If $N = 2$, as in two-photon absorption, then a factor of 2 improvement in optical lithography could be envisaged, without the need of higher photon fluxes. This is a very interesting approach, which needs the proof of concept in order to see the magnitude of the challenges for actual implementation. The other approach with light beams is based on holography for atom lithography as a step towards 3D nanofabrication. It is based on neutral atoms, and relies on the modulation of a light beam in the nanometer scale caused by light forces of a coherent multibeam light field. In turn, this field is reconstructed from a hologram stored in a photorefractive crystal. The light field intensity distribution captures specific atoms, which are deposited on a substrate, can be changed and or repeated along the z-axis, thus leading to the growth of a 3D nanostructure.[40] This approach has been demonstrated with Cesium and gold atoms, the latter in stripes of 426 nm period in a 1:1 pitch as for holography.

Among emerging nanopatterning techniques, stenceling[41] plays a significant role, as it is fairly adaptable to well-established processes with the advantage of bringing in the nanometer scale and large area patterning. Other recent developments in nanopatterning include dip pen lithography[42] and nanoplotting,[43] both based on scanning probe techniques and SAMS, which some potential of becoming parallel as the "Millipede" approach has demonstrated.

1.7. STATUS AND PERSPECTIVES

At an international workshop on NIL and related techniques held in Lund in January 2002 the participants produced a table with the status on some alternative nanofabrication techniques.[44] An updated version in shown in Table 1, which includes the recent report of laser assisted NIL for ultra fast silicon patterning.[45]

TABLE 1.1. Comparison of printing and related techniques. (Modified from Reference 44.)

Technique	Minimum feature Minimum pitch	Combined small and large features	Largest area printed	Overlay Accuracy	Time for alignment, printing, release, cycle time	No of times stamp used
Nanoimprint Ligraphy NIL[9]	6nm-40nm	10nm-100's μm	100mm^2 wafer	0,5 μm [46]	Few mins 10 sec Few mins 10-15 min	50
Step and Stamp Imprint Lithography (SSIL)[48]	10nm-50nm	200nm-10's μ m	150mm^2	1 μ m	1 min 5 min few sec 6min/step	36
Step and Flash Imprint Lithography (SFIL)	10nm-50nm	20nm-~mm	1 inch wafer	1 μm	1min 10-20sec 10sec 5min/flash	>100 with-out clean
Micro Contact Printing (MCP)[1]	60nm-200nm	60nm-60nm	12 inch wafer	0,5-1 μm	1 min 1-30 sec 10 sec 2min	>100
Stenciling [41,49]	10nm	10nm-10'sμm	From 50μm^2 to 1mm^2	-	Few min 15 min Few min 30 min	-
Laser assisted NIL[45]	10nm	140nm	10nm-10's μm	2.3mm^2	- 250 ns - -	-

It can be safely argued that NIL for nanoelectronics needs to demonstrate multilevel capabilities. However, simpler applications, which encompass magnetic arrays, biosensors, passive optical elements, MSM detectors, etc do not have the strong overlay demand. Patterning in 3D, akin to using SU8, will lead to extensions to 3D micro and nanosystems by using multilayers of suitable printable polymers, which are also sensitive to UV and electron beams.

However, in order to realise many of the potentials of NIL, MCP and other alternative nanofabrication techniques, efforts need to be made towards automation and alignment down to 10's of nm and over 200 mm wafers. Much remains to be understood, especially in the viscoelastic properties of very thin polymer layer in order to make progress on standards and critical dimension control, which necessarily pass through adequate design of the stamp features. Nevertheless, the versatility of the NIL and the set accompanying of alternative nanofabrication processes require the qualification of CD control with respect to a particular application. This means that we can expect further developments in an increasing number of nanofabrication approaches, for example, concerning NIL, developments could come in the field of optically active polymers suitable for printing, transparent stamps, etc. for low-cost options needing sub wavelength scale replication of plastic optical elements and circuits. As nanotechnology develops the need for accessible nanofabrication techniques expands. This need is there at both the laboratory scale and the manufacturing level. In this respect, one can be optimistic about the future of novel nanofabrication techniques as these are bound to expand in planar and 3D configurations.

ACKNOWLEDGEMENTS

The author has benefited from participating in several project on nanoimprint lithography including the EU ESPRIT project NANOTECH (project No. 28785), the EU IST FET project CHANIL (IST-1999-13415), EU Growth project MONALISA (GRD-2000-25592), the EU IST project APPTech (IST-2000-29321), the German Research Council (DFG) and the INTAS project 99-009928. The author is indebted to her projects partners for many illuminating discussions and would like to thank Dr Y Cheng (CNRS-LPN), for his help in compiling and discussing Table 1.

REFERENCES

1. B. Michel, A. Bernard, A. Bietsch, E. Delamarche, M. Geissler, D. Juncker, H. Kind, J-P. Renault, H. Rothuizen, H. Schmid, P. Schmidt-Winkel, R. Stutz and H.Wolf, IBM J. Res. & Dev.**45** 697 (2001).
2. Y. Xia, X-M. Zhao and G. M. Whitesides, Microelectr. Eng. **32**, 255-68 (1996).
3. P. Vettiger et al., Microelectronic Engineering **46** (1-4) 101 (1999).
4. H.-C. Scheer, H. Schulz, T. Hoffmann and C. M. Sotomayor Torres, in: H. S. Nalwa (Ed.) Handbook of Thin Film Materials, vol. 5, Academic Press, place, 2002, pp 1-60.
5. R. Feynmann, talk at the California Institute of Technology, 29th December 1959.
6. H. Tan, A. Gilberston and S. Y. Chou, J Vac Sci Technol B **16** 3926 (1998).
7. B. Heidari , I. Maximov, E-L. Sarwe and L. Montelius, J. Vac.Sci.Technol B **18,** 3552-6 (2000).
8. T. Haatainen, J. Ahopelto, G. Gruetzner, M. Fink and K. Pfeiffer, *Proc SPIE* **3997** 874-80 (2000).

9. S. Y. Chou, P. R. Krauss and P. J. Renstrom, Appl. Phys. Lett. **76,** 3114 (1995).
10. S. Y. Chou, P. R. Krauss, W. Zhang, L. Guo, and L. Zhuang, J. Vac, Sci Technol B **15,** 2897 (1997).
11. J. Wang, X.Sun, L. Chen and S. Y. Chou, Appl Phys Lett **75,** 2767 (1999).
12. A. Lebib, Y. Chen, J. Bourneix, F. Carcenac, E. Cambril, L. Couraud and H. Launois, Microelectronic Engeneering **46,** 319 (1999).
13. Z. Yu, S. J. Schablitsky and S. Y. Chou, Appl Phys Lett **74,** 2381 (1999).
14. J. Wang, S. Schablitsky, Z. Yu, W. Wy and S. Y. Chou, J Vac Sci Technol B **17**, 2957 (1999).
15. K. Pfeiffer, G. Bleidiessel, G. Gruetzner, H. Schulz, T. Hoffmann, H.-C. Scheer, C. M. Sotomayor Torres and J. Ahopelto, Microelectronic Engineering **46,** 431 (1999). See also www.microresist.de.
16. H. Schulz, D. Lyebyedyev, H.-C. Scheer, K. Pfeiffer, G. Bleidiessel, G. Gruetzner and J. Ahopelto, J Vac Sci Technol B **18,** 3582 (2000).
17. T. Mäkela, T. Haatainen, J. Ahopelto and H. Isolato, J Vac. Sci. Technol. B **19**, 487 (2001).
18. K. Pfeiffer, M. Fink, G. Aherens, G. Gruetzner, F. Reuther, J. Seekamp, S. Zankovych, C. M. Sotomayor Torres, I. Maximov, M. Beck, M. Grazcyk, L. Montelius, H. Schulz, H-C. Scheer and F. Steingrueber, Microelectronic Engineering **61-62,** 393 (2002).
19. V. Studer, A. Pépin, and Y. Chen, Appl. Phys. Lett. **80** (19), May 13 (2002).
20. M. Beck, M. Graczyk, I. Maximov, E.-L. Sarwe, T. G. I. Ling, M. Keil and L. Montelius, Microelectronic Engineering **61-62,** 441 (2002).
21. C. Finder, M. Beck, J. Seekamp, K. Pfeiffer, P. Carlberg, I. Maximov, F. Reuther, E.-L. Sarwe, S. Zankovych, J. Ahopelto, L. Montelius, C. Mayer and C. M. Sotomayor Torres. To appear in Microelectronic Engineering
22. K. Pfeiffer, F. Reuther, M. Fink, G. Gruetzner, N. Roos, H. Schulz, H.-C. Scheer, J. Seekamp, S. Zankovych, C. M. Sotomayor Torres, I. Maximov, L. Montelius and Ch. Cardinaud. To appear in Microelectronic Engineering.
23. W. Zhang and S. P. Chou, Appl Phys Lett **79** 845 (2001).
24. S. Zankovych, T. Hoffmann, J. Seekamp, J.-U. Bruch and C. M. Sotomayor Torres, Nanotechnology **12,** 91 (2001).
25. C. Gourgon, C. Perret and G. Micounin, Microelectronic Engineering **61-62,** 385 (2002).
26. K. Pfeiffer. M. Fink, G. Bleidissel, G. Gruetzner, H. Schulz, H.-C. Scheer, T. Hoffmann, C. M. Sotomayor Torres, F. Gaboriou and Ch. Cardonaud, Microelectronic Engineering 53, 411 (2000).
27. F. Gaboriau, M. C. Perpignon, A. Barreau, G. Turban, Ch. Cardonaud, K. Pfeiffer, G. Bleidiessel and G. Gruetzner, Microelectronic Engineering 53, 501-505 (2002)
28. M. V. Maximov, J. Seekamp, et al., unpublished data.
29. S. Zankovych, I. Maximov, I. Shorubalko, J. Seekamp, M. Beck, S. Romanov, D. Reuter, P. Schafmeister, A. Wiek, J. Ahopelto, C. M. Sotomayor Torres and L. Montelius, To appear in Microelectronic Engineering
30. J. Seekamp, S. Zankovych, A. H. Helfer, P. Maury, C. M. Sotomayor Torres, G. Boettger, C. Liguda, M. Eich, B. Heidari, L. Montelius and J. Ahopelto, Nanotechnology **13** 1 (2002)
31. C. Clavijo-Cedeño, J. Seekamp, A. P. Kam, T. Hoffmann, S. Zankovych, C. M. Sotomayor Torres, C. Menozzi, M. Cavallini, M. Murgia, G. Ruani, F. Biscarini, M. Behl, R. Zentel and J. Ahopelto, Microelectronic Engineering **61-62** 25 (2002)
32. M. Behl, J. Seekamp, S. Zankovych, C. M. Sotomayor Torres, R. Zentel and J. Ahopelto, Adv Mater **14** 588 (2002)
33. C. Kim, M. Stein and S. R. Forrest, Appl Phys Lett **80** 4051 (2002)
34. M D Austin and S Y Chou, Appl Phys Lett **81** 4431 (2002)
35. A. Lebib, S. P. Li, M Natali and Y. Chen, J. Appl Phys **89**, 3892 (2001).
36. P. R. Krauss and S. P. Chou, Appl Phys Lett **71** 3174 (1997)
37. B. G. Casey, D. R. S. Cumming, I. I. Khandaker, A. S. G. Curtis and C. D. W. Wilkinson, Microelectronic Engineering **46** 125 (1999)
38. J. Gierak, D. Mailly, G. Faini, J. L. Pelouard, P. Denk, F. Pardo, J. Y. Marzin, A. Septier, G. Schmid, J. Ferre, R. Hydman, C. Chappert, F. Flictein, B. Gayral and J.-M. Gerard, Microelectronic Engineering **57** 865 (2002)
39. A. N. Boto, P. Kok, D. S. Abrams, S. L. Braunstein, C. P. Williams and J. P. Dowling, Phus Rev Lett **85** 2733 (2000)
40. M. Muetzel, S. Tandelr, D. Haubrich, D. Meschede, K. Peithmann, M. Flaspoehler and K. Buse, Phys Rev Lett **88**, article no 083601-1 093601-3 (2002).

41. J. Brugger, J. W. Berenschot, S. Kuiper, W. Nijdam, B. Otter and M. Elwenspoek, Microelectronic Engineering **53** 403 (2000)
42. D. Pinner, J. Zhu, F. Xu and S. Hong, Science **283**, 661 (1999)
43. S. Hong and C. A. Mirkin, Science **288** 1808 (2000)
44. for a report see the PHANTOMS Newsletter, May 2002 Issue 6, p23 at http://www.phantomsnet.com
45. S. Y. Chou, C. Keimei and J. Gu, Nature **417** 835 (2002)
46. S. Y. Chou, private communication
47. J. Ahopelto, private communication
48. H. Kurz, private communication
49. M. M. Deshmukh, D. C. Ralph, M. Thomas and J. Silcox. Appl. Phys. Lett. **75** 11 (1999)

2

Nanoimprint Lithography

Stephen Y. Chou
Department of Electrical Engineering, Princeton University, Princeton, USA

2.1. INTRODUCTION

Our ability in patterning nanostructures offers a unique path to discovery and innovation in science and technology. When nanostructures are smaller than a fundamental physical length scale, conventional theory may no longer apply and new phenomena emerge. To fully benefit from the discovery and innovation in nanostructures and commercialize them, a low-cost and high-throughput manufacturing of nanostructures is essential. One biggest challenge today is that we still do not have a mature nanostructure manufacturing technology that has the needed low-cost and high-throughput. Nanoimprint lithography is one of the most promising low-cost, high-throughput technologies for manufacturing nanostructures. Significant developments have been made in recent years.

2.2. NANOIMPRINT LITHOGRAPHY (NIL) PRINCIPLE AND PROCESS

NIL has two steps: imprint and pattern-transfer Figure 1.

In the imprint, a mold with nanostructures on its surface is used to deform a thin resist film (or an active material) deposited on a substrate. In the pattern-transfer, an anisotropic etching process, such as reactive ion etching (RIE), is used to remove the residual resist in the compressed area, transferring the thickness contrast pattern created by the imprint into the entire resist.

The resist can be either a thermal plastic[1] or uv (or thermal) curable polymer[2,3] or other deformable materials. For a thermal plastic resist, the resist is heated above its glass transition temperature during imprint, where the resist becomes viscous liquid and the mold is pressed into the resist. Then the resist is cooled below its glass transition temperature before being separated from the mold. For an uv or thermal curable polymer, a monomer mix, often in a liquid thin film form on a substrate, is

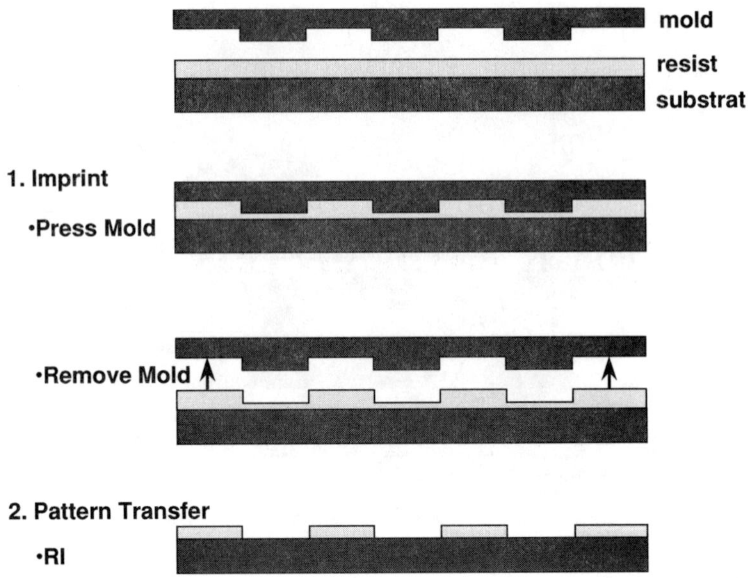

FIGURE 2.1. Schematic of nanoimprint lithography process: (a) imprinting using a moldcreate a thickness contrast in a resist, and (b) pattern transfer using anisotropic etching to remove residue resist in the compressed areas.

imprinted by a mold and then cured by uv light radiation or thermal heating to from polymers (due to polymerization and cross-linkings) before being separated from the mold.

NIL is a primarily a physical deformation process and is fundamentally different from the "stamping" with self-assembly "ink"[4] and other lithographical methods. Such unique principle, although simple, allows NIL to avoid many problems in other lithographical methods and to achieve high resolution and high throughput and low cost at the same time.

2.3. RESOLUTION

The resolution of NIL is determined by the mechanical strength of the mold and polymer. Using silicon-dioxide mold and PMMA resists (the glass transition temperature of 100 C), the holes of 6 nm diameter (60 nm deep) in PMMA and the PMMA pillars of 30 nm diameter (35 nm tall) have been achieved using NIL (Figures 2-4).[5]

Many experiments have indicated that with suitable resist and mold, the resolution of NIL can be below 5 nm. In general, the small holes in a resist are much easier to imprint than the small resist pillars, because the small pillars can easily tear off during the mold separation and are extremely difficult to image (because scanning electron microscopy can easily melt a small polymer pillar).

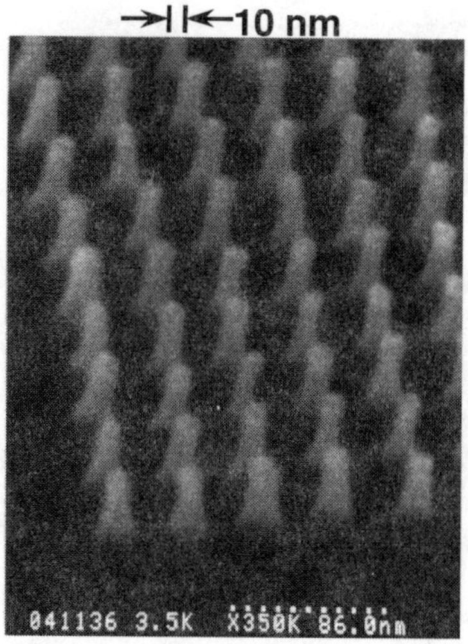

FIGURE 2.2. SEM micrograph of a SiO2 mold with 10 nm minimum diameter pillars with a 40 nm period which are 60 nm tall, after being used 12 times.

2.4. 3-D PATTERNING

Another unique feature of NIL is that it is a 3 dimensional pattern technology, rather than a 2D patterning as in other lithographies. The 3D features are very desirable some applications, such as microwave and MEMS. For example, the "T" gate for microwave transistors has a narrow foot print for high speed, but wide top for lower resistance. The T-gate fabrication often requires two electron beam lithography steps: one for the foot

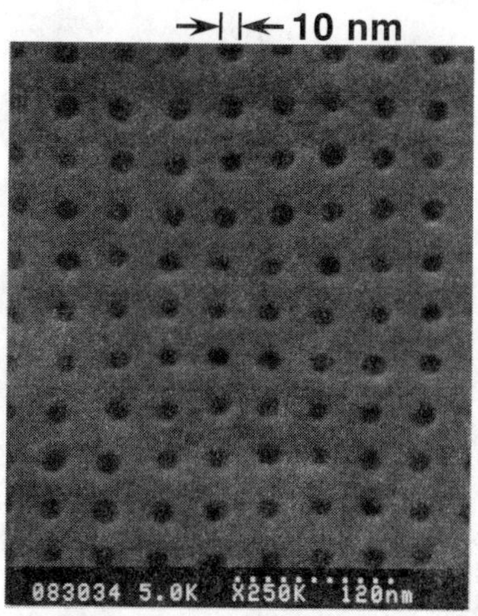

FIGURE 2.3. SEM micrograph of a top view of 60 nm deep holes imprinted into PMMA which have a 10 nm minimum diameter and a period of 40 nm.

FIGURE 2.4. SEM micrograph of a top view of 10 nm minimum diameter metal dots with a period of 40 nm, formed by imprinting into PMMA and a lift-off process.

print and one for the wide top. Each lithography exposure could take over 2 hours for one 4" wafer. With NIL, the entire 4" wafer can be finished in less than 60 seconds. Figure 5 shows examples of 40 nm T-gates fabricated by a single NIL and lift-off of metal.[6]

2.5. IMPRINT OVER NON-FLAT SURFACES

Often, a wafer surface is not flat and already has features on it. To imprint resist patterns on a non-flat surface, multilayer resist methods have been developed and demonstrated.[7] Regardless of using double layer resists or triple layer resists, the first (i.e. bottom) layer of the resists is the planarization layer, which is thicker than the height of the steps on a wafer surface. The planarization layer can result in a flat wafer surface, by (a) a free viscous flow of the resist at an elevated baking temperature or (b) a forced

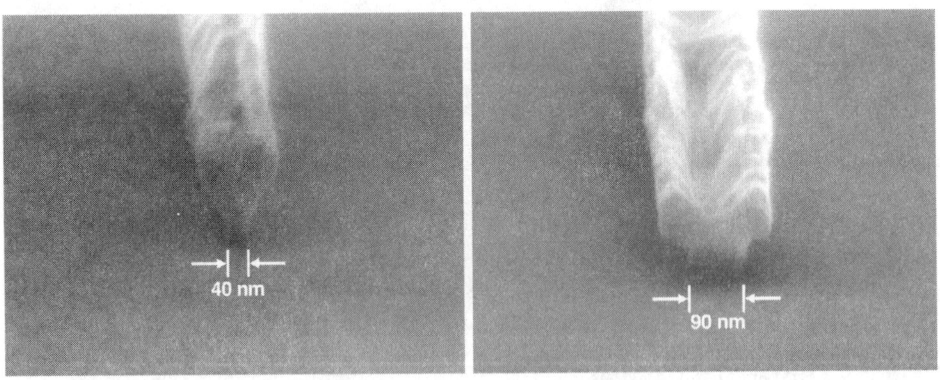

FIGURE 2.5. SEM micrograph of two T-gate of 40 nm and 90 nm foot-print respectively fabricated by a single NIL and a lift-off of metal.

resist flow through pressing a flat mold on the resist surface while heating. The free viscous flow approach works well for PMMA, when heated up to 175 C for 24 hours, but not for novolak resin because it is a thermal-set plastic and has a high viscosity. The forced flow approach works well for both PMMA and novolak resin, when heated at 175 C for 20 min, with a 600 psi pressure on the flat mold. The steps over 200 nm deep have been planarized.

The top layer resist in the multi-layer resist methods is the imaging layer, which will be patterned by imprint. A key difference between the triple layer and the double layer methods is that the triple layer method use a pattern transfer layer in the middle, while in the double layer method, the pattern in the imaging (top) layer will be transferred directly to the planarlization (bottom) layer. The transfer (middle) layer in the triple layer resist scheme is needed, if the top layer resist can dissolve the bottom layer, or has a etching resistance that is too low to be used as etch mask for the bottom layer. In the other words, the middle layer serves as either the barrier to prevent resists mixing or the mask for pattern transfer.

2.6. UNIFORMITY AND SUBMICRON ALIGNMENT OVER 4 INCH WAFERS

To use NIL as a tool to manufacturing integrated circuits (either electronic or photonic), it requires the NIL of nanostructures be uniform over entire wafer and accurate alignment between the different lithography layers. The NIL machine, mask and resist technologies all play significant roles in improve the uniformity and alignment accuracy. Many proprietary technologies have been developed in these areas. Figure 6 shows that uniform 90 nm wide resist gratings imprinted over entire 4 inch wafer.[8]

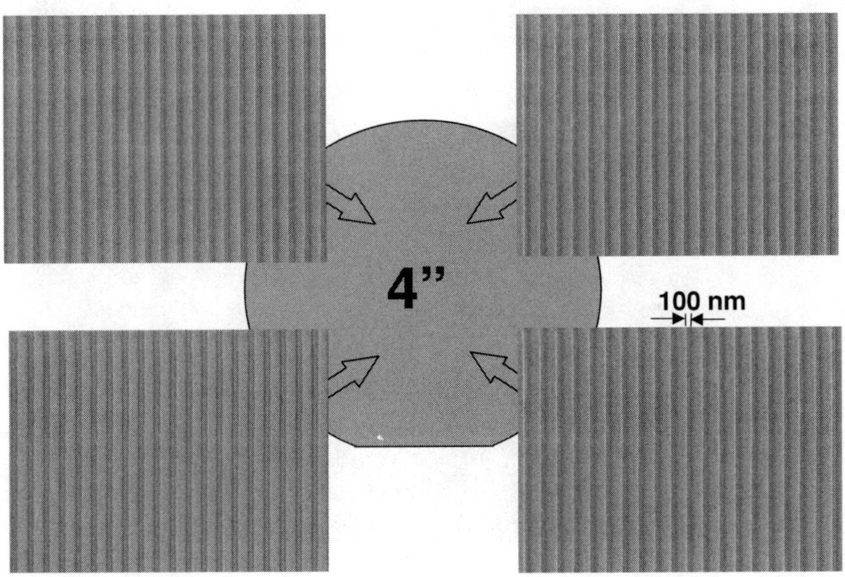

FIGURE 2.6. SEM micrograph of 90 nm wide PMMA grating (200 nm Period) uniform over entire 4 inch wafer by a single layer NIL.

Submicron alignment accuracy over 4 inch wafers has been demonstrated in 10 consecutive runs.[8] Many new NIL resists have been developed to have lower imprint temperature and pressure, better flow properties, smaller shrinkages, and better etching resistance. The NIL machine, mask, and resist technologies developed can be readily scaled to larger wafer size. Better alignment methods as well as local misalignment correction methods are being actively developed.[9]

2.7. DIFFERENT IMPRINT MACHINES

NIL can be carried out using three different types of machines: single-imprint, step-and-repeat, and roller. Single-imprint machine imprints entire wafer at once. Step-and-repeat NIL machine imprints a small area (called a die) of a wafer a time, and then move to a new area of the wafer and imprint. The process repeats until entire wafer gets imprinted. The first advantage of step-and-repeat NIL machine is that it is easier to achieve a higher alignment accuracy in a smaller area than that in a larger area (such as in single-imprint machine). The second advantage and maybe the most significant advantage for many applications is that it allows to use a very small mold to create a large imprint area or a large imprint mold. It might take several days of current electron beam lithography to exposure 10 nm features over entire mold of one centimeter square area. Using step-and-repeat NIL, the one-centimeter-square mold can pattern a much large size of mold (say: 8 inch diameter) in a few minutes - six orders of magnitude faster and perhaps even higher orders of magnitude cheaper! Figure 7 shows 120 nm lines fabricated by step and repeat of NIL on a thermal plastic resist (PMMA).

Step-and-repeat NIL on uv curable resists has also been demonstrated.[3]

Roller nanoimprint (RON) machines have the advantages of better uniformity, less imprint force, simple machine construction and the ability of using a mask continuously on a large substrate.[10] There are two ways to use RON (Figure 8).

FIGURE 2.7. SEM micrograph of 120 nm wide PMMA lines (300 nm period) fabricated by step-and-repeat NIL.

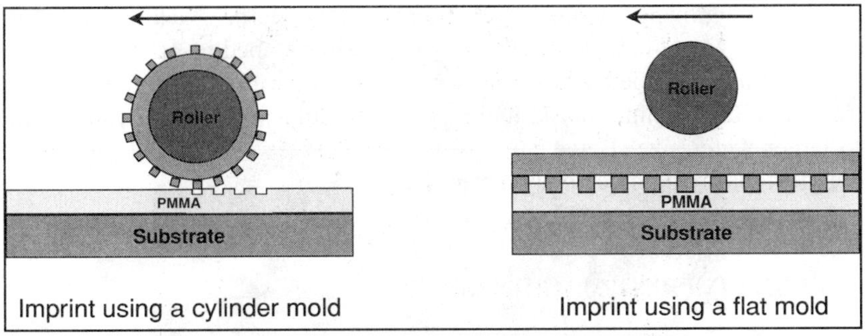

FIGURE 2.8. Schematic of two roller nanoimprint lithographical methods: the mask is cylindrical and the mask is flat.

One uses a cylinder mold, which can be made by bending a thin metal mold around a smooth roller. The other uses a flat mold with a smooth roller rolling over. To keep the shape of the previously imprinted patterns while imprinting other areas, the substrate is held at a temperature below the resist glass transition temperature, but the mold is above the transition temperature. 30 nm wide PMMA lines have been successfully patterned by RON (Figure 9), as well as zone plates (Figure 10).[11]

2.8. APPLICATIONS

NIL have been used to fabricated a variety of electronic, photonic, magnetic, and biological nanodevices, including silicon nanotransistors,[5] high-speed photodetectors,[12] magnetic nanostructures for storages,[13] passive optical elements,[14,15] organic light emitting materials[16] and other materials,[17,18] just name a few. The use of NIL for devices fabrication has become broader.

FIGURE 2.9. SEM micrograph of 30 nm wide PMMA grating (200 nm Period) by roller nanoimprint (RON).

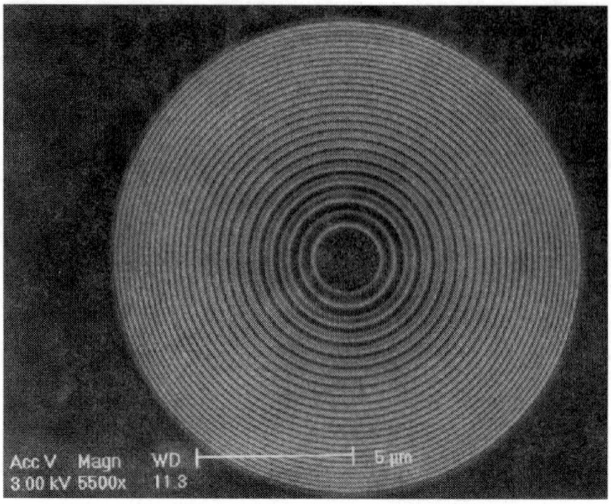

FIGURE 2.10. SEM micrograph of PMMA zone plates with 70 nm min feature size by Roller nanoimprint (RON).

2.9. SUMMARY AND FUTURE

Since its proposal seven year ago, NIL has advanced significantly and begins to be used widely in nanodevice fabrication. Yet, it is just a beginning. Many key issues in NIL machines, masks, resists, and processes still need to be solved. However, it is clear that NIL will be used for manufacture nanostructures. The first step for NIL is not to compete and replace the conventional lithographies, but to find new markets, where the conventional lithography fails to serve. As have said in the beginning of the article, as high-throughput and low cost nanopatterning technologies, NIL will play an important role in future development and manufacturing of nanostructure.

ACKNOWLEDGEMENTS

It is my great pleasure to thank many previous and present members of my group for the their contributions to NIL and LISA and the high lights reviewed here, in particular, P.R. Kraus, P.J. Renstrom, L. Guo, L. Zhuang, W. Zhang, W. Wu, L. Chen, Y. Zhaoning, Mingtao Li, Jian Wang, H. Tan, Paru Deshpande, L. He, B. Cui, X.Y. Sun, S.J. Schablitsky, and A. Gilbertson. The work are supported in part by DARPA, ONR, ARO, and AFOSR. The majority of material in this chapter has taken from the Author's article of Material Research Society Bulletin, Volume 27, No. 7, pp 512-517, July 2001.

REFERENCES

1. S. Y. Chou, P. R. Krauss, and P. J. Renstrom, *Appl. Phys. Lett.* 67, 3114 (1995) and *Science,* (1996), 272, 85.
2. J. Haisma, M. Verheijen, K.van der Heuvel, and J. van der Berg, *J. Vac. Sci. Technol. B,* (1996), 14, 4124.

3. P. Rucjjpeft, M. colburn, B. Choi, Nounu, S. Johnson, C. G. Wilson, et al., *J. Vac. Sci. Technol.,* (1999), *B* 17, 2965.
4. A. Kumar, G. M. Whitesides, *Appl Phys Lett.,* (1993), 63, 2002.
5. S. Y. Chou, P. R. Krauss, W. Zhang, L. Guo and L. Zhuang, *J. Vac. Sci. Technol.,* (1997), *B* 15, 2897.
6. M. Li, L. Chen, W. Zhang, and S. Y. Chou, "3-Dimensional Microfabrication Using Deep Nanoimprint Lithography", EIPBN, Palm Spring, CA, May 30-June 2, 2000.
7. X. Sun, L. Zhuang, W. Zhang and S. Y. Chou, J VAC SCI TECHNOL B, (1998),16, 3922-3925.
8. W. Zhang and S. Y. Chou, "Multilevel Nanoimprint Lithography With Submicron Alignment on 4" Wafers", EIPBN, Palm Spring, CA, May 30-June 2, 2000.
9. F. R.W. Pease, et al., unpublished; and H. I. Smith, unpulished.
10. H. Tan, A. Gilbertson, S. Y. Chou, J VAC SCI TECHNOL, (1999),B 16, 3926.
11. M. T. Li, J. Wang, L. Zhuang, and S. Y. Chou, *Appl Phys Lett.,* (2000), 76, 673.
12. Z. H. Yu, S. J. Schablitsky, S. Y. Chou, *Appl. Phys. Lett.,* (1999), 74, 2381.
13. W. Wu, B. Cui, X. Y. Sun, W. Zhang, L. Zhunag, and S. Y. Chou, J VAC SCI TECHNOL, (1998), B 16, 3825.
14. J. Wang, S. Schablitsky, Z. Yu, W. Wu, S. Y. Chou, *J Vac Sci Tech,* (1999), *B* 17, 2957.
15. Z. H. Yu, P.u Deshpande, W. Wu, J. Wang, and S. Y. Chou, *Appl Phys Lett.,* (2000), 77, 927.
16. J. Wang, X. Y. Sun, L. Chen, S. Y. Chou, *Appl. Phys. Lett.,* (1999), 75, 2767-2769.
17. T. I. Kamins, D. A. A. Ohlberg, R. Stanley Williams, W. Zhang, and S. Y. Chou, *Appl. Phys. Lett.*, (1999), 74, 1773.
18. J. Wang, X. Y. Sun, L. Chen, L. Zhuang, S. Y. Chou "Molecular alignment in submicron patterned polymer matrix using nanoimprint lithography", *Appl. Phys. Lett.,* (1999), 76, 166-168.

3

Viscoelastic Properties of Polymers

Relevance for Hot Embossing Lithography

Thomas Hoffmann
IMEC, B-3100 Leuven, Belgium

3.1. INTRODUCTION

The attractiveness of hot embossing lithography (HEL) also known as nanoimprint lithography (NIL) as nanofabrication technique is mainly founded on its simplicity. This became already apparent in the works of S. Chou,[1-3] who demonstrated that features of a few ten's of nanometers can be imprinted with a standard laboratory press. The principles of embossing reach back into history and are applied today in the production of numerous products, in particular a broad variety of materials is presently micro-moulded in micro-system technology.[4] Although earlier attempts to modify polymer surfaces on a nanometer scale were reported,[5-8] the work of S. Chou certainly has stimulated the interest of the nanofabrication community, since expectations were raised, that a low cost nanofabrication technique was at hand. Since then, a number of technical realisations and variations to the process have been developed, which are reviewed in considerable detail.[9] Most of them share a common principle: a pattern is transferred into a fluid layer by embossing the mould into it. Usually a thin polymer layer spun on a silicon substrate is moulded by a sequence of steps, which are schematically depicted in Figure 1.

The polymer is heated above its glass transition temperature, T_g, to obtain a melt with the desired flow properties. Applying pressure to the mould and the substrate squeezes the polymer into the mould cavities. After completing the filling of the cavities, the system has to be cooled below T_g to guarantee sufficient mechanical stability of the polymer to withstand the separation from the mould. These are the essential steps of the embossing process to which further steps may be added according to the aimed at application. For instance, if it is desired to pattern the substrate, an additional etching step has to be added to open the remaining layer at the bottom of the moulded features (Figure 1).

Alternative Lithography: Unleashing the Potentials of Nanotechnology
Edited by Sotomayor Torres, Kluwer Academic/Plenum Publishers, 2003

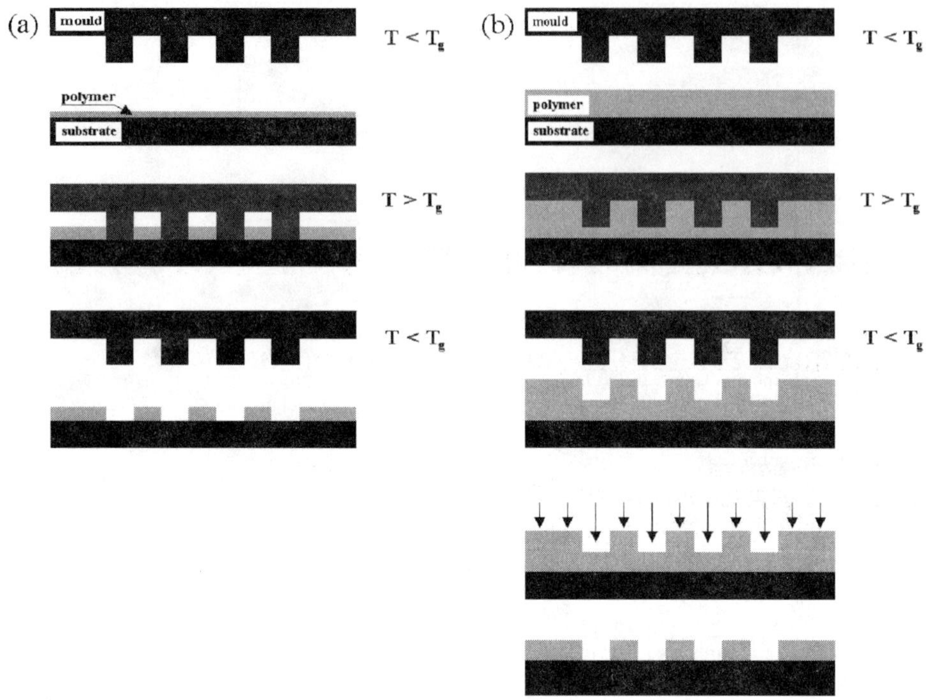

FIGURE 3.1. Scheme of HEL showing heating, embossing, cooling and separation, additional etching step to open remaining layer, additional cross-linking step.

It has to be emphasized that the latter example represents an application with a rather particular functionality of the polymer as etching mask. The variety of properties, which are provided by polymers and organic compounds, can be explored for many applications, in which the optical, electrical, chemical, biological and surface properties of the compounds are of interest. Hence, it is useful to make abstraction from any application at thin stage and to consider the general requirements to be fulfilled by the materials with respect to pattern transfer. Obviously, their flow properties should be adjustable within a certain range to ensure that the pattern transfer is completed within a reasonable regime of processing times. On the other hand, their mechanical properties must guarantee the stability of the moulded features with respect to the aimed functionality. This difference in mechanical behaviour is usually adjusted by application of appropriate temperatures at different stages of the embossing process. Alternatively, the mechanical stability after embossing can be obtained by cross-linking of a low molecular weight compound. Finally, the actual interesting properties of the material must withstand the temperature and pressures applied during embossing.

The latter condition is certainly an issue for polymers or organic layers, because the material flow is achieved by heating to a temperature at which the material shows liquid or viscoelastic behaviour. Limits for a suitable range of processing temperatures are set by the thermal stability of the polymer or organic compound. A chemical degradation of the compound leads usually to the loss of the interesting properties, but in some cases chemical reactions are desired. Examples are low molecular weight compounds or

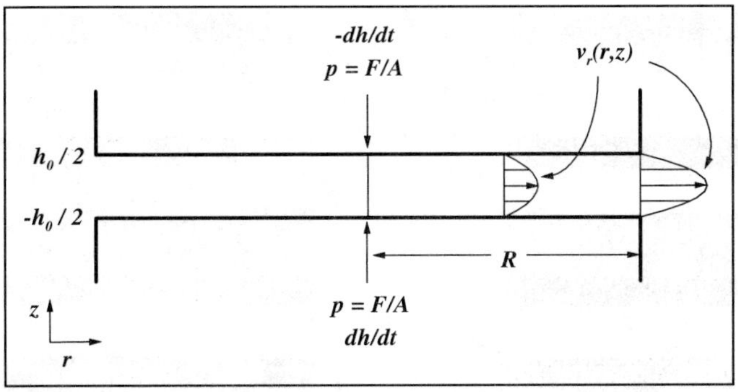

FIGURE 3.2. Scheme to illustrate different flow distances for different feature arrangements/Schematic views of HEL process displaying the flow behaviour of a fluid between two parallel disks of the same radius R.

pre-polymers, which are cross-linked by supplying the required energy either in form of UV light[10-13] or application of a thermal treatment.[1-3,14-39]

Although other materials like metals have been imprinted, the particular interest in polymers and organic compounds is based on the experience that they are easily processed at relatively low temperatures and many of their physical and chemical properties can be tailored by synthesis. The processing temperatures rarely exceed 250°C, which gives raise to the expectation that HEL matches important demands of production, namely high throughput and low cost. The throughput is directly related to the required processing time, which is the sum of the times needed for the heating and cooling cycle and complete filling of the mould cavities. Both times are closely linked to the flow properties of the polymer through the temperature at which a sufficiently low viscosity is obtained.

Other important contributions arise from the applied pressure, the thickness of the material layer and the distances over which the material has to flow to fill the cavities. A first insight into the role of transport distances was provided by experimental results, which indicated a remarkable difference in the filling of cavities according to their size and arrangement. The shorter the distances the faster the filling of the mould cavities under identical embossing conditions. The order of magnitude of these distances can be roughly estimated by summing half the width and the spacing of the features as shown in Figure 2 and 3.

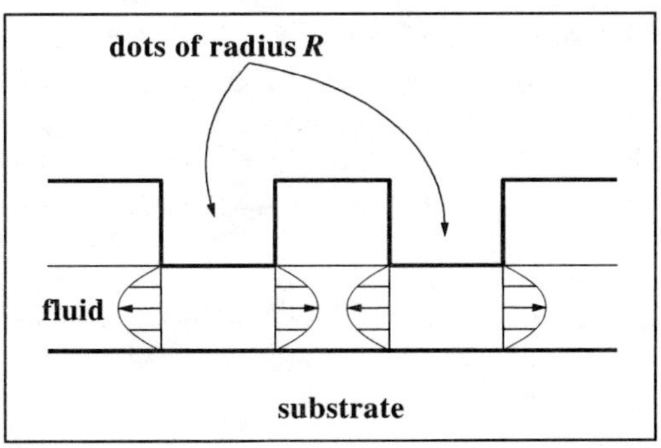

FIGURE 3.3. Scheme to illustrate different flow distances for different feature arrangements/ Schematic views of helprocess squeezing flow in geometry alike to that encountered in nanoimprinting.

Large, isolated features surrounded by large spacing are very challenging with respect to a complete filling. This is true for either elevated features or cavities. A much easier arrangement consists of small periodic structures with a ratio of 1:1 between width and spacing. In addition to the short flow distances, the displaced volume equals those to be filled (Figure 2 and 3). These basic considerations illustrate that certain nanometer-sized patterns are relatively easy to transfer.

A further important issue is the applied pressure to obtain complete pattern transfer in the shortest possible time. Ideally, the forces acting on the mould features should be low enough to ensure that they withstand a large number of embossing steps. The lifetime of the mould is certainly an issue in cost reduction, because so far the fabrication of the mould itself is rather cost intensive. Although not studied or reported in detail, it can be expected that particularly nanometer-sized features with high aspect ratio may set limits to the forces. Other damages that may be introduced by the applied pressure are changes to the properties of the substrate and the imprint layer, which may be avoided by reducing the pressure at the expense of increased flow times. In general, such trade-offs need to be addressed to define process windows, which compromise between low cost, high throughput, high yield, good repeatability and reliability of a process.

The brief summary above given of some of the mayor aspects of HEL should have made clear, that a good understanding of the interdependence of the material viscosity, pressure, layer thickness and arrangement and size of the features is needed. Such an understanding will certainly contribute to the definition of the potential and limitations of HEL. In this chapter, an attempt will be made to summarize those aspects of polymer flow properties, which are closely related to the embossing process. The discussion begins with the squeezing flow of a Newtonian liquid as the simplest approach, which can be adapted to HEL. Although the conclusions that can be drawn are rather qualitative, because the model is too simple to account properly for the flow properties of polymers and the complexity of the flow fields, some considerable insight into the flow process can be gained. In particular, the dependence of the mould motion or flow time on the transport distances, fluid layer thickness and pressure will become apparent. An interesting adaptation of the squeezing flow problem is derived and discussed in considerable detail in Chapter 4.

Even in the limit of linear flow, a description of the viscoelastic properties of polymers needs to take into account both the reversible and irreversible contributions. Both contributions determine a characteristic time beyond which the irreversible deformations dominate the overall deformation of the polymer. Additionally, this characteristic time provides a measure for the relaxation of build up stress in the polymer melt after cessation of the flow process. The most important aspects of this behaviour will be discussed qualitatively in terms of the temporary network of entangled polymer chains and the resulting constraints on the chain motions. Finally, a few arguments are given which point to the relevance of non-linear flow behaviour in certain embossing conditions.

3.2. SQUEEZING FLOW OF A NEWTONIAN LIQUID IN HEL

3.2.1. General Considerations

The squeezing flow of a fluid between parallel disks is a problem considered in fluid dynamics, which comes closest to the situation encountered in HEL (Figure 4).

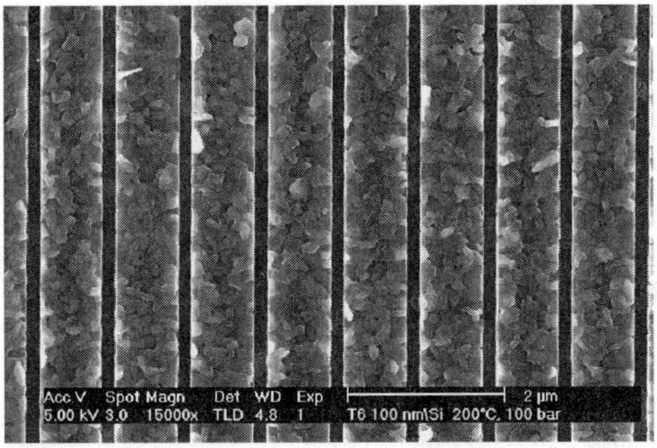

FIGURE 3.4. Scanning electron micrograph of 100 nm thick imprinted alpha sexithiophen layer (T6) on silicon substrate with parallel lines.

A Newtonian liquid is placed in the gap between two parallel disks separated by a distance $2h_0$ corresponding to the thickness of the fluid layer. The application of a force F leads to the squeezing flow of the fluid layer, which is maintained against the atmospheric pressure p_a. A quasi steady state solution of the problem is given by the Stefan equation,[40] which expresses the force in terms of the disk radius R, the motion of both disks of same magnitude but opposite sign ($\pm\, dh/dt$) and the viscosity η_0 of the fluid as:

$$F = -\frac{3\pi R^4}{4h_0^3}\frac{dh}{dt}\eta_0 \tag{1}$$

The strong dependence of the force on the disk radius, $F \propto R^4$, and the fluid thickness, $F \propto h_0^{-3}$, indicates that very large forces are required to obtain a fast fluid motion in thin films and over large transport distances[1]. The fluid viscosity enters only linearly into equation1, but for polymers it usually decreases by various orders of magnitude with increasing temperature. The viscosity of a Newtonian fluid does not depend on the shear rate, which also applies to more realistic fluids in the limit of small shear rates. At moderate or high shear rates non-linear flow behaviour can lead to a further decrease of the viscosity by various orders of magnitude. Thus, in certain conditions an extremely reduced viscosity may positively contribute to a reduction of the pressure or processing times.

To adapt the squeezing flow model to the situation encountered in nanoimprinting, it is necessary to understand – at least qualitatively - the different contributions to the force, which arise from the pressure and the shear stress related to the fluid motion. The area of contact between the mould and the fluid determines the required force to obtain the pressure p-p_a. The contribution arising from the shear stress is related to the flow velocity, which depends on the transport distances as defined by the mould design (Figure 5). In the case of a flat disk, both the area and the transport distances are identical and given by the disk radius R. The flow of the fluid results from the transfer of the

[1] The discrepancy between experimentally applied and theoretically computed pressures was mentioned as early as 1999 by V. Sirotkin, A. Svintsov, and S. Zaitsev, "Deformation and viscous flow in nano-imprinting", presented at the 4th MEL-ARI/NID Workshop, July1999, Duisburg, Germany.

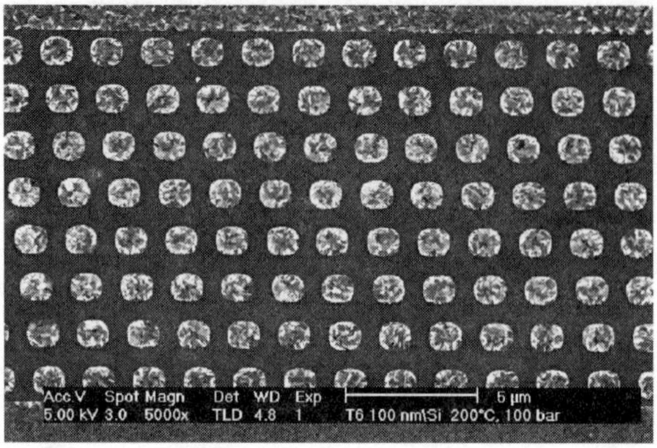

FIGURE 3.5. Scanning electron micrograph of 100 nm thick imprinted alpha sexithiophen (T6) layer on silicon substrate with a hole array.

mould motion, which is transferred into a radial flow of the fluid, $v_r(r, z)$, from the disk centre outwards to its circumference. At a given radial distance r from the centre, the fluid motion of a volume element depends not only on the amount of material displaced by the disks, but also on the amount of material moving from the centre outwards. To fulfil the conditions of continuity and conversation of momentum for an incompressible liquid, the velocity has to increase with the radial distance as:

$$v_r(r, z) = -\frac{3}{4}\frac{dh}{dt}r\left[1 - \left(\frac{z}{h_0}\right)^2\right] \tag{2}$$

The velocity profile causes shear stresses acting between the different fluid layers. Integration over these shear stresses leads to a quadratic dependence of the force on the transport distances. The velocity profile in z – direction is of the well known parabolic shape, which takes its maximum velocity at the central layer ($z = 0$) and zero velocities at both interfaces ($z = \pm h_0/2$). It is assumed that no slippage between the fluid and the mould or substrate surfaces occurs. Again, the mould motion adds an additional contribution to the force-thickness dependence leading to the $F \propto h_0^{-3}$ relation.

The usual HEL set-up is characterised by the application of a constant pressure and the quantity of interest is the time dependent penetration depth as given by the motion of the mould and substrate, $2\ \mathrm{d}h/\mathrm{d}t$, which can be computed from Equation (1). Two different moulds are considered: (i) a flat disk with a radius of $R = 10^{-2}$ m and (ii) a disk of the same radius covered to 50% by dots, each dot having a radius of $R = 10^{-6}$ m (Figure 4-7). Both, the disk and the dot radius define the corresponding transport distances. The area of contact between the mould and the fluid is given by $A_{\text{flat}} = \pi \times 10^{-4}\ \text{m}^y$ and $A_{\text{patt}} = 0.5 \times \pi \times 10^{-4}\ \text{m}^2$, respectively. For a typical pressure of 10^7 Pa (100 bar) the force acting on the flat mould amounts to $\pi \times 10^3$ N. Furthermore, two typical initial thicknesses of the fluid layer are considered. Reported values of the viscosity of PMMA at different temperatures agree reasonably well.[41-43] The computed values are summarised in Table 1.

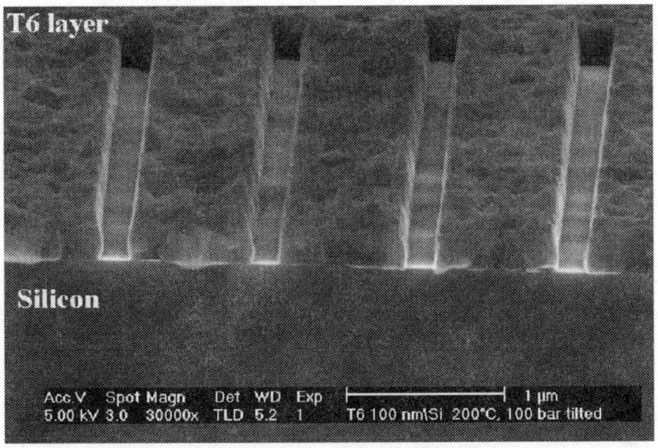

FIGURE 3.6. Scanning electron micrograph of 100 nm thickimprinted alpha sexithiophen (T6) on silicon substrate in different geometric arrays. Detail of SEM micrograph of Figure 4.

Values for the speed, which are not beyond any range of practical interest, are only obtained with the patterned mould. Depending on the thickness of the initial layer and the temperature speeds larger than 10^{-9} nm s^{-1} are achieved, which would lead to embossing times of less than 100 s for an imprint depth of 100 nm. Regarding the elapsed time during embossing another quantity, the time required for thinning the fluid layer to a thickness $h_1 < h_0$, can be derived by integrating the Stefan equation with respect to dh/dt, which leads to:

$$\frac{1}{h_1^2} - \frac{1}{h_0^2} = \frac{16Ft}{3\pi R^4 \eta_0} \tag{3}$$

This equation may be further simplified to express the time $t_{1/2}$ required to reduce the thickness of the fluid by 50% as:

$$t_{1/2} = \frac{9\pi R^4 \eta_0}{16F h_0^2} \tag{4}$$

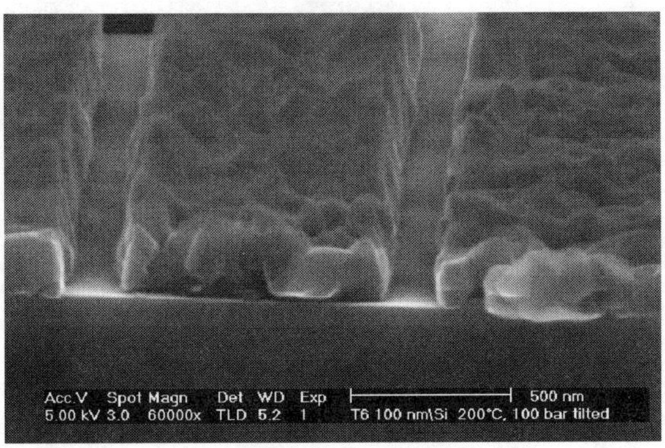

FIGURE 3.7. Scanning electron micrographies of 100 nm thick imprinted alpha sexithiophen (T6) on silicon substrate in different geometric arrays. Magnified details of SEM micrograph of Figure 4.

TABLE 3.1. Computed motion of mould and substrate ($2dh/dt$) for an applied pressure of 10^7 Pa, different viscosity of PMMA and two different initial fluid thicknesses ($2h_0$). Both the flat and the patterned mould have a radius of $R = 10^{-2}$ m. The patterned mould consists of single dots of radius $R = 10^{-6}$ m, which cover the surface to 50%. Values for the viscosity are taken from reference.[43]

PMMA $M_w \approx 1.1 \times 10^5$ g mol^{-1}		Flat mould layer thickness $2h_0$		Patterned mould Layer thickness $2h_0$	
		500 nm	100 nm	500 nm	100 nm
T(°C)	η_0 (Pa s)	Motion 2dh/dt (m s^{-1})		Motion 2dh/dt (m s^{-1})	
140	4.6×10^9	2×10^{-18}	1×10^{-20}	2×10^{-10}	1×10^{-12}
160	7.1×10^7	1×10^{-16}	1×10^{-18}	1×10^{-8}	1×10^{-10}
200	1.5×10^6	6×10^{-15}	4×10^{-17}	6×10^{-7}	4×10^{-9}

Table 2 summarises the half times computed for the patterned mould and the two initial thicknesses. The values compare reasonably well to those usually experienced in HEL experiments. Note that the times for the flat disk increase by a factor of 10^8, since the transport distance is 4 orders of magnitude larger. Again, the predictions of the squeezing flow model for the flat disk arrangement are beyond any practical use.

Both examples can be considered as the two limiting cases encountered in HEL. Since the cavities are increasingly filled with increasing imprint depth, the fluid has to flow over larger distances, which finally reaches the size of the mould. Accordingly, a slowing down of the mould motion with increasing imprint depth can be expected. This slowing down during embossing is considered in more detail also in Chapter 4.

The predicted times in Table 2 compare reasonably well to those regimes typically encountered in HEL. It seems that the role of the viscosity and the transport distances is reasonably well described. For instance, it has been reported that periodically arranged lines of 500 nm width and spacing are completely filled by application of 100 bar even at temperatures as low as 140-150°C during a typical imprint time of 5 minutes. In

TABLE 3.2. Computed half times for a patterned stamp covered to 50% by dots of radius $R = 10^{-6}$ m, an applied pressure of 10^7 Pa and two different initial fluid thicknesses. Values for the viscosity of PMMA at the given temperatures are taken from Table 1.

PMMA	$t_{1/2}$	
T(°C)	5×10^{-7} m	1×10^{-7} m
140	1h 10 min	28 h 45 min
160	64 s	27 min
200	1 s	35 s

contrast, a complete filling of features and spacing with sizes of 100 μm and more could be achieved only at temperatures of 200°C. These results were already interpreted in terms of different transport distances.[35] Other examples, which support the idea that flow distances play a very important role for the embossing process, are reported in the literature.[27,28] An interesting approach to access the dependence of flow times on the transport distances and compare them to theory is described in Chapter 4.

The imprint into a thin layer of the low molecular compound alpha sexithiophen (T6) provides another example, which illustrates the interplay between localised flow and viscosity.[39] The layer thickness of 100 nm is smaller than the feature height on the mould, which amounts to 400 nm. The imprint was carried out at 200°C. The much lower viscosity of this compound and the fact that the stamp cavities are not completely filled allows printing through the layer down to the substrate (Figure 8).

The example shows a possible approach to avoid the change in the transport distances and possible long imprint times. The approach is likely to be limited to materials with lower viscosity than polymers and smaller mould feature, because otherwise it is barely impossible to thin the remaining layer down to the substrate. Additionally, the complex flow patterns observed with polymers make it more difficult to obtain a uniform level in the filling of the cavities.

It remains to discuss some of the limitations of the previously discussed approach. The assumption that the fluid flow caused by the elevated features is independent from the feature surroundings is certainly not realistic. From Figures 4-7 it can be already inferred that the velocity profiles emerging from adjacent features in opposite direction must interact. Therefore, the flow patterns are more complex. Additionally, build-up of stresses and non-linear flow properties of polymers may contribute to this complexity. Although the squeezing flow model is too simple to account correctly for the polymer flow properties and the complex interactions between the flow fields, it identifies those quantities, which contribute mainly to the embossing time, namely transport distances, viscosity and thickness of the fluid layer. The transport distances depend on the particular arrangement and size of the features on the mould, which is certainly an advantage for

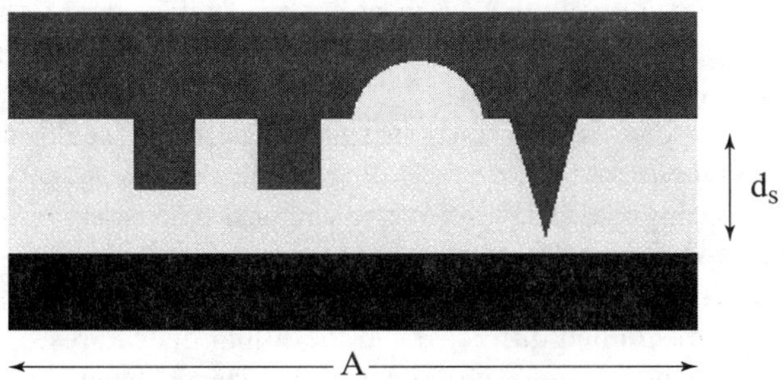

FIGURE 3.8. Definition of polymer thickness and effective volume (Vfill) to be filled during the imprint process. In this example, ds is defined as the distance between the tip of the triangle which is the most elevated structure in the mould, and the stamp baseline. The triangle and the two squares are positive structures while the semicircle is a negative one.

the embossing of nano-structures. On the other hand, for many application moulds are applied, which contain a variety of features differing in size by orders of magnitude. The largest transport distances on the mould define the minimum time, required to complete the filling of all mould cavities. Reduction of transport distances by intelligent mould design is possible within limits set by the functionality of the target device. Even for moulds containing only nano-sized features, a change in transport distances over various orders of magnitude usually takes place during embossing. By comparison, the initial thickness of the fluid varies within narrower limits, which usually lie between 10^{-7} and 10^{-6} m and the thickness is usually not reduced much below 10^{-7} m by the embossing step. Again, the range of the remaining thickness is likely to be limited by the application. Therefore, it is mainly the viscosity, which can be changed with relative easiness by adjusting the temperature or by the choice of fluid material.

3.2.2. Process Control and Slowing down of Embossing

As previously discussed, two extreme situations are encountered in HEL, which differ largely by their respective transport distances. Localised flow over distances of the order of the feature size and spacing takes place only in the beginning of the embossing process, when the mould comes into contact with the fluid layer. Accordingly, flow can proceed either by application of a lower pressure or a faster mould motion is achieved at constant pressure. This situation is dramatically changed when the filling of the mould cavities is completed. The transport distances are determined by the size of the mould and additionally, the imprint can proceed only by thinning the remaining layer.

Such a decrease of the mould motion or correspondingly an increase of the pressure could be advantageous for the control of the imprint process. It could provide a useful measure in automated processing to determine whether complete filling has been achieved or not. Up to now force-displacement curves have not been measured in nanoimprinting, thus it remains open whether such an on-line control of the pattern transfer can be efficiently implemented. It is desirable in terms of both, process control and understanding of the flow to collect more data on how the pressure and the mould motion depend on the size and arrangement of the mould features. An interesting approach to access quantitatively the filling of cavities is described in Chapter 4. The approach is not suitable for on-line control, but may help to understand the flow and the force-displacement curves.

In practical terms one expects a gradual transition from localised flow to large transport distances. One reason is that the filling of the cavities by polymers does not simply proceed by raising the fluid level in the cavities, but rather starts at the border of the features, thus changing the transport distances already in the early stages of embossing. This is a change on the scale of the feature sizes and cavities. Another reason is related to the aforementioned differences in the filling of features of different size. Structures of small size and spacing are filled at earlier stages than larger cavities. After completing the filling of a whole array of small structures, the entire array contributes to the flow process with a transport distance comparable to the array size itself. Examples, which illustrate nicely how single cavities or entire arrays are filled, are discussed in Chapter 4.

3.2.3. Appropriate Fluid Thickness for Arbitrary Mould Patterns

For a number of applications it is desirable to obtain complete filling of the cavities and a as thin as possible remaining layer. A frequently mentioned example makes use of the polymer as a mask for a subsequent etching process. This requires to open first the remaining layer and then to transfer the pattern into the substrate. For the former step the remaining layer should be as thin as possible, for the latter a maximum thickness contrast allows to etch as deep as possible into the substrate. The thickness contrast is defined by the height of the imprinted patterns measured with respect to the remaining layer and is only obtained by complete filling of the mould cavities.

Complete filling is also of mayor importance for applications, which do not require a subsequent etching step, e.g., all applications that require solely the structured modification of a polymer surface. For instance, flow borders, which show up, when cavities are incompletely filled by the polymer material, may deteriorate the periodicity and therefore the quality of an optical grating or photonic band gap structure. For optical waveguides the remaining layer should be as thin as possible, since leaking of the optical modes depends on its thickness. For some applications this condition maybe more relaxed and the imprint can be carried out either in thicker initial layers to reduce the processing time or flow borders are acceptable as long as the functionality of the desired device is not affected. This may well be the case for applications where the definition of a capillary in the polymer is the most important issue as, for instance, in the fabrication of nano- and micro-sized chemical or bio-chemical reactors. However, only a complete transfer usually achieves the desired pattern definition.

The previously discussed time ranges show that it is barely impossible to thin the remaining layer after complete filling is achieved and any attempt to do so will require practically unreasonable processing times. Therefore, a definition of the thickness of the initial and the remaining layer is needed. A simple approach to calculate the initial polymer thickness for arbitrary but given feature sizes and shapes is readily obtained. Without considering the complex pattern of filling, the volume of the mould cavities can be divided into two contribution. First, the cavities from their lowest level up to the level of the most elevated structure on the stamp are filled. This corresponds to the filling from the bottom of the groove to the tip of the triangle in Figure 9.

Without any structure present, the volume would simply be Ad_s, where A corresponds to the area of the mould and d_s is the height of the most elevated structure with respect to the stamp baseline. From this volume either the volume of each elevated or positive structure V_i has to be subtracted or the volume of each hollow or negative structure V_j has to be added to obtain the total volume:

$$V_{fill} = Ad_s - \sum V_i + \sum V_j \tag{5}$$

The volume V_{fill} has a particular interesting meaning, since it corresponds to the minimum volume of resist needed to fill the structures on the stamp up to its most elevated structure with a remaining thickness of zero. Thus, only the volume of an additional layer defined by the desired remaining thickness, Ad_{rem} has to be added to obtain the overall

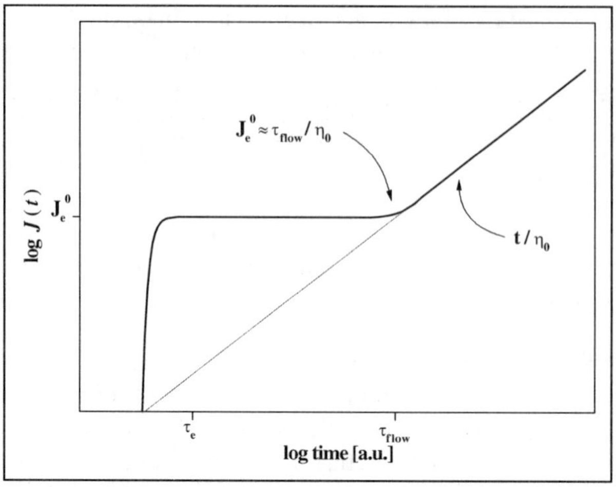

FIGURE 3.9. Deformation (J(t)) behaviour as a function of the time in a polymer layer under the application of a constant stress. Initially, the deformation – proportional to J0e –does not show any dependence on the time. This constant value is reached after a characteristic time te and persists until a second characteristic time tflow has been elapsed.

volume of the initial polymer layer V_0 :

$$V_0 = Ad_{rem} + V_{fill} \tag{6}$$

From the last equation it follows that the thickness of the initial layer d_0 is given by:

$$d_0 = d_{rem} + \frac{V_{fill}}{A} \tag{7}$$

If an optimum value for the remaining thickness is determined by considering the desired functionality and the flow behaviour, the appropriate thickness of the polymer layer can be calculated by Equation (7). In terms of processing this can be easily done during stamp design by using all the structural information to compute the appropriate thickness.

3.3. VISCOELASTIC PROPERTIES OF POLYMERS

3.3.1. Linear Flow Properties of Polymer Melts

The embossing process depends in many ways on the flow properties of polymers. As already discussed, the viscosity can change by 3 to 4 orders of magnitude in the temperature range of interest for HEL (Table 1). The glass transition temperature T_g of the polymer in question naturally defines a point of reference for the temperature range. In view of the flow properties, T_g can be defined by the temperature at which the viscosity reaches a value of 10^{12} Pa s. A few degrees Kelvin below T_g the flow processes in the fluid become extremely slow. The polymer responds as a solid to external forces or deformations except for very long times.

The temperature dependence of the viscosity can be described as a thermally activated process. An empirical relation, which holds for most of the polymers[44,45] is given by the Vogel-Fulcher law:

$$\eta_0(T) = \eta_c \exp\left(-\frac{T_A}{T - T_V}\right) \tag{8}$$

where T_A refers to the activation temperature. The Vogel temperature T_V lies approximately 50 K below T_g and refers to the temperature at which the flow processes actually cease. Note that the pragmatic definition of T_g is based on a large, but still measurable viscosity. The parameters η_c, T_A and T_V have to be determined from experimental data. A convenient way to do so is based on the Williams-Landel-Ferry equation,[44-46] which is an equivalent formulation of Equation (8). An example for the application of the WLF-equation is discussed in Chapter 4. Although, the validity of the empirical Vogel-Fulcher law has been questioned[47,48] the WLF-equation has proven to be a powerful mean for analysing the mechanical properties of amorphous polymers. The empirical relation for the viscosity is usually valid between T_g and 100 degrees above it. For the particular interesting case of PMMA with a T_g values lying between 95 to 105°C, it was shown that this relation holds up to temperatures of 240°C.[41,42] Knowledge of the value of T_g is certainly advantageous for the introduction of new polymers into HEL.

So far the discussion of the squeezing flow model has focussed on Newtonian liquids whose flow properties are completely described by knowledge of $\eta_0(T)$. For such a liquid the *time dependent shear compliance* $J(t)$ has the simple form:

$$J(t) = \frac{t}{\eta_0} \tag{9}$$

and the deformation $e_{zx}(t)$ under a constant shear stress σ_{zx}^o is given by

$$e_{zx}(t) = J(t)\,\sigma_{zx}^0 \tag{10}$$

In contrast, polymer melts display an additional elastic behaviour even at higher temperatures. To account for this reversible contribution to the overall deformation a further quantity, the steady state shear compliance J_e^0 , is required. Then the time dependent deformation can be described as:

$$J(t) = J_e^0 + \frac{t}{\eta_0} \tag{11}$$

It has to be noted that both quantities η_0 and J_e^0 are material parameters only in the limit of small shear stresses or rates. For relatively moderate values of the shear rate non-linear response of a polymer melt is usually observed. The application of a constant

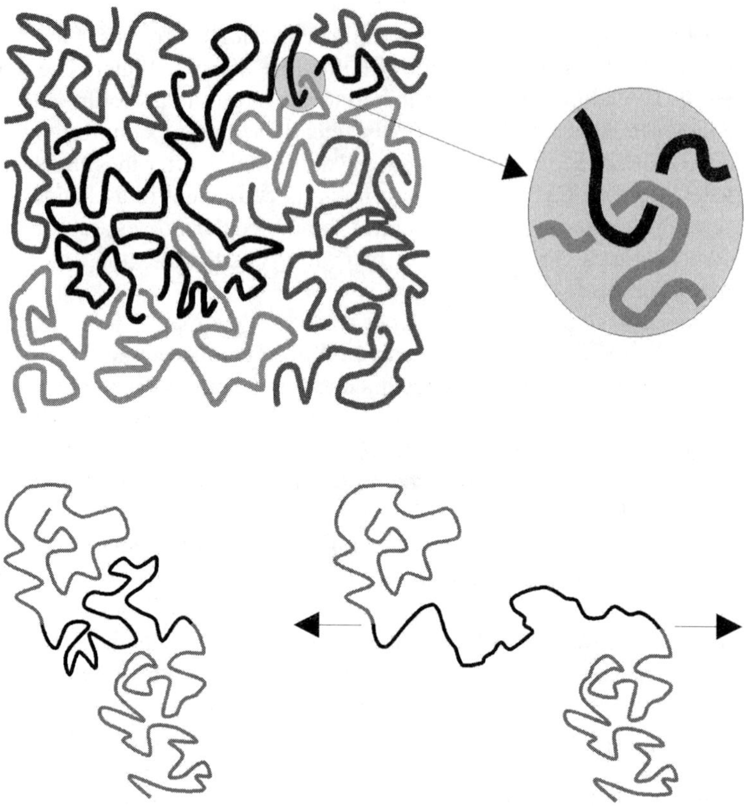

FIGURE 3.10. Schema of temporary network of entanglements, which is displayed in the virtual magnification of the figure (top). The movements of the polymer chains are constrained by the presence of neighbouring chains. The time needed for those movements are much longer than the ones observed during stretching and relaxation of the segments between entanglements. External force acting (arrows) on a single chain (bottom).

shear stress at time t_0 leads to a time dependent deformation as illustrated schematically in Figure 10.

When the stress is applied a constant deformation $J_e^0 \sigma_{zx}^w$ is reached after a characteristic time τ_e, which persists until a second characteristic time τ_{flow} has elapsed. Beyond this time, the deformation shows the linear increase of a Newtonian fluid. For polymer melts, J_e^0 is usually of the order of 10^{-5} to 10^{-6} m^2 N^{-1} and therefore about three orders of magnitude larger than the value of an amorphous polymer in the solid state.[41-43,49-52] Accordingly, the deformations are large as it is typically the case for rubber-elastic behaviour. It should be noted that this part of the deformation is recovered after cessation of the stress. The flow time characterises the situation where the viscous part of the deformation τ_{flow}/η_0 equals the elastic contribution J_e^0. Thus, a simple relation can be derived which permits to estimate the times beyond which flow processes begin to dominate the deformation:[53-56]

$$t_{fill} \approx J_e^0 \eta_0 \tag{12}$$

Similarly, t_{flow} defines the typical times for relaxation of stresses in a polymer melt. The importance of stress relaxation for HEL will be discussed later. The temperature dependence of $J(t)$ can be mainly attributed to the strong temperature dependence of η_0 (Equation 8). In contrast, J_e^0 varies only linearly with temperature. At higher temperatures viscous flow begins to dominate the deformation after the elapse of a shorter time. In general, a change of temperature finds its equivalent in a change of the time scale. Increasing the temperature accelerates the molecular processes and accordingly the characteristic times are shorter. This behaviour is known as the time-temperature superposition principle.[44-46,57,58]

Theoretical descriptions of viscoelastic behaviour are mainly founded on two concepts: (i) a single polymer chain owes its flexibility to a large number of conformations which it can take by rotation around the bond axis along the chain backbone and (ii) the interactions in a multi-chain system impose constraints on the motion of the macromolecules. The interactions between chains are usually pictured as a temporary network of entangled chains, which permits motion only by slippage along the entanglements.

Naturally, the average entanglement lifetime is related to the aforementioned flow time. If an external force acts on a melt over times longer than this average lifetime, the motion of chains leads to irreversible macroscopic flow. On the other hand, if the force persists for shorter times, slippage of entanglement is not very likely to occur. On this time scale the melt responds by stretching of the flexible segments fixed between the entanglements until the external force is counterbalanced. On cessation of the force, the segments re-coil into equilibrium conformations. This large reversible response of the melt is related to the conformational changes which the chain segments between the entanglements undergo. Rotation around the backbone bonds permits the macromolecule to change its conformation continuously by thermally activated jumps across the rotational energy barriers. This leads to a large number of conformations, which a polymer chain may take. The conformations differ by the set of rotational angles and energies related to each rotational state, and their likelihood. In the simplest approach the conformations can be modelled by a random walk of constant step length.[45,59-62] The most likely conformations are those in which the chain takes the form of a coil, whereas extended chains are very unlikely, since only a few extended conformations exist. Therefore, an extended chain or a segment re-coils due to thermally activated rotational jumps in the absence of an external force. This relaxation is driven by entropy. At higher temperatures chains will stretch out faster, thus establishing equilibrium at J_e^0 within shorter times. In addition, the equilibrium value of the deformation is lower, since the entropy driven forces, which counteract the external force, are becoming stronger. As a consequence, the steady state shear compliance J_e^0 decreases linearly with increasing temperature.

It remains to discuss briefly how the quantities η_0 and J_e^0 depend on the molecular weight M. Both quantities display different behaviour in two regions, which are separated by a certain critical molecular weight M_c. In terms of the entanglement model, M_c can be interpreted as the molecular weight at which a temporary network of entanglements spans over macroscopic dimensions. Below M_c entanglements may be present, but their number is too small to lead to a sufficiently connected temporary network, which would impose constraints on the chain motion. Accordingly, the polymer melt resembles the flow properties of a low molecular weight compound, namely, the viscosity increases linearly with the molecular weight. However, the action of flow fields causes the molecules to take conformations away from their equilibrium. This shows up in a linear increase of the recoverable shear compliance J_e^0 with increasing molecular weight. Above M_c the segment length between entanglements appears to be constant in accordance with the experimental observation that J_e^0 remains unchanged with increasing M. Consistent with the assumption of a constant segment length, the viscosity shows a strong dependence on M, since the number of entanglements per chain has to increases with M for a constant segment length. The relations for η_0, J_e^0 and τ_{flow} are summarised by the following equations:[45,63,64]

$$M < M_c : \eta_0 \propto M, \quad J_e^0 \propto M, \quad \tau_{flow} \propto M^2 \tag{13a}$$

$$M > M_c : \eta_0 \propto M^{3.4\pm0.2}, \quad J_e^0 \propto const, \quad \tau_{flow} \propto M^{3.4} \tag{13b}$$

The critical molecular weight of PMMA is approximately 3×10^3 g mol^{-1}.[65,66] So far the relations have been discussed only for the ideal case of linear polymer chains with a uniform molecular weight. Deviations from the predicted behaviour are frequently observed in experiments and ascribed to the molecular weight distribution and the presence of branches. In the case of PMMA, it is assumed that the presence of polar groups leads to deviations from the predicted behaviour.[41]

3.3.2. Implications of linear visco-elastic flow properties to HEL

From the brief discussion of the linear mechanical properties of polymer melts it becomes apparent, that the flow time is the quantity of interest for HEL, since it determines the time beyond which the irreversible deformations dominate over the elastic contributions. However, even beyond τ_{flow} one has still to account for the reversible elastic contributions. For a polymer of given molecular weight $M > M_c$, the flow time decreases strongly with increasing temperature due to the exponential decrease of the viscosity. Limits for the applications of high embossing temperatures are set by (i) the thermal stability of the polymer and (ii) the time required for the entire heating and cooling cycle, which may compensate the gain in the embossing time. Low molecular weight compounds with $M < M_c$ can be imprinted at lower temperatures, lower pressures or within shorter times. In some cases, the absence of a network of entanglements may lead to more brittle behaviour. In connection with nano-sized features, this may lead to a fracture of the features during separation of the mould from the imprinted layer.[35] A certainly interesting approach consists in the use of cross-linkable monomers or pre-polymers, which are cross-linked after completion of the pattern transfer to achieve higher mechanical stability.

What remains to be discussed is whether the recoverable part of the deformation bears any importance for HEL. During embossing, the chain segments are deformed by the shear stresses established by the velocity gradient. When the flow slows down, the chain segments attempt to recover their equilibrium conformations. As a result, the polymer may exert forces on the mould by attempting to recover partly its original shape. If these stresses do not relax before the polymer is cooled down below T_g, the forces may be frozen in and contribute to the sticking of the mould. To proof this assumption, the question has to be answered whether the stresses can relax during the time needed for cooling the polymer from the embossing temperature to a temperature below T_g. The time for the cooling cycle is of the order of 3 to 6×10^2 s depending on the embossing temperature. The flow time can be computed from Equation (12) with the viscosity values given in Tab. 1. For the same PMMA, a value of approximately 2×10^{-5} Pa^{-1} is reported for J_e^0.[14] For an imprint temperature of 140°C one obtains a value of approximately 9×10^4 s for τ_{flow}, which is two orders of magnitude larger than the typical time for cooling below T_g. In contrast, at 200°C, the computed flow time is approximately 3×10^1 s and any build up stress in the polymer relaxes before the cooling cycle is completed.

The assumption that a lack of stress relaxation may contribute to the sticking phenomenon is supported by a few experimental observations. For a number of different polymers, best quality of pattern transfer defined by completely filled and intact features was achieved at imprint temperatures 80 to 90°C above T_g. For imprints carried out 40 to 50° above T_g, severe damage of the moulded polymer was easily identified, because the polymer was peeled away from the substrate at many places.[35] It has to be emphasised, that a similar behaviour was obtained even with moulded amorphous Teflon®, for which it is difficult to imagine that adhesive forces contribute to the peel off of the polymer. The presence of stress in moulded polymers is nicely demonstrated by the changes, which the moulded features undergo after finishing the embossing process. The changes are proceeding slowly over weeks, since the polymer is kept below T_g.[6]

3.3.3. Non-linear flow behaviour

At low shear rates both η_0 and J_e^0 are material parameters, the values of which depend solely on the temperature for a polymer of a given molecular weight. Deviations from this linear behaviour are commonly observed in processing of polymers even for moderate shear rates or stresses.[44,45] The practically motivated interest in polymer processing as well as the growing interest in non-linear fluid dynamics has lead to an increased amount of work in this area of research. It is beyond the scope of this chapter to discuss matters in an extent appropriate to this topic. Some very nice illustrations of non-linear flow patterns as observed in HEL are discussed in Chapter 3. In this section the discussion focuses on two well known effects described in literature, namely shear thinning and extrudate swelling, which are of immense importance in the processing of polymers.

Shear thinning is characterised by a decrease of the viscosity with increasing shear rate ($\partial\gamma / \partial t$), which sets in at relatively moderate shear rates of 10^{-2} to 10^0 s^{-1}.[44,45,67] Depending on the polymer, the processing temperature and shear rate may decrease by 5 or 6 order of magnitude, which is a considerable amount in view of the processing times predicted by the squeezing flow model. An empirical description of shear thinning can be given, by, for instance, the relation

$$\eta = m \left(\frac{\partial \gamma}{\partial t} \right)^{n-1} \tag{14}$$

TABLE 3.3. Shear rates calculated for different temperature and two layer thickness.

T	$2h_0 = 5 \times 10^{-7}$ m		$2h_0 = 1 \times 10^{-7}$ m	
(°C)	$2h/dt$ (nm)	$\partial\gamma/\partial t$ (s^{-1})	$2h/dt$ (nm)	$\partial\gamma/\partial t$ (s^{-1})
140	2×10^{-10}	5×10^{-3}	1×10^{-12}	6×10^{-4}
160	1×10^{-8}	2×10^{-1}	1×10^{-10}	6×10^{-2}
200	6×10^{-7}	1×10^{1}	4×10^{-9}	2×10^{0}

For polymers the exponent n takes values between 0.1 and 0.6, which corresponds to a decrease of the viscosity at different rates. Whether the effect of shear thinning plays a role in embossing or not is not an easy question to answer. However, it is worth to estimate the shear rates on grounds of the squeezing flow model. The shear rate is obtained from the velocity profile in radial direction $v_r(r, z)$ as given in Equation (2) by derivation with respect to ∂z:

$$\frac{\partial\gamma}{\partial t} = \frac{3}{2}\frac{dh}{dt}\frac{rz}{h_0^3} \tag{15}$$

For the mould motion, the values given in Tab. 1 are used, which are computed for the patterned mould and different temperatures for an applied pressure of 10^7 Pa. The maximum shear rate is obtained at the contact planes between fluid and either mould or substrate ($z = \pm h_0/2$) and for r equal to the dot radius ($R = 10^{-6}$ m). Table 3 summarises the computed shear rates for two different layer thicknesses $2h_0$.

The obtained values of the shear rate fall partly into the regime where shear thinning may be expected. Even at a temperature of 160°C non-linear shear thinning may play a role at least in the early stages of the embossing process. At later stages, the mould motion becomes too slow to obtain sufficient shear rates for the shear thinning effect to occur (cf. entries in Tab. 1 for the flat mould). Interestingly, shear thinning is expected to be more importance for smaller feature size, thicker layers and higher temperatures (Equation 15).

Another commonly observed phenomenon in polymer flow is the swelling of the extrudate, which occurs after a polymer fluid has been forced through a confined volume like, for instance, a capillary. After leaving the capillary, the confinement by the capillary walls is released and the melt can relax. The shear flow stretches the chains in direction of the flow into non-equilibrium conformations. After leaving the zone of confinement, the polymer chains re-coil into equilibrium positions, which results in a swelling of the melt outside the capillary. The swelling can be interpreted in terms of normal stresses acting perpendicular to the flow direction. In HEL this may occur, when the polymer material passes the border between feature and cavity, where the chains can relax and swell only in direction upwards into the cavity, because of presence of the substrate. The amount of swelling, which can be expressed as the ratio between the diameter of the capillary and the diameter of the extrudate increases with increasing shear rate and decreasing temperature. For polystyrene, which has a T_g comparable to PMMA, the diameter is almost doubled at shear rates of approximately 10^2 s^{-1} and a temperature of 160°C.[45,68]

It is difficult to judge, whether swelling leads to the typical flow patterns and borders, which result from polymer that flows the cavity walls upwards. Other contributions may arise from the interaction of the flow fields emerging from the elevated mould features, surface energies and electrical charge of the polymer. It is likely that more than one process is contributing to the formation of the complex flow pattern.

ACKNOWLEDGEMENTS

The partial support of the EU–MEL ARI Project Nanotech is gratefully acknowledged. The T6 samplers were kindly provided by Dr F. Biscarini.

CONCLUSIONS

The relevance of linear and nonlinear processes has been illustrated with estimates of flow and mould motions for PMMA. In particular, the role of the rate at which pressure and temperature are varied has been highlighted vis-à-vis polymer properties and starting thickness and mould design. It is shown that, assuming a better knowledge of the viscoelastic properties of thin polymer films used in HEL suitable windows of parameters may be found to bring out the potentials of HEL in nano fabrication.

REFERENCES

1. S.Y. Chou, P.R. Krauss, and P.J. Renstorm, Appl. Phys. Lett. **67**, 3114 (1995).
2. S.Y. Chou, P.R. Krauss, and P.J. Renstorm, Science 272, **85** (1996).
3. S.Y. Chou, P.R. Krauss, and P.J. Renstorm, J. Vac. Sci. Technol. B **14**, 4129 (1996).
4. W. Ehrfeld, V. Hessel, H. Löwe, C. Schulz, and L. Weber, Microsystem Technol. **5**, 105 (1999).
5. R. Bartolini, W. Hannan, D. Karlsons, and M. Lurie, Applied Optics **9**, 2283 (1970).
6. R. Ulrich, H.R. Weber, E.A. Chandross, W.J. Tomlinson, and E.A. Franke, Appl. Phys. Lett. **20**, 213 (1972).
7. C. Puech, Opt. Commun. **7**, 135 (1973).
8. G.D. Aumiller, E.A. Chandross, W.J. Tomlinson, and H.P. Weber, J. Appl. Phys. **45**, 4557 (1974).
9. H.-C. Scheer, H. Schulz, T. Hoffmann, and C.M. Sotomayor Torres, *Nanoimprint Techniques* in Handbook of Thin Film Materials, Vol. 5: "Nanomaterials and Magnetic Thin Films," Ed. H.S. Nalwa, Academic Press, pp 1-60, 2002.
10. J. Haisma, M. Verheijen, K. van den Heuvel, and J. van den Berg, J. Vac. Sci. Technol. B. **14**, 4124 (1996).
11. B.J. Choi, S. Johnson, M. Colburn, S.V. Sreenivasan, and C.G. Willson, Proc. of ASPE, 1999 Annual Meeting
12. M. Colburn, S. Johnson, M. Stewart, S. Damle, T. Bailey, B.J. Choi, M. Wedlake, T. Michaelson, S.V. Sreenivasan, J. Ekerdt, and C.G. Willson, Proc. SPIE: Emerging Lithographic Technologies III, Vol. 3676, 379 (1999).
13. M. Colburn, A. Grot, M. Amistoso, B.J. Choi, T. Bailey, J. Ekerdt, S.V. Sreenivasan, J. Hollenhorst, and C.G. Willson, Proc. SPIE: Emerging Lithographic Techniques IV, Vol. 3997, p. 453 (2000).
14. S.Y. Chou and P.R. Krauss, Microelectron. Eng. **35**, 237 (1997).
15. S.Y. Chou, P.R. Krauss, W. Zhang, L. Guo, and L. Zhuang, J. Vac. Sci. Technol. B. **15**, 2897 (1997).
16. P.R. Krauss and S.Y. Chou, Appl. Phys. Lett. **71**, 3174 (1997).
17. L. Guo, P.R. Krauss, and S.Y. Chou, Appl. Phys. Lett. **71**, 1881 (1997).
18. H. Tan, A. Gilbertson, and S.Y. Chou, J. Vac. Sci. Technol. B. **16**, 3926 (1998).

19. W. Wu, B. Cui, X. Sun, W. Zhang, L. Zhuang, L. Kong, and S.Y. Chou, J. Vac. Sci. Technol. B. **16**, 3825 (1998).
20. J. Wang, X. Sun, L. Chen, and S.Y. Chou, Appl. Phys. Lett. **75**, 2767 (1999).
21. D. Eisert, W. Braun, S. Kuhn, J. Koeth, and A. Forchel, Microelectron. Eng. **46**, 179 (1999).
22. D.-Y. Khang and H.H. Lee, Appl. Phys. Lett. **75**, 2599 (1999).
23. B.G. Casey, W. Monaghan, and C.D.W. Wilkinson, Microelectron. Eng. **35**, 393 (1997).
24. B.G. Casey, D.R.S. Cumming, I.I. Khandaker,, A.S.G. Curtis, and C.D.W. Wilkinson, Microelectron. Eng. **46**, 125 (1999).
25. R.W. Jaszewski, H. Schift, P. Gröning, and G. Margaritondo, Microelectron. Eng. **35**, 381 (1997).
26. R.W. Jaszewski, H. Schift, J. Gobrecht, and P. Smith, Microelectron. Eng. **41/42**, 575 (1998).
27. H. Schift, R.W. Jaszewski, C. David, and J. Gobrecht, Microelectron. Eng. **46**, 121 (1999).
28. L.J. Heyderman, H. Schift, C. David, J. Gobrecht, and T. Schweizer, Microelectron. Eng. **54**, 229 (2000).
29. H. Schift, C. David, J. Gobrecht, A. D'Amore, D. Simoneta, W. Kaiser, M. Gabriel, J. Vac. Sci. Technol. B **18**, 3564 (2000).
30. H. Schift, L.J. Heyderman, M. auf der Maur, and J. Gobrecht, Nanotechnology **12**, 1 (2001).
31. S.W. Pang, T. Tamamura, M. Nakao, A. Ozawa, and H. Masuda, J. Vac. Sci. Technol. B **16**, 1145 (1998).
32. B. Faircloth, T. Kowalewski, H. Rohrs, K. Wooley, R.R. Krchnavek, A. Mueller, and R. Ruoff, "Hyperbranched Fluoropolymer Nanoimprint Lithography", unpublished.
33. H.-C. Scheer, H. Schulz, T. Hoffmann, and C.M. Sotomayor Torres, J. Vac. Sci. Technol. B **16**, 3917 (1998).
34. K. Pfeiffer, G. Bleidiessel, G. Gruetzner, H. Schulz, T. Hoffmann, H.-C. Scheer, C.M. Sotomayor Torres, and J. Ahopelto, Microelectron. Eng. **46**, 431 (1999).
35. F. Gottschalch, T. Hoffmann, C.M. Sotomayor Torres, H. Schulz, and H.-C. Scheer, Solid-State Electronics **43**, 1079 (1999).
36. H. Schulz, H.-C. Scheer, T. Hoffmann, C.M. Sotomayor Torres, K. Pfeiffer, G. Bleidiessel, G. Grützner, Ch. Cardinaud, F. Gaboriau, M.-C. Peignon, J. Ahopelto, and B. Heidari, J. Vac. Sci. Technol. B. **18**, 1861 (2000).
37. K. Pfeiffer, M. Fink, G. Bleidiessel, G. Gruetzner, H. Schulz, H.-C. Scheer, T. Hoffmann, C.M. Sotomayor Torres, F. Gaboriau, and Ch. Cardinaud, Microelectron. Eng. **53**, 411 (2000).
38. S. Zankovych, T. Hoffmann, J. Seekamp, J.-U. Bruch, and C.M. Sotomayor Torres, Nanotechnology **12**, 91 (2001).
39. C. Clavijo Cedeño, J. Seekamp, A.P. Kam, T. Hoffmann, S. Zankovych, C.M. Sotomayor Torres, C. Menozzi, M. Cavallini, M. Murgia, G. Ruani, F. Biscarini, M. Behl, R. Zentel, and J. Ahopelto, Microelectronic Engineering **61-62,** 25 (2002)
40. R. B. Bird, R. C. Armstrong and O. Hassager, *Dynamics of Polymeric Liquids,* Vol. 1 "Fluid Mechanics", John Wiley & Sons, 1977.
41. S. Onogi, T. Masuda and T. Ibaragi, Kolloid Z. - Z. Polym. **222**, 110 (1967).
42. T. Masuda, K. Kitagawa and S. Onogi, Polymer J. **1**, 418 (1970).
43. J. L. Halary, A. K. Oultache, J. F. Louyot, B. Jasse, T. Sarraf, and R. Muller, J. Polym. Sci. (B) Polym. Phys. **29**, 933 (1991).
44. J. D. Ferry, *Viscoelastic Properties of Polymers,* 3rd ed., John Wiley & Sons, 1980.
45. G. R. Strobl, The Physics of Polymers (2nd Edition), Springer, 1997.
46. M. L. Williams, R. F. Landel and J. D. Ferry, J. Am. Chem. Soc. **77**, 3701 (1955).
47. S. S. N. Murthy, J. Polym. Sci., Part C., Polym. Lett. **26**, 361 (1988).
48. S. S. N. Murthy, J. Polym. Sci. Part B: Polym. Phys. **31**, 475 (1993).
49. J. R. McLoughlin and A. V. Tobolsky, J. Colloid. Sci. **7**, 555 (1952).
50. A. Gourari, M. Bendaoud, C. Lacabanne, and R. F. Boyer, J. Polym. Sci., Polym. Phys. Ed. **23**, 889 (1985).
51. N. Ouali, M.B.M. Mangion, and J. Perez, Philosophical Magazine A **67**, 827 (1993).
52. A. Dufresne, S. Etienne, J. Perez, P. Demont, M. Diffalah, C. Lacabanne and J. J. Martinez, Polymer **37**, 2359 (1996).
53. F. Bueche, J. Chem. Phys. **20**, 1959 (1952).
54. P.E. Rouse, .J. Chem. Phys. **21**, 1272 (1953).
55. W.W. Graesseley, J. Chem. Phys. **43**, 2696 (1965).
56. W.W. Graesseley, J. Chem. Phys. **47**, 1942 (1967).
57. H. L. Williams, Polymer Engineering, pp. 86-99, Elsevier, 1975.

58. C. N. Robinson, Basic Concepts of Polymer Mechanical Behaviour in "Polymers for Electronic and Photonic Applications," pp. 633-50, Ed. C. P. Wong, Academic Press, 1993.
59. P. J. Flory, Statistical Mechanics of Chain Molecules, Interscience Publishers, 1969.
60. M. Doi and S. F. Edwards, The Theory of Polymer Dynamics, Clarendon Press, 1986.
61. M. Doi, Introduction to Polymer Physics, Clarendon Press, 1996.
62. P. G. de Gennes, Introduction to Polymer Dynamics, Cambridge Univ. Press, 1990.
63. G. C. Berry and T. G. Fox, Adv. Polym. Sci. 5, 261 (1968).
64. T. P. Lodge, N. A. Rotstein and S. Prager, Adv. Chem. Phys. 79, 1 (1990).
65. V. P. Privalko and Y. S. Lipatov, Makromol. Chem. 175, 641 (1974).
66. S. M. Aharoni, Macromolecules 16, 1722 (1983).
67. J. Meisser, Kunststoff 61, 576 (1971).
68. J.J. Burke and V. Weiss, *Characterisation of Materials in Research*, Syracuse University Press, 1975.

4

Nanorheology

Squeeze Flow in Hot Embossing of Thin Films

Helmut Schift and Laura J. Heyderman
Laboratory for Micro- and Nanotechnology, Paul Scherrer Institut, Villigen PSI, Switzerland

4.1. INTRODUCTION

Molding techniques based on imprint processes make use of the differences between the mechanical properties of a structured stamp and a molding material. The viscous molding material is shaped by either pressing the hard stamp into it (embossing or imprinting) or by filling of a cavity containing a structured mold insert (casting or injection molding). In order to achieve a reasonable process time and yield, this is usually carried out under pressure and using molding materials with relatively low viscosities.

As a simple example in daily life we can take a look at the waffle baking process in which soft dough is processed into a surface structured disc-shaped waffle. A quantity of dough is placed at the center of the lower plate and the upper plate is pressed onto it. This causes the dough to spread to the borders of the mold, so continuously filling the mold cavities. After forming and cooking these delicious waffles, which have a finely structured surface relief providing a large surface area for shorter cooking times, they can be released from the mold. The quality of the waffle is mostly dependent on the ingredients of the dough, and the temperature and time used for processing. The size and final thickness of the waffle is predominantly dependent on the initial quantity of dough. A perfect waffle is the result of optimization of all of these parameters, including optimization of the pattern used as a mold. For example, the more dough we place at the center of the mold, the easier it is to fill all the mold cavities. Taking less dough would result in more crusty waffles which may be incomplete and inhomogeneous, even if we press for a long time. This simple waffle model has an important shortcoming because the waffle dough is a thermoset-type polymer which changes its chemical properties irreversibly during the thermal treatment. In hot embossing processes we mostly deal with thermoplastic materials such as PMMA whose mechanical properties can be repeatedly and reversibly changed from a solid into a viscous state by simply varying the temperature. The rheological processes described here for thermoplastic materials can be considered to be similar for thermoset or UV-curable materials as long as the thermomechanical properties can be changed without affecting the chemical ones.

In this chapter we want to take a closer look at the rheology of molding of thin thermoplastic films as used for hot embossing lithography. This discussion can be transferred with some restrictions to other types of polymers and processes. Before we derive formulas for the description of squeeze flow in Section 4.2, we will briefly present some rules of thumb for hot embossing lithography and finish by comparing this with industrial molding processes. This introduction will enable the reader to understand rheology in hot embossing lithography from a practical point of view bringing in information from other molding processes for completion. We then address the theoretical description which gives a better insight into the physical phenomena present in squeeze flow. In Section 4.3 we present experimental results and in Section 4.4 we include some additional aspects which should be considered for future development of the hot embossing process. More fundamental questions of non-linear flow and treatment of special polymers will be presented in other chapters of this book (see Chapter 2 by S.Y. Chou or Chapter 3 by T. Hoffmann).

4.1.1. Rules of Thumb for Hot Embossing

The quality of molding depends much on the purpose for which the generated relief is used for, e.g. in *Hot Embossing Lithography* (HEL, also referred as *Nano-Imprint Lithography*) reactive ion etching (RIE) is used to open substrate windows in the resist for subsequent pattern transfer processes. Therefore the residual height of polymer at the bottom of the resist relief has to be reduced to an extent that RIE does not change significantly the topography and shape of single structures. As we will see, reducing the thickness of polymer layers is directly linked with an increase of process times needed for embossing. Process times of half an hour or more are not acceptable for mass fabrication and in fact, it would be desirable to reduce them to a few seconds as is the case in modern Compact Disc (CD) injection molding processes. There are different areas for optimization:

1. *Initial film thickness:* Process times to mold thin polymer layers can be reduced to a few seconds by simply using a high initial layer thickness. The reason for this is easy to understand: for thicker films the squeeze polymer can flow more freely in the central plane of the film unaffected by the friction at the boundaries. Care must be taken in HEL where a high initial layer thickness usually results in an unacceptable residual thickness, i.e. too high for further pattern transfer.
2. *Molding temperatures:* High molding temperatures can lower polymer viscosities and reduce times to fill the mold cavities. However in current embossing setups, the relatively high heat capacities and the low heating and cooling rates often limit embossing times. This is because the stack of mold and polymer, as well as the substrate and heating plates of the press have to be heated and cooled. Furthermore, unwanted temperature effects may occur such as a decomposition of the polymer or changes in the electronic, chemical or mechanical properties of the underlying substrate or stamp (e.g. the deterioration of the anti-adhesion layer on the stamp[1] or a loss of mechanical hardness and stiffness in electroplated

Nickel stamps).[2] An example where heat capacities are kept to a minimum is AFM (atomic force microscope) embossing[3,4]. A short current flash heats the tip of a scanning cantilever over the melting point of the polymer, causing the AFM tip to make a local indentation by sinking locally into the polymer. Using this technique, molding processes at 100 kHz rates can be realized. For a good replication fidelity, often processes are preferred where all the stack elements are kept in thermal equilibrium during the temperature cycle, consequently leading to long embossing times. In conclusion, in order to reduce embossing times, either one should keep the heat capacity of the system low, use efficient heating and cooling devices, take a process where the polymer is only locally heated and reduce the polymer viscosity by using higher temperatures or low molecular weight polymers.

3. *Stamp geometry:* The distances of lateral polymer flow should be kept as small as possible. Stamp cavities and protrusions (elevated structures) should be distributed homogeneously over the entire stamp area i.e. the fill factor should be kept constant in order to limit the displacement of polymer in the vicinity of each structure. This reduces the tendency for incomplete filling and variation of residual layer thickness over the surface area. Often an equilibration mechanism such as a thin silicone mattress (e.g. PDMS) is used to equilibrate pressure inhomogeneities due to thickness variations in the mold or the press setup. Another possibility is to reduce wedges and thickness variations to a minimum using an entirely stiff setup which is not controlled by pressure but exact positioning of the stamp in the vertical (z-)direction, as used for high aspect ratio LIGA structures.[5] *Demolding:* It seems likely that any reasonable structure present on a stamp (with sizes down to 10 nm or aspect ratios up to 5) can be molded, by using sufficient heat (assuring low viscosity), pressure (assuring sufficient flow) and time. For high aspect ratio structures, demolding becomes a critical step in the replication process. High aspect ratio structures can be realized if molds with vertical sidewalls are used. Here, adhesion, imperfections of the walls and roughness in conjunction with lateral deformation due to the different thermal expansion of mold, polymer and substrate might lead to interlocking or high friction during demolding (see Section 4.3.8). Anti-adhesion layers on the mold can reduce friction forces, but will become less effective as the cavities become very small and only a few polymer chains have to be extracted from a cavity. Many interesting developments have been made for microscale molding in LIGA technology[2,5] where slow directional demolding was realized in special mechanical setups, air was pressed into the cavities for easier demolding, and multilayer polymer films with high anti-adhesive properties at the top surface were successfully employed.

In the following sections, many of these points will be substantiated by examination of theoretical considerations and experimental investigations although, with such a complex subject, we do not claim to give a fully conclusive summary of all effects and parameters used in present hot embossing processes. In order to transform hot embossing into a widely applicable process, further investigations will be a prerequisite for understanding and optimization of new structures. The possible areas for investigation are discussed in the concluding section of this chapter.

4.1.2. Nanorheology in Other Replication Processes

The hot embossing process is suited to imprint a pattern onto a very thin layer of thermoplastic polymer for lithographic applications. There are several molding processes currently used in industry for thick films or bulk materials and it is of benefit to compare these processes.

Most prominent in industry are roll embossing and thermal injection molding. In the first case a polymer foil is continuously embossed while being squeezed through two heated rolls, one of them being covered with a surface structured metal shim. The foils are normally several 10's of μm thick. This concept has already been used for hot embossing lithography.[6] Using this process very rapid embossing times are achieved, filling the mold cavities in a matter of milliseconds. Here the bending of the shim helps to reduce the contact area and process times, and even might ease demolding of low aspect ratio structures. However, if heat conductance of the substrate is high, already molded structures released from the stamp cavities can be damaged by the ongoing molding of neighboring structures.

In injection molding, a melt with a relatively low viscosity is injected with high pressure through a nozzle into a closed cavity containing a structured mold insert. This cavity is large in comparison to the cavities on the mold. For high aspect ratio microstructures, variotherm processes have to be used where the mold temperature is kept above the glass transition temperature during injection. This ensures that the melt is sufficiently viscous when it moves into the structure cavities of the mold. When molding is completed, the mold is cooled down and the melt freezes. In non-variotherm processes, the mold has a lower temperature than the melt, resulting in an immediate local freezing of the polymer when the melt comes into contact with the surface structures. Meanwhile the injection of the melt continues and fills the rest of the cavity. Surface nanostructures can be molded this way at very high speeds (10 sec for the replication of a compact disc (CD) with 25 nm structures).[7,8] The height of the CD-mold cavity (1.2 mm) is much larger than the film thickness used in hot embossing lithography (100 to 1000 nm). However a constant molding fidelity over the whole disc diameter is only achieved if high pressures and polymers with low viscosities are used. Hot spots in the polymer melt and even decomposition of polymer may occur due to the high speed of injection and the high shearing forces present. In addition, flow freezing, surface tension and other rheological effects determine whether structures are molded or not. The main development e.g. in CD molding (as needed for the 600 μm thick DVD with nanostructures with dimensions around 400 nm) is currently directed towards a combination of injection molding with a successive compression of the cavity (so-called compression molding). Here the compression of the cavity is used to compensate for the shrinkage of the disc during processing. Although the thickness of a DVD is still much larger than the film thickness used in HEL, lateral flow between mold boundaries plays an important role in the filling of the thin disc-like cavity.

4.2. BASICS OF THIN FILM RHEOLOGY

For a deeper insight into the embossing process and the problems occurring during processing, knowledge about the mechanical properties of the material is indispensable.

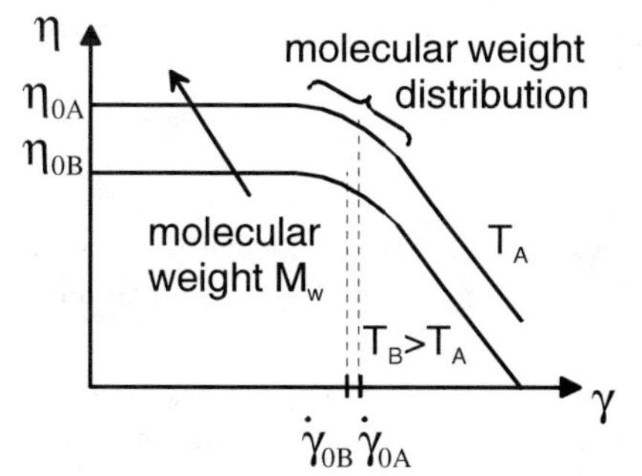

FIGURE 4.1. Viscosity shear curve (shear velocity $\dot{\gamma}$, shear viscosity η) schematically shown for two temperatures T_A and T_B.

Rheological aspects are of the highest importance for the processing of the thermoplastic materials, for example frozen stress can be avoided if the processing of the material is adapted to the relaxation time of the material.

In this contribution we cannot dwell on the details of the linear visco-elasticity of polymers and we assume the reader has a basic understanding. In the following paragraphs we will nevertheless give a short introduction into squeeze flow of polymers followed by a derivation of the theory. This is helpful for the basic understanding of important parameters required for the optimization of hot embossing processes and demonstrates how rheological measurements coincide with the theoretical investigations. As a further introduction into rheology the reader can find excellent textbooks.[9,10] Other aspects of rheology such as non-linear flow are described in Chapter 3.

4.2.1. Shear Viscosity and Velocity

The reduced viscosity of polymers at higher temperatures is a result of the increasing ability of the chains to move freely, while entanglements and Van der Waals interaction of the chains are reduced. Materials such as PMMA (polymethylmethacrylate) or PC (polycarbonate) are generally considered as amorphous, although on a microscopic scale PC has semicrystalline properties. For most of the applications considered here, effects caused by orientation of chains (such as birefringence effects in Compact Disc manufacturing) are of minor importance. However taking a closer look at the properties and processes, especially in the nanorange, regimes can easily occur where these polymers can no longer be considered as amorphous and processes are no longer reversible. Most of our considerations here are valid for a range of practical process parameters, as used in current hot embossing processes, where linear behavior can be assumed.

PMMA, the classical resist material used in HEL, can be considered to a good approximation as purely viscous and can be described by a truncated power law model (Figure 1).

$$\breve{\eta}(\dot{\gamma}) = \begin{cases} \eta_0 , & \dot{\gamma} < \dot{\gamma}_0 \\ \eta_0(\dot{\gamma}/\dot{\gamma}_0)^{(1-k)/k} , & \dot{\gamma} > \dot{\gamma}_0 \end{cases} \tag{1}$$

η_0 is the zero shear viscosity, with the limiting value

$$\eta_0 = \lim_{\dot{\gamma}\to 0} \breve{\eta}(\dot{\gamma}_0) \tag{2}$$

$\dot{\gamma}_0$ is the limiting shear velocity, at which the approximated stationary shear viscosity changes over from the constant value η_0 to a power law model, and k is a constant.

In order to determine viscosities of the polymers used in hot embossing, thin polymer films have to be prepared and measured using a standard rotation rheometer setup. The values obtained at one temperature can be extrapolated to an entire temperature range, which is often sufficient for theoretical modeling of processes.

We obtained rheology measurements for 25 kg/mol and 75 kg/mol molecular weight PMMA (from Polysciences Ltd) at different temperatures. From these measurements, the relationship between zero shear viscosity and temperature could be determined using the Williams-Landel-Ferry (WLF) equation:

$$\log a_T = -\frac{C_1(T - T_0)}{C_2 + T - T_0} = \log \frac{\eta_0(T)}{\eta_0(T_0)} \tag{3}$$

The best fit to the experimental data was obtained with the following parameters T_0= 393.1 K, $C_1 = 12.21$, $C_2 = 70.1$ K (see Figure 2).[11] This dependence of the zero shear viscosities will be used for the experiments described in Section 4.3.

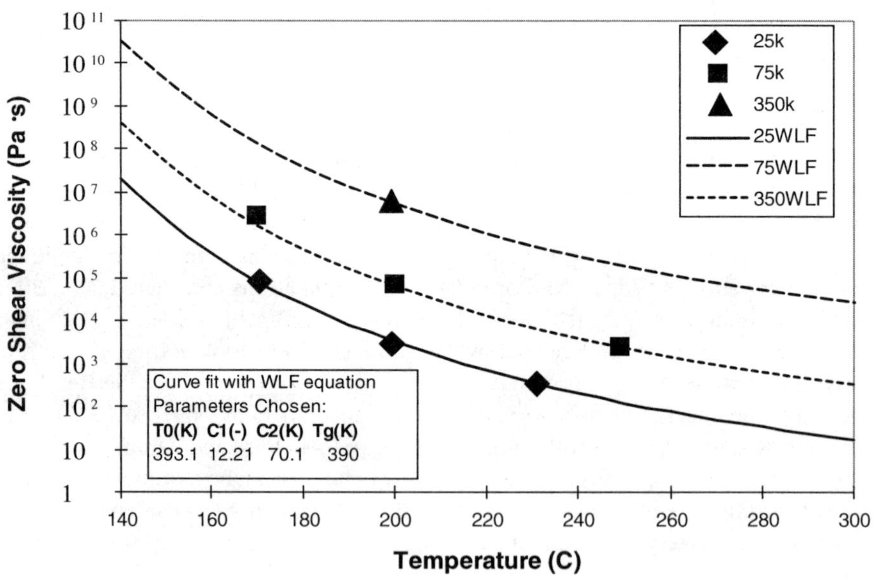

FIGURE 4.2. WLF plots of zero shear viscosities of PMMA for three molecular weights Mw (rheology measurements carried out at ETH Zürich by T. Schweizer).

4.2.2. Theory of Squeeze Flow

We now describe the squeeze flow theory in detail. In hot embossing the PMMA-coated substrate is heated to a temperature $T > T_g$ (glass transition temperature) and kept at this temperature during a time t_f. A stamp with a nanorelief is placed onto the substrate and a constant force is applied during the embossing process. The viscous polymer is forced to flow into the cavities of the mold so that it conforms exactly to the surface relief of the stamp. At the moment when the molten PMMA has filled the cavities (recessed stamp areas) of the mold and has assumed the shape of stamp's microrelief, it is cooled down to the demolding temperature $T_f < T_g$ and the embossing force removed. The embossed sample can then be released from the stamp relief.

A simple model for the viscous flow of a polymer melt is a sandwich of two discs which squeeze a polymer layer between them. This flow, referred to as *squeeze* or*squeezing flow,*[12] is found in the parallel discs rheometer setup.[13] Instead of the cylindrical symmetry normally cited[12], we take rectangular plates with long narrow line cavities and consider the two-dimensional case in the plane perpendicular to the direction of the lines (see Figure 3). The derivation for a circular setup, as used for the description of non-linear flow in Chapter 3, can be found in reference 12 and the resulting formula is given in Eq.2. Let us assume that the embossing stamp relief consists of N parallel, straight line stamps (protrusions) of length L, of widths s_i with distances w_i (recessed areas or cavities), which can be repeated several times (each iteration with $s_N = s_0$). The stamp s_i corresponds to the areas where the residual layer should be as thin as possible to ensure window opening for the subsequent pattern transfer process and the distances w_i correspond to the parts of the substrate protected by resist areas after removal of residual layer.

The main part of the derivation is taken from[14] where it was applied to the embossing of long straight micrometer sized polymer waveguides on glass substrates for integrated optics, and is now adapted to nanomolding. The process described in reference 14 is similar to hot embossing lithography because a small residual layer thickness was also required for the embossed waveguides to avoid an intensity loss of the guided optical mode due to leakage.

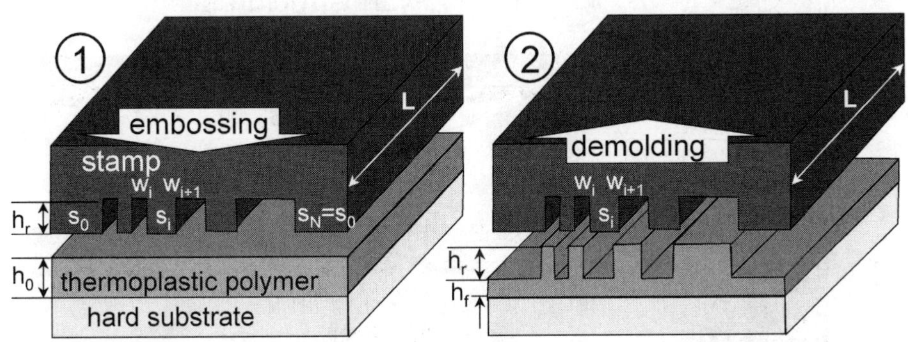

FIGURE 4.3. Geometrical definitions used for the description of the flow process. (1) before molding, (2) after demolding.

Before embossing, the polymer film has an initial thickness h_0 and the depth of the micro relief is h_r. For a fully inserted stamp, the film thickness under the single stamp protrusions with width s_i is $h_f \equiv h(t_f)$. Applying the continuity equation with the assumption that the polymer melt is incompressible (conservation of polymer volume), we get:

$$h_0 \sum_{i=1}^{N} (s_i + w_i) = h_f \sum_{i=1}^{N} (s_i + w_i) + h_r \sum_{i=1}^{N} w_i \tag{4}$$

The initial height, h_0, should be chosen so that it the final thickness, h_f, of the residual layer is that required for window opening. From Eq.4 we derive h_0 to be:

$$h_0 = h_f + \frac{h_r}{\Lambda} \sum_{i=1}^{N} w_i, \quad \Lambda \equiv \sum_{i=1}^{N} (s_i + w_i). \tag{5}$$

It is evident that as long as we assume that there is no local bending of the stamp, the same relation has to be valid for small stamp areas and for every protrusion of the stamp:

$$\frac{1}{2} s_{i-1}(h_0 - h_f) + \frac{1}{2} s_i (h_0 - h_f) = w_i (h_f + h_r - h_0) \tag{6}$$

This results in the condition for s_i:

$$s_i = \frac{2 w_i h_r}{h_0 - h_f} - 2 w_i - s_{i-1}. \tag{7}$$

It will be shown in this section that at identical pressure, small single stamp protrusions will sink faster than wide ones. The stamp geometry can therefore be optimized by reducing the dimensions of the protrusions. In Figure 4a, a stamp which contains an

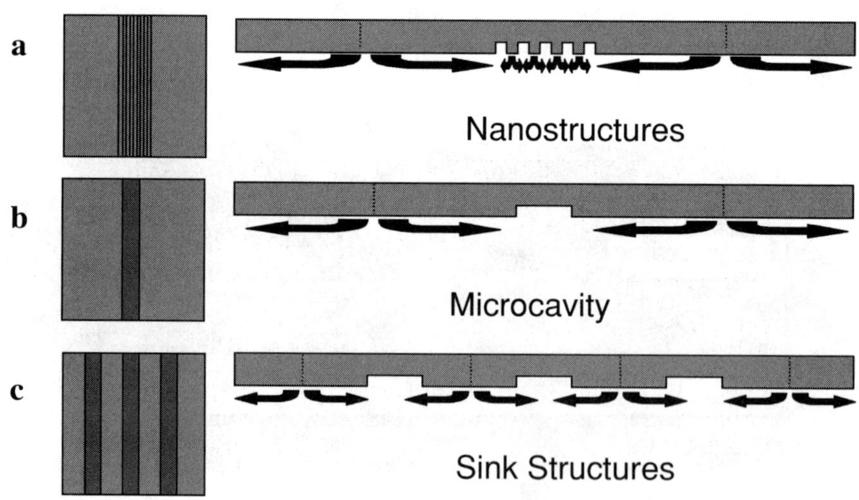

FIGURE 4.4. Molding of nanocavities in comparison to microstructures and possible introduction of sink microstructures for faster and more homogeneous molding. Left side: top view. Right side: side view.

array of small structures in the center is shown where the large single stamp protrusions surrounding the array dominate the sinking velocity (larges_i, see also Figure 9). The array in Figure 4a is equivalent to the microcavity in Figure 4b which has the same volume as the added volume of the cavity array. As will be demonstrated in Section 4.3, this simplification can be used for the calculation of the embossing times. The fill factor, i.e. ratio of area covered by cavities to the total stamp area, should be kept constant, both locally (at length scales corresponding to the cavity dimensions) and also across the wafer, i.e., for large stamp protrusions, to ensure better flow of the polymer and shorter embossing times. For this purpose, additional protrusions or cavities can be placed in intermediate areas not needed for the device function (Figure 4c). We would also like to draw the reader's attention to the fact that the different sinking rates of protrusions of different sizes means that the stamp, which is normally backed by an elastic silicone mattress, can bend locally. This will result in a residual layer height that is not uniform over the entire embossing area. This case is discussed with an experimental example in Section 4.3.5.

For the description of the flow process we consider the polymer being squeezed between two parallel plates as shown in Figure 5. Two stamp protrusions with widths s contribute symmetrically to the flow which fills the cavity between them. Half the polymer under each stamp protrusion flows towards the stamp cavity and half to the borders. Consequently only the volume between the two dotted lines contributes to the molding of the cavity. The origin is located in the center of the left stamp protrusion and of the gap $h(t)$ (depicted as ⊙). The melt should stick ideally at the plate surfaces and be able to move freely both into the cavity and away from the cavity beyond of the borders of the regarded volume. Consistent with the condition of Eq.7, the embossing time t_f is simply the time which is reached when the distance between the two plates is h_f. To solve the present *squeeze flow* problem we make the following assumptions:

- ideal viscous liquid
- incompressibility of the polymer melt

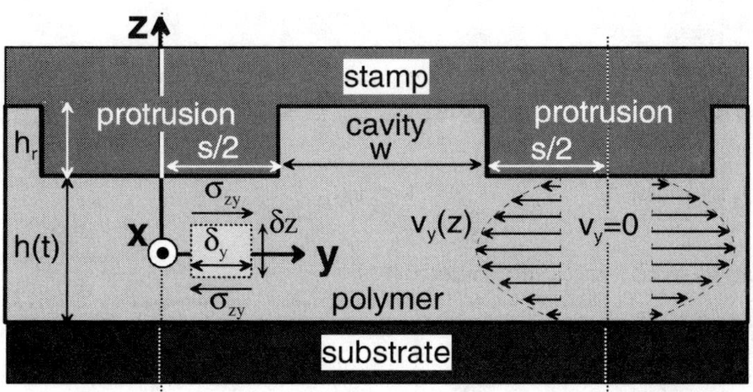

FIGURE 4.5. Model describing the flow process. Left side: Flow element. Right side: Velocity distribution in a Newtonian flow. Only the flow of polymer between the centers of the two stamp protrusions contribute to the filling of the cavity.

–$s \gg h(t)$
–no air trapped in the cavities
–no stick-slip effect at the interfaces ($v_y(t, z = \pm h/2) \equiv 0$)

In a hot embossing setup, the adhesion of the polymer film to the stamp is normally reduced by applying an anti-adhesion layer to the stamp surface, while the adhesion to the substrate should be as good as possible. As shown in Figure 2, at temperatures of more than 170°C, PMMA can be regarded as a purely viscous liquid. Therefore the difference in the interfacial energies of substrate (good adhesion of polymer) and the stamp (no sticking during demolding) are considered to be negligible above this temperature.

For the calculation of $h(t)$ we have to make an assumption about the (stationary) shear velocity $\dot{\gamma}_{zy}$. The viscosity function can be described very well by a truncated power law model (see Figure 1). For the calculation of a *squeeze flow,* the truncated power law model does not give a solution. We have either to consider the shear viscosity as constant ($\breve{\eta}(\dot{\gamma}_0) = \eta_0$, left side of the graph) or describe it by a power law model (right side of the graph). In the first case, the shear velocity $\dot{\gamma}$ is below the limiting shear velocity $\dot{\gamma}_0$, and we can consider the flow as a Newtonian flow with a parabolic velocity distribution between the flow borders. Above $\dot{\gamma}_0$, Newton's law no longer applies. For $z = 0$ the shear velocity $\dot{\gamma}_{zy}$ is zero. For many practical cases, as we will see later, the maximum shear velocity occurring during embossing is smaller than the limiting shear velocity $\dot{\gamma}_0$, therefore we conduct our calculations using $\breve{\eta}(\dot{\gamma}_0) = \eta_0$.

The two-dimensional description gives rise to $\partial/\partial x = 0$. Therefore in the x-direction, both plates should have the length $L \gg s$. For an infinitesimally small cuboid of the size $L \cdot \delta y \cdot 2\delta z$ at the axis $z = 0$ (see Figure 5), from the motion equation (for quasi-stationary flow):

$$-\partial_j p + \nabla_i \sigma_{ij} = 0 \tag{8}$$

the shear stress in y-direction can be derived:

$$\sigma_{zy} = \frac{dp(y)}{dy} z \tag{9}$$

As already mentioned, here the pressure$p$ is considered as independent from z. Hence Newton's friction law results in:

$$\frac{dv_y}{dz} = \dot{\gamma}_{zy} = \frac{1}{\eta_0}\sigma_{zy} = \frac{1}{\eta_0}\frac{dp}{dy} z \tag{10}$$

Integration over z gives the velocity field:

$$v_y(y, z, t) = \frac{1}{2\eta_0} \cdot \left\{ z^2 - \left(\frac{h(t)}{2}\right)^2 \right\} \cdot \frac{dp(y, t)}{dy} \tag{11}$$

with the boundary conditions $v_y = 0$ for $|z| = h(t)/2$.By integrating with boundaries $\pm h(t)/2$, we introduce the dependence on the time.

The volume flow $q(y, t)$ through the area $y = const$ results in an integration of Eq.11 over z and of the continuous equation:

$$q(y,t) = 2L \cdot \int_0^{h(t)/2} v_y(y,z,t)dz = \cdots = -\frac{L}{12\eta_0} \cdot h^3(t) \cdot \frac{dp(y,t)}{dy} = \\ = -Ly \cdot \frac{dh(t)}{dt} \tag{12}$$

Integrating over y results in the hydrostatic pressure as a function of the film thickness $h(t)$:

$$p(y,t) = \frac{6\eta_0}{h^3(t)} \cdot \frac{dh(t)}{dt} \cdot \left\{ y^2 - \left(\frac{s}{2}\right)^2 \right\} \tag{13}$$

with the boundary condition $p(s/2, t) = 0$.

Integration of Equation 13 over the width of a single stamp results in the exerted force:

$$F = L \cdot \int_{-s/2}^{s/2} p(y,t)dy = -L\eta_0 \cdot \left\{ \frac{s}{h(t)} \right\}^3 \tag{14}$$

Finally from Eq.14 we can derive the film thickness as a function of time using a constant force F:

$$\frac{1}{h^2(t)} = \frac{1}{h_0^2} + \frac{2F}{\eta_0 L s^3} \cdot t \tag{15}$$

Inserting the final thickness $h_f \equiv h(t_f)$in Eq.15 gives the embossing time:

$$t_f = \frac{\eta_0 L s^3}{2F} \left(\frac{1}{h_f^2} - \frac{1}{h_0^2} \right) \tag{16}$$

For many practical cases, where a constant pressure under each stamp protrusion $p = F/(s \cdot L)$ is assumed, this formula gives:

$$t_f = \frac{\eta_0 s^2}{2p} \left(\frac{1}{h_f^2} - \frac{1}{h_0^2} \right) \tag{17}$$

From this expression we can see how strongly s influences the embossing time ($t_f \sim s^3$using a constant force F). It is therefore worthwhile reducing the size of the stamp protrusions as much as possible. The strong dependence of the embossing time on the pressing area has the consequence that for a fully inserted stamp relief (full contact

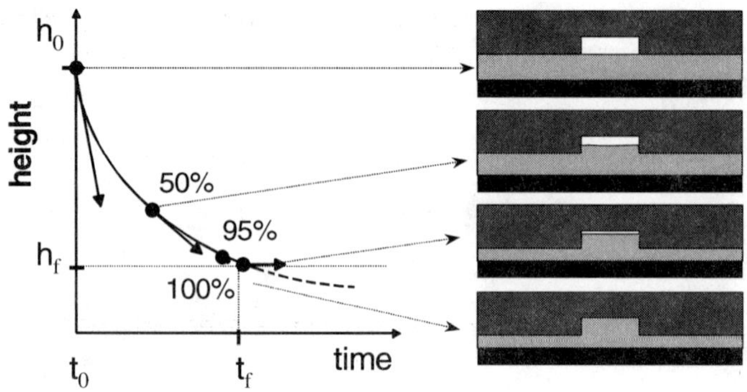

FIGURE 4.6. Squeeze flow of polymer into one cavity. Once the cavity is filled the stamp continues to sink but at a much slower rate.

over the total stamp area), the flow practically stops (schematically shown in Figure 6). For this case, s becomes large and the flow continues only towards the stamp borders. It is also evident that there is only weak influence of the embossing force ($t_f \sim 1/F$). At first sight there is a similar weak influence for η_0. However, the viscosity can be changed significantly by varying the temperature (see Figure 2). We can draw curves showing fill time versus stamp width, temperature, pressure, thickness, examples of which are given in Figure 7. Here, standard embossing parameters are used.

From Eqs.10,12 and 14 the following shear velocity relation can be derived:

$$\frac{dv_y}{dz} = \dot{\gamma}_{zy}(y, z) = \frac{12F}{\eta_0 L s^3} y \cdot z \tag{18}$$

Note that the shear velocity is stationary and does not depend on $h(t)$.

Finally, we will show that the assumption of a constant shear viscosity η_0 was justified:

For this purpose we calculate the curve $z(y)$ for which $|\dot{\gamma}_{zy}(y, z)| = \dot{\gamma}_0$ applies. The finite shear velocity, $\dot{\gamma}_0$, limits the valid range of Newton's law for the viscosity function. Solving Eq.17 with respect to z results in the desired curve:

$$|z(y)|_{\dot{\gamma}_{zy}=\dot{\gamma}_0} = \dot{\gamma}_0 \cdot \frac{\eta_0 L s^3}{12F} \cdot |y^{-1}| \tag{19}$$

The shear velocity $\dot{\gamma}$ is at a maximum at the position of the stamp corners occurring at the interfaces of stamp protrusion and substrate ($y = s/2$ at $z = \pm h/2$). If we take as typical embossing parameters $\dot{\gamma}_0 = 2\ \mathrm{s}^{-1}$, $\eta_0 = 70717$ Pa·s (75k PMMA at 200°C), $y_{max} = s/2 = 50$ µm, $p = 3.6 \cdot 10^6$ Nm^{-2} ($F = 5.4$ N for a single stamp protrusion), we obtain $z(y_{max} = 50\ \mu\mathrm{m}) = 650$ nm $= h/2$ as the maximum film thickness to avoid exceeding the limiting shear velocity $\dot{\gamma}_0$. This is well above the film thickness generally used for hot embossing lithography. Therefore for this stamp relief and with a film thickness $h(t) \leq 1.3$ µm, Newton's approach is a very good approximation. However already at 250°C ($\eta_0 = 2648$ Pa·s), we get $z(y_{max} = 50\ \mu\mathrm{m}) = 12$ nm, which is below the

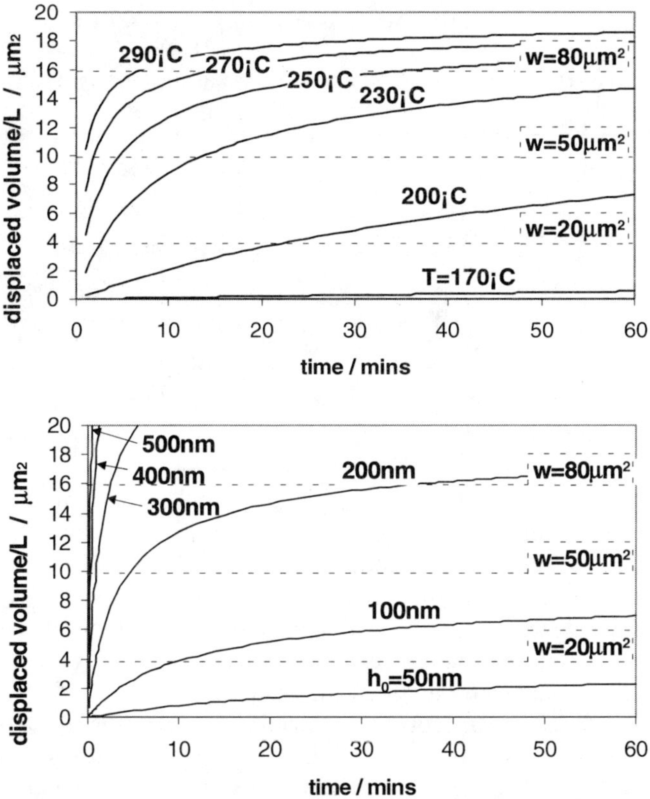

FIGURE 4.7. Graphs describing the dependence of the displaced polymer volume $(h_o\text{-}h(t))\times s$ on time for a line cavity with length L. For different cavity widths w, the time at which the cavity is filled can be determined when $(h_o\text{-}h_f)\times s = h_r\times w$ corresponding to the point at which hashed lines cross the curves. The graphs show the displaced volume as a function of temperature (above) and the initial film thickness (below). Similar graphs can be obtained for pressure and stamp width. Example parameter values taken are $s = 100\mu m$, $h_0 = 200$ nm, $h_r = 200$ nm, $p = 3.6\times 10^6$ Nm^{-2}, $h_0 = 2648$ Pa$\times$s.

film thickness generally used in HEL. For this latter case, the calculated shear velocity is higher than the limiting shear velocity $\dot{\gamma}_0$ and the viscosity follows a power law model (see Figure 1). Consequently the shear velocity would vary locally and using Eqs.16 or 17 only gives the calculation of an upper limit of embossing times. Care should be taken with this issue due to some uncertainty about the measured values $\dot{\gamma}_0$ and η_0 at higher temperatures. We can conclude that, although the typical experimental situation of hot embossing lithography is often well inside the Newtonian flow regime, for specific experimental setups non-linear conditions have to be considered. Note that for the polymers used here (Table 1 and Figure 2), effects such as shear thinning and swelling might occur at shear viscosities already in the range of $\dot{\gamma}_0$. Such effects are described in Chapter 3 in more detail.

The values given here are valid for the case of long straight line cavities, where the flow in y-direction is determined by the size of structure protrusions s and the negligible flow in x-direction by the stamp length L. When complex stamp geometries have to

TABLE 4.1. Experimental conditions required to fill a 20 mm wide cavity. Trials 1 to 4 used stamps with the Series 1 structures, and Trials 5 to 8 used stamps with Series 2 structures. Embossing forces of 15 and 30 kN correspond to embossing pressures, p, of $3.6 \cdot 10^6$ Nm^{-2} and $7.2 \cdot 10^6$ Nm^{-2}, respectively.

Trial No.	Mol. Wt. M_W (kg/mol)	PMMA Thickness (nm)	Force (kN)	Cavity Depth (nm)	Temp. (°C)	Zero Shear Viscosity (Pa·s)	Time of Embossing (mins)	Cooling Time (mins)
1	25	200	15	175	230	363	2	12.0
2	75	200	15	175	250	2648	2	12.5
3	350	200	15	175	290	-	2	14.0
4	350	200	15	175	270	-	10	14.5
5	25	100	15	70	200	3244	10	12.0
6	75	100	15	70	230	7915	10	13.5
7	75	385	15	70	200	70717	2	12.0
8	350	100	30	70	250	-	10	12.5

be embossed, such as checkerboards or 2D-gratings where the flow in the x-direction can not be neglected, the use of complex methods and detailed flow calculations is required. For this case, if stamp sizes s differ in x and y (even slightly because of the s^3-dependence in (15)), the smaller s will dominate the direction of the main flow and so the direction of the filling of the cavities. Once single cavities are filled, they will contribute to a larger stamp size to fill the yet unfilled cavities.

A final word should be said about local bending of stamps during embossing. Local differences in the stamp widths play a crucial role because, as stated in formula (16), smaller stamp widths will sink faster than large ones. For example, when a large structured area (typically a grating consisting of an array of line cavities) is surrounded by a large unstructured area, the stamp sinks down quickly in the center of the grating whereas due to the slow squeeze flow in the surrounding area the polymer thickness stays almost constant. In Section 4.3.5 we describe such a case where the stamp bending results in an inhomogeneous residual layer. Local bending of some nanometers will always occur due to small inhomogeneities and variations of stamp geometry, but in most cases will not affect the final quality of the embossed structures. According to the theory, only the geometry of the stamp is responsible for bending and therefore the most important way to reduce bending is by an optimization of the structure design and the arrangement of cavities and protrusions.

Another source of bending may be the effect of different flow regimes at different places of the same stamp. Exceeding the maximum shear velocity in the vicinity of cavity corners may result in a lowering of viscosity and may enhance the tendency of the corners to sink in faster than the center of stamp protrusions (see Figure 5). A thick polymer may thin down quickly at these places and may soon reach a layer thickness where Newtonian flow conditions are reached and further bending slows down. The lower as expected "effective" stamp widths derived in Section 4.3.3 are an indication for this behavior which may lead to bending. If embossing continues for some time, the stamp could even bend back to its original form. Bending may cause distortions in the final embossed structure, particularly for densely packed high aspect ratio structures, but until now this has not been proven by experiment.

For completion we now give the expression similar to Eq.15, but derived by taking two parallel discs of radius R and distance $h(t)$.

$$\frac{1}{h^2(t)} = \frac{1}{h_0^2} + \frac{4F}{3\pi\eta_0 R^4} \cdot t \tag{20}$$

4.3. HOT EMBOSSING IN PRACTICE

4.3.1. Embossing of Large Nanostructured Arrays

The hot embossing process is a low cost, fast method for the parallel replication of structures at the micro- and nanoscale. With a single master or stamp, identical structures can be produced as required over large surfaces.[15] Applications pursued by our group include large area nanoscale electrode structures,[16] sieves for biological applications and calibration structures for Scanning Probe Microscopes.[7]In addition, diverse structures (dots, lines and meanders) with lateral sizes down to 25 nm and aspect ratios of up to 3 have been replicated.[17]

The fabrication of interdigitated electrode arrays (IDEAs) is shown in Figure 8.[16] The meander like cavities of the stamp, here with a period of 400 nm, are replicated into a thin layer of polymer. The polymer lines serve as resist and by using RIE and lift-off an array of thin electrode lines is obtained. For such devices, surrounding connection lines and pads must be included so that a mixture of both micro- and nanostructures is required in the stamp.

It has already been reported that material transport and stamp geometries should be considered when embossing polymer materials.[18-21] With a better understanding of the

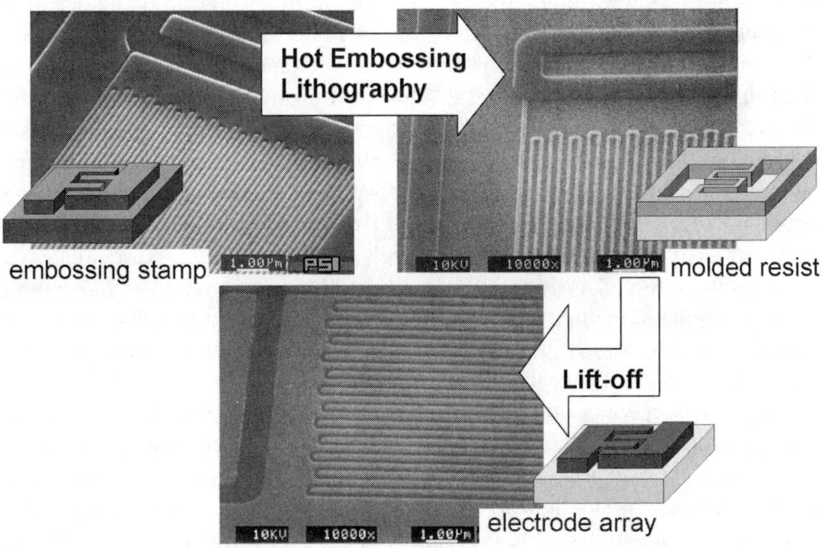

FIGURE 4.8. Hot embossing lithography steps for the fabrication of interdigitated electrode structures (IDEAs).

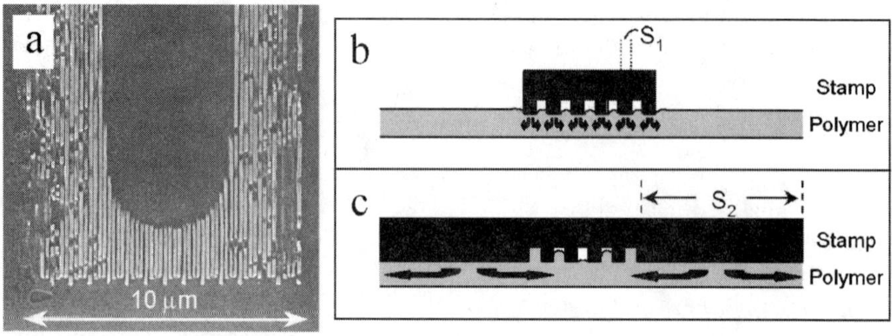

FIGURE 4.9. (a) SEM image of PMMA nanostructure intended for the fabrication of interdigitated electrode arrays (IDEAs). Typical defects such as incomplete molding and abrasion which occur during hot embossing can be seen. (b) and (c) Polymer flow during hot embossing for different stamp geometries.

polymer behavior, both the embossing conditions and stamp design can be optimized, and a good replication fidelity for the fabrication of micro- and nanodevices can be ensured. In Figure 9a, a case can be seen where the molding of the meander like structures was incomplete and resulted in an unstructured area in the center. The polymer appears to have flown from the array boundaries and filled the nanostructures as it moved towards the center. Similar effects have also been reported in.[18,19] Note also that there are local discontinuities in the individual lines which are a result of incomplete filling at the nanoscale and breakage during demolding. The observed flow of the polymer can be explained with the schematic representations given in Figure 9b and c. The thin polymer film undergoes a squeeze flow when it is compressed between the silicon stamp and substrate. In the case of large area periodic structures such as gratings, the specific area which contributes to the filling of a single cavity is limited by the structure density or the distances between the cavities given by s_1 in Figure 9. The distance s_1 corresponds to half the size of the two protrusions with width s adjacent to the cavity in Figure 5 as described in the theory in Section 4.2.2. The local flow of the polymer is indicated by the arrows in Figure 9 and results in simultaneous filling of the cavities. For the IDEA shown in Figure 9, the nanostructured region is surrounded by a large unstructured area and the majority of material flows from the boundaries. This is schematically shown in Figure 9c and results in a sequential filling of the cavities. For this case, the squeeze flow in the intercavity regions can be neglected and the effective stamp area is then the area between the edge of the structured region and the stamp border, indicated by the stamp width s_2 and results in the polymer flow given by the arrows. The time to fill such a nanostructure array should be the same for an equivalent microcavity with the same stamp area and cavity volume.

We have carried out a series of embossing trials in thin PMMA films in order to elucidate the embossing process and provide information for the optimization of HEL. We have looked in detail at the filling of micro- rather than nanoscale cavities in order to facilitate the observation and interpretation of the polymer behavior. In Sections 4.3.3 and 4.3.6 we describe two different observed filling modes of the microcavities and in Section 4.3.4 we compare the experimental embossing times with simple theory. Successful demolding is dependent on the stamp quality and in Section 4.3.8 we describe how non-vertical side walls can result in a plastic deformation of the polymer at temperatures below the glass transition temperature, T_g.

4.3.2. Experimental Setup

We used thin films of PMMA as the resist material with different molecular weights, M_w, of 25, 75 and 350 kg/mol providing a range of polymer viscosities. The thin films were produced by dissolving PMMA powder (from Polysciences Ltd) in ethyllactate and spinning these solutions onto silicon wafer substrates. Following spincoating, the films were baked on a hotplate for 1 minute at 170°C. The PMMA films were produced with various thicknesses ranging from 50 nm to 3 μm. The glass transition temperatures for PMMA given by the manufacturer are 105°C, 98°C and 95°C for molecular weights of 25, 75 and 350 kg/mol, respectively.

The masters were produced from silicon wafers and the structures were manufactured using either an electron beam or a laser to write the required pattern into a resist, depending on the structure sizes needed. In both cases, RIE was used to transfer the resist pattern into silicon dioxide providing the necessary hard embossing stamp. The stamps themselves have surface dimensions of 15 by 15 mm^2 and the area containing structures has dimensions of a few mm^2.

An anti-adhesive layer applied to the surface of the stamp lowers its surface energy, for example using fluorinated Teflon®-like polymer films[1] or fluorinated self-assembled monolayers.[23] Another possibility is the use of mold release agents.[24] This prevents adhesion of the embossed PMMA film to the stamp surface, so enabling several replications to be made with a single stamp. With an anti-adhesive layer, we have embossed over 50 replica structures and have not yet observed adhesion of the PMMA to the stamp.

Prior to embossing, the substrate was heated to the required temperature above T_g. The silicon stamp was then pressed against the polymer using embossing forces ranging from 10 to 40 kN. Following the specified embossing time, the substrate was cooled to 70°C and the polymer demolded. The application of a vacuum during embossing[14,25] was not found to be necessary.

4.3.3. Filling of Stamp Mold Cavities

The first stamp used in this section (referred to as Series 1 structures) provided long microcavities with different widths of 20 μm, 10 μm and 6 μm and periods of 80, 40 and 24 μm, respectively. The cavity depth was 175 nm. A region of 10 by 10 square cavities with dimensions 20 μm by 20 μm was also created. The squares had rounded corners due to the resolution limit of the laser writer used.

The first step was to determine the temperature required for the PMMA to completely fill the stamp cavities using a short embossing time of 2 minutes. Ideally, we would like reduce this time to a few seconds but this does not make any sense when the cooling step also requires several minutes. The results for a press force of 15 kN are given in Table 1, Trials 1 to 4. The 350 kg/mol polymer required an embossing temperature of 290°C to reach the 2 minute embossing times. However, in order to avoid any polymer degradation at high temperatures, an embossing was also carried out at a lower

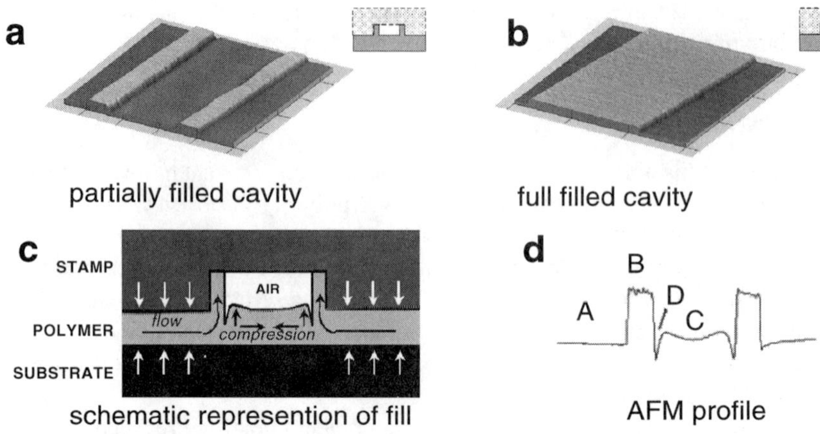

FIGURE 4.10. Filling of cavity: (a,b) AFM micrographs of replicas of a 20 μm line stamp cavity in a 200 nm polymer thin film. (c) schematic representation showing filling of a stamp cavity. (d) AFM height profile of a partially filled 20 μm line cavity.

temperature of 270°C. Here a longer embossing time of 10 minutes was necessary to give a complete fill.

The way in which the mold is filled can be observed by embossing at lower temperatures, and choosing a short enough embossing time so that the polymer is hardened and demolded when the cavity is only partially filled. The filling was observed as flow of the polymer from the borders into the stamp cavity. The start of the filling process was investigated in the 20 μm and 10 μm line cavities. The distance that the PMMA moves into the stamp cavity appears to be independent of the cavity width and periodicity. A constant volume flow per cavity border would imply that a slight local bending of the stamp occurs. A height profile of a partially filled 20 μm groove measured by AFM and a schematic diagram illustrating the flow behavior of the polymer during embossing are given in Figure 10. The PMMA which has moved into the cavity from the cavity borders is clearly seen as raised regions, B. It seems likely that the polymer expands into the cavity as if it has passed through a nozzle and climbs up the cavity walls. The polymer inside the cavity region prior to embossing does not mix with the polymer flowing in from the borders and instead is pushed by the moving polymer. This causes a compression of the central region, C, so that it has a convex profile. The central region is separated from the polymer flowing into the cavity by a dip at D. It is well known in polymer injection molding that internal friction or shearing can cause the polymer to heat up drastically. It therefore seems likely that the polymer which is squeezed into the cavity undergoes a greater shear and therefore has a lower viscosity resulting in the flow illustrated in the schematic of Figure 10. This would be in accordance with the calculations described at the end of Section 4.2.2, where the highest shearing velocities would be expected at the borders of a cavity. Apart from effects due to local differences in viscosity, electrostatic effects have been found to play a crucial role during the filling of cavities (see Section 4.3.6). Note that the experimental polymer profile is very different to that shown schematically in Figure 6.

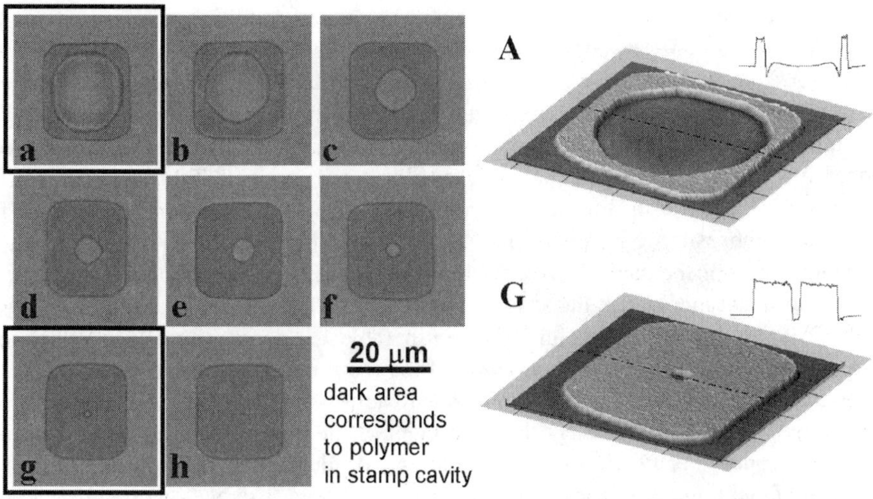

FIGURE 4.11. Left side: Optical images of the filling of a 20 μm square cavity. Right side: Three dimensional AFM images of different stages (A and G corresponding to a and g) of fill of 20 μm square cavities. The structures are 175 nm high corresponding to the cavity depth. For each image a height profile taken across the center of the structure, indicated by the dashed line, is given.

It is possible to observe how a 20 μm square cavity is filled in one embossing as small local differences in the pressure result in small differences in the time to fill each cavity in the array. The step-by-step filling of the square cavity is given in Figure 11. The filling starts from the edge as seen in Figure 11a and b. Here, the PMMA flows further at the corners due to the combination of pressure from two edges and this results in the formation a square shaped hole at 45° to the original cavity shape (Figure 11c and d). As the square cavity continues to fill, the shape of the hole is no longer dictated by the original cavity shape. The surface tension is now more important and the hole becomes circular to minimize the surface energy (Figure 11e and f). In the final stages of filling, the circular hole contracts (Figure 11g and h) until it eventually disappears. It is likely that the air can not escape here and is compressed to a very small fraction of its original volume. The polymer structures during filling are well defined as seen in the AFM images and profiles given in the right side of Figure 11. Each image corresponds to a different stage of fill of the 20 μm square cavities and it can be seen that at each stage, the side walls of the moving boundary are of the same quality as the embossed walls. In addition, there is good definition of features much smaller than the size of the cavities; the raised areas in Figure 11A are 3 μm wide and the hole in Figure 11G has a diameter of 2 μm.

A second set of structures (Series 2) has been used for further investigations. The stamps consist of two identical arrays of rectangular cavities with dimensions 20 μm by 80 μm, a depth of 70 nm and a distance between rectangles of 130 μm. The two identical arrays are separated by 1650 μm and between the two arrays is a 500 by 500 μm area containing nanostructure arrays (10's μm wide) of lines, grids and dots with periods down to 120 nm.[26] For an embossing force of 15 kN and embossing time of 2 minutes, the optimum embossing temperatures required for complete filling were the same as for the

Series 1 structures (Table 1, Trials 1 to 4). The embossing conditions (time, temperature and pressure) required for replication on the microscale also result in a good replication at the nanoscale. This agrees with the initial assertion that the time to fill a nanostructure array with a large unstructured surrounding area is the same for a microcavity with the same surrounding area and cavity volume. Additional results obtained using lower temperatures, increased pressure and thicker films were obtained (Table 1, Trials 5 to 8). As a general rule, a drop in temperature of 20°C from the optimum temperature for a 2 minute embossing required an increase in embossing time to 10 minutes to allow complete filling of the cavities. Decreasing the thickness of the polymer was found to give a significant increase in the required embossing temperature. An example is given for the 75 kg/mol molecular weight polymer in Table 1; a decrease in PMMA thickness from 385 nm to 100 nm gave an increase in the required embossing temperature from 200 to 250°C. The increase is likely to be due to adhesion of the PMMA at the substrate and stamp surface which hinders the movement of the polymer chains. This retarding effect on the polymer flowing into the cavities will increase as the polymer thickness decreases. Doubling the embossing force from 15 kN to 30 kN (a value close to the maximum force available with the current press setup), gave a reduction by 20°C in the required embossing temperature.

4.3.4. Comparing Model and Reality

Following on from the work of reference 14 we attempted to use the squeeze flow theory to model the filling behavior of the stamps described in Sections 4.3.2 and 4.3.3. However in comparison to reference 14 our stamp geometries are more complex with a variety of cavity lengths ranging from 10's μm up to several mm, and we used smaller film thicknesses and cavity depths of the order of 100 nm. In the experimental setup, the embossing forces of 15 and 30 kN correspond to embossing pressures, p, of $3.6 \cdot 10^6$ Nm^{-2} and $7.2 \cdot 10^6$ Nm^{-2}, respectively. The relationship between zero shear viscosity and temperature was determined using equation (3) for the 25 kg/mol and 75 kg/mol molecular weight PMMA. The zero shear viscosities of the PMMA are included in Table 1 and it can be seen that with the minimum embossing time of 2 minutes and embossing force of 15 kN (Trials 1 and 2), complete fill of the cavity occurs when the PMMA has a viscosity in the range 300 to 3000 Pa·s.

An effective stamp width, s, can then be calculated from the theory using the experimental embossing times. The exact time to fill the cavities was not known so, with the inclusion of data from incomplete embossings, it was only possible to determine the range of values in which s lies. For the stamp used with the Series 1 structures in Section 4.3.3, s was found to lie in the range 280 to 350 μm and for the Series 2 structures, in the range 70 to 175 μm. In order to pinpoint the exact fill time, many more embossing trials are required or, even better, real time observations would allow direct identification of the time at which the cavities fill. For simplicity, the flow occurring during the cooling period (given in Table 1) has not been included and should also be taken account in future calculations for improved accuracy. The stamp widths obtained are of the order of 100 μm but in reality the total area of compression, i.e. the stamp size, is much greater and the distances over which polymer flow contributes to the filling of a cavity can be much larger. It therefore appears that the area of polymer flow has been restricted. This may reflect the distance between the stamp cavities, also of the order of 100 μm, or may be a result of local inhomogeneity in embossing pressure, i.e., local bending of the stamp,[30] which will limit the area of polymer flow.

The theory described in this section does not take into account the more complex flow which would occur for a more complicated stamp geometry. In addition, effects such as capillary action, surface tension, the elastic nature of the polymer, local shear of the polymer, distortion of the stamp and complex filling are not considered. Nevertheless, this simple theory has served to highlight the important factors which determine the filling of a stamp cavity and, with a knowledge of the polymer viscosity at given temperatures, can be used to predict the required embossing times from the calculated effective stamp width.

4.3.5. Bending of Stamps

In hot embossing lithography, the stamp is often considered as a hard tool which is inflexible over millimeter distances. However, this is only true for special cases, e.g. when density and sizes of stamp protrusions are homogeneous over the whole stamp surface. Significant bending may occur when dust particles of some μm diameter are present between the stamp and substrate. Because of the high Young's modulus of the stamp, the equilibration mechanism (often a thin layer of silicone behind the stamp) is not able to achieve a conformal contact of the stamp with the polymer and consequently structures are not embossed in the vicinity of these particles. Whereas contamination can be avoided largely under clean room conditions, local bending of some nanometers occurring due to small local variations of the stamp geometry has to be considered as the general case during hot embossing of thin films.

For many applications, when pattern transfer has to be achieved after the embossing, it is of high interest to know the final residual thickness h_f of the polymer layer (see Figure 3). Furthermore it is important to know how homogeneous the thickness is over the embossed area. Here stamp bending must be kept to a minimum in critical areas. Otherwise, parts of the structure will be lost during pattern transfer. This is especially critical when the thin resist is used as a masking layer in a subsequent reactive ion etching step.

For the ideal case where no significant bending would occur, e.g. by using a thick rigid stamp or small grating areas, the overall sinking of the stamp would be determined by the largest stamp protrusion, in this case the surrounding area (see also Figure 9). In the ideal case of a grating covering the whole stamp surface, the molding process would be determined by the grating protrusions. The residual layer thickness for these ideal cases can be simply calculated applying the continuity equation (4). For example, using a stamp covered with a regular grating structure with a fill factor of 50% ($s = w$) and a structure depth $h_r = 160$ nm gives a vertical movement $\Delta h = 80$ nm. An initial PMMA thickness $h_0 = 320$ nm would therefore result in a final residual thickness $h_f = 240$ nm and, with the fully molded structures on the top, in an overall thickness $h_f + h_r = 400$ nm.

For a typical case where the grating is surrounded by a large unstructured area, stamp bending results in an inhomogeneous residual layer. Figure 12 shows such a case, with a stamp area of some square mm surrounded by a non-structured area. The final thickness values achieved after embossing, in the center of the grating and at the border, are almost similar to those calculated by the simple theory (the values correspond to those taken in

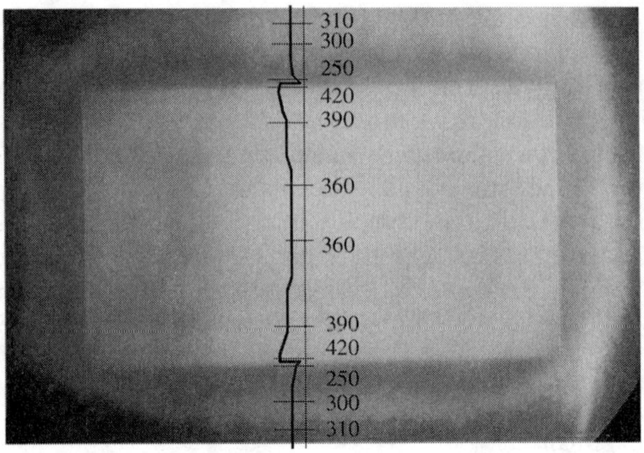

FIGURE 4.12. Micrograph of a grating (area 5x3mm2) embossed in resist: lighter area corresponds to the grating, higher than the surroundings. Profilometric height measurement of a cross-section schematically represented in nm indicate areas of large resist thickness variations near and within the grating.

the example above). Note that the profilometer does not resolve the grating structures. Slight differences can be attributed to variations in fill factor and structure depth.

The profile of the transition areas between the center and border regions are due to bending. For typical stamps made from 500 μm thick silicon wafers (100 mm diameter), transition distances of several 100gμm can be measured. In Figure 13, the effect of

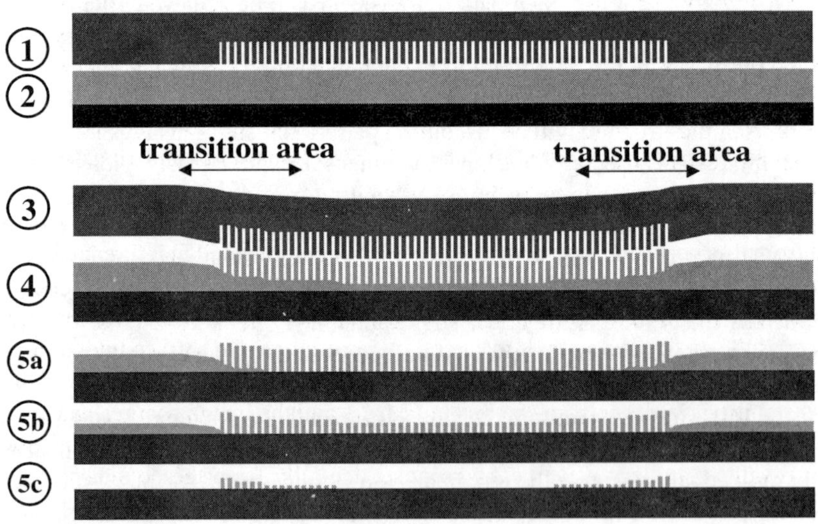

FIGURE 4.13. Schematic view of the molding of a large array of nanocavities surrounded by large stamp structures. Stamp (1) and resist (2) before embossing. (3) Local bending of stamp during molding. (4) Resist after embossing with inhomogeneous residual layer. (5a) to (5c) show different states of resist thinning. (5b) is the state where complete pattern transfer is possible.

transition areas on the quality of pattern transfer is schematically shown. Care must be taken with the amount of resist thinned down during the removal of the residual layer in order to ensure complete transfer of the grating (section 5b). Under- and over-thinning will result in a loss of the grating structure (section 5a and 5c). For large differences in height in the transition region, i.e. large bending, complete transfer of the structure is impossible. For every new design, particularly for unevenly distributed cavities, such transition areas between structured and unstructured areas have to be carefully considered. A few additional protrusions or cavities can be added in order to keep the transition area out of the grating structure. These should be placed in the vicinity of nanostructures, with intercavity distances smaller than typical transition distances. The additional cavities also ensure better flow of the polymer and shorter embossing times (see also Figure 4).

4.3.6. Squeeze Flow and Self Assembly

The topic of self-assembly of molten polymers has been addressed by other groups. They have shown how the self assembled polymer structures can be used for nanoscale lithography and have already developed theories for the structure generation.[27,28] Our process differs from the Lithographically Induced Self Assembly (LISA[27]) process, as we have a squeeze flow of polymer in addition to the formation of mounds.[26,29] In this section we describe how self-assembly occurs during a squeeze flow process.

During hot embossing self-assembled polymer mounds can be created the form of which depends on the local nucleation and compressive effects. We have found that it is possible to tailor the shape and size of the polymer pillars by defining the stamp size and shape. In addition, as the cavity size is decreased, it is possible to isolate single polymer mounds. Different polymer patterns resulting from different stamp geometries are given in Figure 14. In Figure 14a, it can be seen that the larger scale grid cavities have some influence on the polymer lines which tend to align themselves with the structure boundaries. In Figures 14b, c and d, the smaller grids and square cavities result in the creation of some interesting mound shapes: line sections, dots and rings have been isolated.

The way in which these regular structures may form in our squeeze flow setup is adapted from the theory in Chapter 4, and described in more detail in reference 29. There are two main effects that should be considered. Firstly, the polymer melt is squeezed into the cavity, flows up the walls of the cavity and compresses the polymer film in the center of the cavity. As a result of lateral compressions, the surface undulations are intensified, causing a buckling of the polymer film. Secondly, the electrohydrodynamic (EHD) instability, resulting from charges in PMMA and the image charges on the mask, favors the formation of periodic capillary waves in the PMMA melt with boundaries at the cavity walls. Capillary wave peaks grow sequentially from the boundary touching the cavity top one at a time. In Figures 14b-d the cavity size coincides with the capillary wave period and only one mound can form. If the embossing process is continued, the polymer melt flows into the cavity until it is fully molded. The squeezing of the polymer may help to form the mounds but does not seem to be prerequisite. In any case, no sign has been found that the self-assembly process affects the final quality of the molded

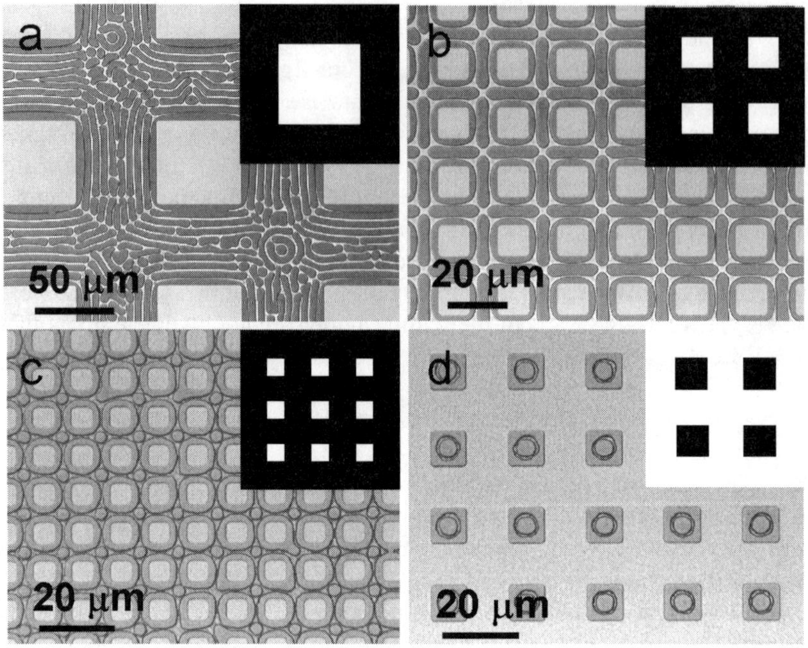

FIGURE 4.14. Effect of stamp geometry on mound structures. All structures are 75 nm high and inserts show original stamp designs where black corresponds to cavities. (a) Line pattern within cavity with dimensions larger than characteristic wavelength. (b,c,d) Isolated mounds forming in cavities with dimensions comparable to characteristic wavelength.

structure as long as embossing times are chosen to be long enough to achieve a complete fill of the master cavities. Indeed it may improve the filling of nanocavities.

4.3.7. Air Inclusions and Viscous Fingering

An important question has to be asked: Where does the air in the cavities go during hot embossing? Figure 11 indicates a possible solution. Due to the high pressure present in the molding process, the air is compressed in the center of the cavity to a fraction of its initial volume. However, in the final molded structure no indication of a remaining air bubble was found. Therefore it appears that a large part of the compressed air diffuses into the polymer. During demolding the expansion of this air might help to separate the stamp from the molded structures. However, it can also cause deformation of the final structure when the structure is demolded before the polymer reaches its solid state on cooling. For this case, viscous fingering patterns result from explosive expansion of air.[29] It is not clear at his stage whether the expansion is solely a result of trapped air or whether there is also trapped moisture present. In addition, overheating and decomposition of polymer due to the high internal or interfacial friction during the lateral shear flow may result in an outgassing of the polymer. Future experiments should determine the exact cause.

4.3.8. *Phenomena Observed During Demolding*

Up to now we have only considered the viscous flow of the liquid polymer during embossing. However, deformation can also occur while the polymer is in its solid state and, as we will describe in this section, we have observed that distorsions can occur due to forces present during the demolding.

During demolding, the mold is separated from the embossed polymer structure by a vertical movement of the stamp. Distortion or damaging of the molded structure during this movement can occur as a result of different effects such as adhesion at the surface, friction due to surface roughness and trapping of the polymer due to negative slopes of cavity sidewalls. If the side walls of the stamp cavities are not completely flat, then the polymer will locally interlock with the mold. As a result the polymer structures or parts of the wall profile are either ripped away or deformed during demolding and for small width, high aspect ratio structures the shear results in a stretching of the total structure when the stamp is pulled away.

The stamps employed for this section were fabricated using a specially developed highly anisotropic and highly selective RIE process.[36,37] They consisted of an array of grooves with linewidth varying continuously from 2 μm down to 100 nm and with a depth of 1 μm (see Figure 15a and b). The stamp was used to emboss a 2 μm thick PMMA film. In general, the stamp grooves have flat side walls and it was possible to replicate fairly high aspect ratio structures, up to 6, in the polymer.

Small changes in processing of the stamp produced non-flat side walls and resulted in stretching of the polymer structures. An example of this phenomena is given in Figure 15c. Images of both the stamp and demolded polymer are shown. There is a burr located on the side walls of the stamp grooves (indicated by arrows) which interlocks with the polymer during embossing. This has little effect on the larger polymer structures greater

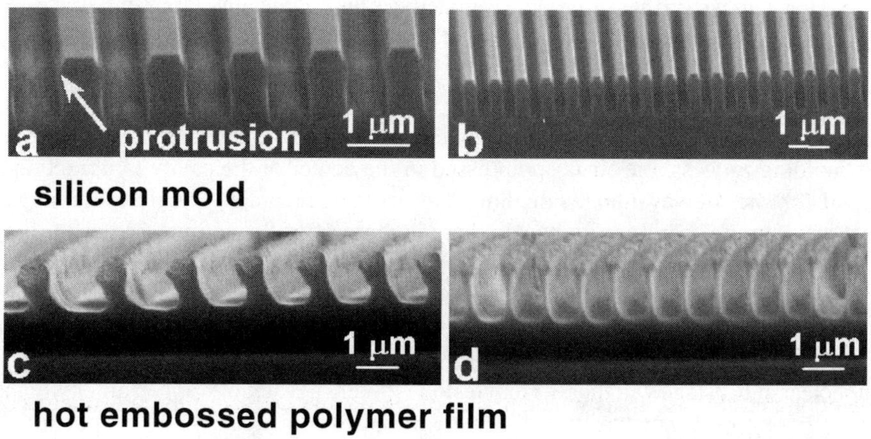

FIGURE 4.15. Interlocking of polymer structures due to non-flat sidewalls leads to deformation during demolding. SEM images of a stamp with 1 μm deep line cavities and a width of (a) 600 nm and (b) 250nm. The associated embossed polymer structures have been (c) distorted or (d) stretched during demolding.

than 1 μm wide but as the linewidth of the stamp decreases, the polymer undergoes a plastic deformation as the stamp is removed. The stretching is significant at linewidths below 300 nm as seen in Figure 15d. The stretched polymer is 1.5 μm high which is 50% higher than the original cavity depth giving an increase in aspect ratio to 6 compared with 4 in the original stamp. The polymer thins in the middle during stretching, decreasing in width from 250 nm to 125 nm. This thinning effect allows the polymer to eventually pull free from the mold. When the linewidth is smaller than 250 nm, the polymer structures are too weak and break on demolding.

The deformation or stretching of the polymer should generally be avoided by ensuring a good quality of the cavity side walls with no undercuts and very smooth surfaces. A reduction in the friction during demolding is also possible by adding a demolding agent to the polymer or by use of an improved anti-adhesive layer on the stamp. However, the stretched structures may be of interest for two reasons: firstly to enhance the aspect ratio of nanostructures and secondly to provide a novel form of molded part needed, for example, for improved lift-off or to provide narrow gaps for electronic applications. Therefore, control of the plastic deformation of the polymer can provide a technique for the production of additional functional structures at the micro- and nanoscale.

4.4. LOOKING AHEAD

In this final section we address the areas in hot embossing that require further development and give suggestions for future work. We have divided this into two main themes that can be represented by the following questions:

- How far have we come in our understanding of polymer rheology at the nanoscale and what do we need to investigate further?
- How can we further improve our understanding of the hot embossing process and use this knowledge for the design of new processes?

Answering both questions will help us to optimize the hot embossing process and solve quality issues when process problems arise.

4.4.1. Rheology at the Nanoscale

Thin films used in HEL have a thickness of the order of 100 nm. Can the properties of these films still be considered as macroscopic, i.e. comparable to properties at the mm-scale? Many of the phenomena observed and described in this contribution can be related to macroscopic effects but there are phenomena which are specifically related to micro- or even nanoscopic interactions. In this section we discuss the ultimate limits of replication and discuss areas where an improved understanding of the sub-microscopic processes are required.

4.4.1.1. Limits of Replication. Nanoscale polymer structures can be produced by irradiation induced chain scission or be molded with topographic details down to a few nanometers. Holes in polymer films down to 6 nm have been fabricated[38] by hot embossing. However, the ultimate limits of nanorheology for structuring purposes are not yet known.

It seems likely that the ultimate resolution of a polymer is given by the extension or the space covered by a single linear polymer chain or a column of aligned polymer chains. For a hydrocarbon chain, the minimum bending radius is determined by the distance (0.154 nm) and the angle (109.5°) between adjacent carbon backbone atoms, and lies below 0.2 nm. However, amorphous polymers are normally present as entangled entities. The size of the macromolecule present as a statistical coil aggregate is then defined by the radius of gyration, which can be calculated from the volume of a sphere containing the coil, e.g. for a PMMA macromolecule with a molecular weight of 25000 (M_W in g/mol) the mass is $4.2 \cdot 10^{-20}$ g and taking the density given for PMMA as 1.19 g/cm^3, the radius of gyration is calculated to be 2 nm. The chain contains 250 MMA monomer elements ($C_5H_8O_2$) with a weight of 100 g/mol each and has a total length of about 80 nm. When the details of a molded structure are smaller than that given by the radius of gyration, the structures have to be generated by overcoming the mechanical force which corresponds to the entanglement and Van der Waals forces present between the chains. For many practical applications, the already achieved resolutions down to 6 nm may be sufficient and it seems that down to these resolutions, the molded feature size is determined by the minimum mold cavity size rather than the chain conformation of the polymer.

4.4.1.2. Areas for Improved Understanding. Some of the important areas for further understanding of micro- and nanorheology include:

- *Nanoscale Interactions:* We described in Section 4.3.6 the formation of periodic patterns with a characteristic wavelength which is dependent on the balance of the Van der Waals interaction, the electrostatic force exerted on the film and the surface tension contribution (Laplace pressure).[31-34]The effect of these interactions at the micro- and nanoscopic level is presently unknown. Roughness or irregularities on the stamp surface due to processing or contamination, and structure boundaries and protrusions may serve as nucleation sites and facilitate forming of self-assembled structures and filling of cavities.
- *Rheology at the Nanoscale:* While standard rheological equipment can be used to analyze bulk polymers, our knowledge about the rheological behavior of polymer thin films is only modest. High pressure embossings of thin polymer layers may result in non-Newtonian flow situations with a non-uniform shear velocity distribution below the microscale. Differences of interfacial energy and stick-slip movement may also lead to flow and filling behaviors different from macroscopic observations. Smaller structures and film thickness down to a few nm are investigated by theories which take into account single layers of polymer chains. For example, single chain movement of compressed lubricants can be simulated and measured in-situ with sophisticated X-ray diffraction methods.[35] A flow theory specifically needs to be developed for hot embossing, i.e. for polymer films with thicknesses in the 100 nm range and nanostructured molds which result in complex flow fields. A step in the right direction would be a model which finds a middle road between the macroscopic flow models (the surface and thin film

effects would need to be taken into account) and the single chain model (the thickness of interest would need to be increased from several molecules thick to several 10's of molecules).

- *Nano-engineering of Materials:* Molding of polymers with layers of different molecular weights, using internal release agents or adhesion promoters provide a method for nano-engineering the thermoplastic polymers used in the HEL process. These areas have already been addressed in the molding of microstructures fabricated by the well-known LIGA process which, by tailoring the polymer, achieves structure heights of up to several 100 μm and aspect ratios of up to 20.[5]
- *Additional Nanoscale Features:* Many of the additional features described in this chapter, for example polymer deformation, dewetting and self-assembly, are unwanted but can be totally avoided or overcome if the process parameters (e.g. embossing times, demolding temperatures, pressure homogeneity) are optimized. Nevertheless a better understanding of these phenomena will provide an invaluable insight for both researchers and process engineers, particularly when the HEL process is driven to its limits.

4.4.2. *Improving the Hot Embossing Process*

For the observation of rheological phenomena, improved setups for both processing and analysis have to be built. Hot embossing equipment with uniform and precise temperature and pressure control is needed. Reduced heat capacities will enable faster embossings by reducing heating and cooling times and by only locally heating a stamp or a substrate, rheological processes would be restricted to the areas which are covered by nanostructures. Stamps with defined structure sizes (micro- and nanoscale) and a uniform fill factor (e.g. a stamp completely covered with a periodic grating) are prerequisite for comparison and modeling. In addition, improved analytical methods have to be developed. A microscope system enabling direct 2-dimensional observation of the cavities during embossing, as already developed in reference 14 would provide a means to observe the polymer fill of microstructures in real time and confirm the conclusions drawn from the static observations. However, for the observation of the 3-dimensional filling of nanostructures, real time measurement methods have still to be developed.

Computer simulations of fill processes, such as those already developed for injection molding of bulk materials, will be very important for future applications. Although the injection molding process appears to be more complex than that of hot embossing, computer simulation tools are still not available for the 3-dimensional case of hot embossing (mainly because of the moving stamps and consequently non-stationary boundary conditions). It would be very useful to simulate embossing for complex stamp designs, and determine the flow direction, the displaced volume and shear velocity locally. These computer simulations should also indicate where the maximum shear velocity is exceeded, stress is present in the molded structure and distortions may occur. Furthermore, local bending of stamps due to inhomogeneous distribution of structures or unevenness due to particles on the mold may be taken into account. The effect of the addition of auxiliary cavity patterns and the variation of initial layer thickness could be calculated to optimize the process. Finally, regimes could be found for non-variotherm processes where structures can be sequentially embossed without damaging neighboring structures (e.g. for processes with a pre-structured polymer such as step-and-repeat processes).

All these simulations need a further knowledge about the materials used for hot embossing. The molecular weight distributions of common polymers are often not given by the manufactures and for detailed analysis, pure materials may be preferred but difficult to manufacture and expensive. Materials have to be characterized using advanced rheological characterization methods and the limits need to be found where macroscopic rheological behavior can no longer be applied to nanoscale hot embossing processes.

ACKNOWLEDGEMENTS

The authors wish to thank M. Auf der Maur, D. Bächle, C. David, F. Glaus, J. Gobrecht, B. Haas, P. Häberling, T. Neiger, S. Park, K. Vogelsang (PSI Villigen) and T. Schweizer (ETH Zürich) for their help and valuable contributions. This work was supported by the Swiss National Science Foundation (Priority program *NFP 36 Nanosciences).*

REFERENCES

1. R.W. Jaszewski, H. Schift, B. Schnyder, A. Schneuwly and P. Gröning, Applied Surface Science **143,** 301-308 (1999).
2. P. Schlossmacher, K. Bade, W.Bacher, K. Ehrlich and T. Yamasaki, FZKA-Bericht **6080**, FZK Karlsruhe, D, 103-108 (1998).
3. H.J. Mamin and D. Rugar, Appl. Phys. Lett. **61,** 1003-1005 (1992).
4. B.W. Chui, T.D. Stowe, S.J. Yongho, K.E. Goodson, T.W. Kenny, H.J. Mamin, B.D. Terris, R.P. Ried and D. Rugar, J. Micro-Electromechanical Systems (MEMS), Vol. **7**, No. 1 69-78 (March 1998).
5. A. Michel, R. Ruprecht, M. Harmening and W. Bacher, KfK-Bericht **5171**, FZK Karlsruhe, D, (1993).
6. H. Tan, A. Gilbertson, S.Y. Chou, J. Vac. Sci. Technol. B **16**(6), 3926-3928 (1998).
7. H. Schift, C. David, M. Gabriel, J. Gobrecht, L.J. Heyderman, W. Kaiser, S. Köppel and L. Scandella, Microelectronic Eng. **53,** 171-174 (2000).
8. H. Schift, A. D'Amore, C. David, M. Gabriel, J. Gobrecht, W. Kaiser and D. Simoneta, J. of Vac. Sci. Technol. B **18**(6), 3564-3568 (2000).
9. F.R. Schwarzl, Polymermechanik, Springer Verlag, Berlin, (1990).
10. J.D. Ferry, Viscoelastic Properties of Polymers, 2nd ed., John Wiley & Sons, Inc. New York, (1970).
11. Physical Properties of Polymers Handbook, J.E. Mark, ed., American Institute of Physics, Woodbury, New York, (1996).
12. R.B. Bird, R. C. Armstrong and O. Hassager, Dynamics of Polymeric Liquids, Volume 1, Fluid Mechanics, John Wiley & Sons, Inc. (1987).
13. J.R. Scott, Theory and application if the parallel-plate plastimeter. I.R.I. Trans. **7,** 169-186 (1931).
14. L.G. Baraldi, Ph.D. Thesis, Diss. ETH Zurich Nr. **10762** (1994).
15. P.R. Krauss and S.Y. Chou , Appl. Phys. Lett. **71**(21), 3174-3176 (1997).
16. H. Schift, R.W. Jaszewski, C. David and J. Gobrecht, Microelectronic Eng. **46,** 121-124 (1999).
17. R.W. Jaszewski, H. Schift, J. Gobrecht and P. Smith, Microelectronic Engineering **41/42,** 575-578 (1998).
18. H.-C. Scheer, H. Schulz, T. Hoffmann and C. M. Sotomayor Torres, J. Vac. Sci. Technol. B **16**(6), 3917-3921 (1998).
19. F. Gottschalch, T. Hoffmann, C. M. Sotomayor Torres, H. Schulz and H.-C. Scheer, Solid-State Electronics **43,** 1079-1083 (1999).

20. A. Lebib, Y. Chen, J. Bourneix, F. Carcenac, E. Cambril, L. Couraud and H. Launois, Microelectronic Eng. **46,** 319-322 (1999).
21. K. Pfeiffer, G. Bleidiessel, G. Gruetzner, H. Schulz, T. Hoffmann, H. -C. Scheer, C. M. Sotomayor Torres and J. Aholpelto, Microelectronic Engineering **46,** 431-434 (1999).
23. N. Roos, T. Luxbacher, T. Glinsner, K. Pfeiffer, H. Schulz, H.-C. Scheer, Microlithography SPIE **4343,** 33-41 (2001).
24. S.Y. Chou and P.R. Krauss, Microelectronic Eng. **35,** 237-240 (1997).
25. S.Y. Chou, P.R. Krauss and P.J. Renstrom, J. Vac. Sci. Technol. B **14**(6), 4129-4133 (1996).
26. L.J. Heyderman, H. Schift, C. David, J. Gobrecht and T. Schweizer, Microelectronic Eng. **54,** 229-245 (2000).
27. S.Y. Chou and L. Zhuang, J. Vac. Sci. Technol. B **17**(6), 3197-3202 (1999).
28. E. Schaeffer, T. Thurn-Albrecht, T. P. Russell and U. Steiner, Nature **403,** 874-877 (2000).
29. H. Schift, L.J. Heyderman, M. Auf der Maur and J. Gobrecht, Nanotechnology **12,** 173-177 (2001).
30. S.Y. Chou, P. R. Krauss and P. J. Renstrom, Applied. Phys. Lett. **67**(21), 3114-3116 (1995).
31. R. Maboudian and R.T. Howe, J. Vac. Sci.Technol. B **15**(1), 1-20 (1997).
32. S. Herminghaus, Phys. Rev. Letts. **83**(12), 2359-61 (1999).
33. S.Y. Chou, L. Zhuang and L.Guo, Appl. Phys. Lett. **75**(7), 1004-1006 (1999).
34. S.Y. Chou, P. Deshpande, L. Chen and L. Zhuang, oral contribution to EIPBN'2000, abstract book (2000).
35. S.T. Cui, P.T. Cummings, and H.D. Cochran, J. Chem. Phys. **111,** 1273-1280 (1999).
36. C. David and D. Hambach, Microelectr. Eng. **46,** 219-222 (1999).
37. C. David and A. Souvorov, Review of Scientific Instruments **70**(11), 1-5 (1999).
38. S.Y. Chou, P.R. Krauss. W. Zhang, L. Guo and L. Zhuang, J. Vac. Sci. Technol. B **15**(6), 2897-2903 (1997).

5

Wafer Scale Nanoimprint Lithography

Lars Montelius and Babak Heidari
Lund University, Div. of Solid StatePhysics & The Nanometer Consortium, Lund, Sweden

5.1. INTRODUCTION

Nanoimprint lithography (NIL) got large attention when it was introduced[1,2] since it described the possibility for a low cost parallel nanotechnology. It got even larger attention when it was reported that sub 10 nm features on a Si stamp could be replicated into a resist layer, which in turn had a profile allowing a lift-off process to be possible.[3,4] More reports about NIL technology followed and soon also devices were reported made with NIL.[5,6,7]This development inspired a few research groups to initiate projects around NIL technology. Also related techniques such as micro contact printing and its relatives were further developed at this time.[8,9]Several different nanoimprint machines, set-ups and processes were reported. Examples of this rich development are for instance the roller imprint,[10]Step and flash imprint lithography,[11] single- and multi-layer resist methods.[12] An interesting observation is that in most of the works reported so far (except the early works in ref. 2,3 above) the feature sizes have rather been closer to the 100 nm domain than to the 10 nm domain. Furthermore in most of the NIL experiments reported the experiments were made with stamps having areas of only one or a few square centimetre(s).[1-7] In order to meet the demand from industries, there was thus a need to develop a large area NIL process.[13] Here two alternatives exists, either use a full wafer single printing step procedure[14-16] or to utilize a step & repeat procedure.[17] Still up to now, more than seven years ago after its introduction, very few reports of real large wafer scale NIL exists. The largest size of the wafers successfully printed is 6 inch[14,15] and for these experiments two kinds of stamps has been used. The stamp was either a metal CD-master stamp or an e-beam defined electroplated metal stamp having regions with lines at the nanometer scale distributed at few locations over the wafer. The printed pattern definition in the resist

Alternative Lithography: Unleashing the Potentials of Nanotechnology
Edited by Sotomayor Torres, Kluwer Academic/Plenum Publishers, 2003

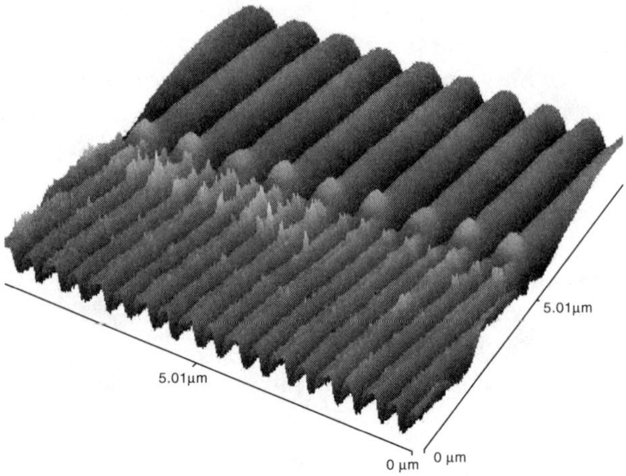

FIGURE 5.1. AFM picture shows a typical example of an imprinted interdigitated pattern in resist from the 6 "wafer NIL process. (reprinted from ref 15; © AVS The Science & Technology Society 2000).

to be accurate. In Figure 1 below is shown as an example an atomic force micrograph (raw data) of the imprinted interdigitated pattern in the employed resist system clearly resolving 100 nm features being reproduced in the 6" wafer NIL process. The stamp used is shown in Figure 19 below.

In Figure 2 an optical micrograph is shown of the resulting pattern in the employed multi layer resist system (for details see reference 15) using the other stamp having features over the whole 6 inch wafer area, i.e a pattern similar to an ordinary CD.

Since on these large wafer NIL experiments reported the stamp has had either 100 nm structures at a few discrete places or structures with sizes around 100's of nm over

FIGURE 5.2. AFM micrograph (50µm ×50µm) showing a typical view of the imprinted CD pattern in a tri-layer resist system on a 6" Si-wafer.

the whole stamp, it would be interesting to investigate NIL having more challenging stamps, such as evenly distributed sub-10 nm patterns over a large wafer. Printing using up to 4 inch wafers with fully covered stamps having features being made on the stamp with holography methods has been reported.[16,18] In this case the periodic features on the stamp had dimensions around 190 nm and 100 nm, respectively. Thus there is still no real wafer scale NIL experiments being reported having a stamp fully covered with demanding nanosized patterns with individual features having dimensions well below 100 nm, e.g. in the 10 nm domain.

In order to implement a large wafer scale NIL process, several issues have to be understood and developed. Within the two European projects NANOTECH and CHANIL such systematic work has been performed. It encompasses to address the stamp fabrication procedure,[19-21] the actual printing on the wafer scale defining a pattern in a resist system,[22,23] development of specially formulated NIL resists having performance superior to normal lithographic resists,[24,25]development of a quality control system giving numbers regarding the fidelity of the printing process,[26,27] the pattern transfer into the underlying substrate material,[28,29] development of lift-off processes and investigation if imprinting induce electronically active defects in the substrate material.[30] Many of these issues are related not only to wafer scale NIL but also generally to NIL and will be reported elsewhere (references 19-31).

In this chapter we will define and describe the problems associated with a large wafer scale NIL and discuss how these problems may be circumvented by design when making a NIL apparatus. Furthermore we will discuss and show recipes for an efficient anti-stick treatment of stamps as well as point to different ways of producing stamps for large wafer scale NIL.

5.2. SPECIAL REQUIREMENTS FOR LARGE WAFER SCALE NIL

In the imprint (or hot embossing) lithography process a thin polymer film is mechanically deformed. The substrate and stamp is heated and pressed together. The temperature, which determines the viscosity of the polymer, the time of embossing and the applied pressure, must be chosen so that the polymer completely fills the cavities of the stamp.[31,32] Furthermore, after the printing cycle the resist has to be cooled below its glass transition temperature so that it is sufficiently hard and stable to enable release of the stamp from the polymer surface. Polymer properties such as viscosity, molecular weight, chain dimensions, glass-transition-temperature, shrinkage, brake limit, adhesion property against substrate and stamp material, etch behaviour in dry and wet processes and other material conditions are all very important for the imprint process and its result.

The polymer used in imprint lithography has fundamentally different requirements than the polymers used in other lithographies. The critical properties of the polymer used in imprint lithography are mechanical, whereas in other lithographies the chemical properties are important such as radiation sensitivity, resolution and contrast. The ideal polymer used in imprint has low viscosity above its glass transition temperature. The glass transition temperature, Tg, is the temperature needed to overcome the chemical forces holding the chain segment of an amorphous polymer together. Above this temperature, the viscosity is many orders of magnitudes lower.[33]

As the structures become smaller and stamp geometries become more complex, understanding of the behaviour of the thin polymer film during both moulding and releasing will be essential. In developing imprint lithography, it is important to understand the relation between the polymer properties[34] and other imprint variables like pressure and imprint time (reference 32).

Adhesion of polymer to stamp during the stamp release can cause damage to the replica in the form of abrasion and distortion. In addition the stamp can be mechanically damaged or residual polymer structures can contaminate its surface. The importance of these effects increases with increasing wafer area and their effects can be significantly reduced by covering the stamp surface with an anti-adhesive monolayer.

In all imprint lithography two surfaces have to come entirely in contact with each other. One problem is surface flatness of the embossing equipment stage and substrates, which affects the pressure uniformity over the printing field and may introduce both relative rotation and shift between the stamp and the substrate during printing (reference 16). Another problem is related to differences in thermal expansion coefficients between e.g. the stamp and the substrate materials.

From the discussion above it ought to be clear that several aspects and problems have to be considered in order to have a successful imprint process. Especially, for imprint with large area wafers these problems are emphasized and have to be carefully and fully understood and controlled or alternatively circumvented. Below we will present our ideas how to solve this complex problem.

5.3. FABRICATION OF A NANOIMPRINT LITHOGRAPHY SYSTEM

We have to consider different parameters with respect to imprint on a large area. In the following the most important parts will be described in some details. These include: homogenous heating and fast cooling; parallelism and flat surfaces on large area; pressure control; stamp manufacturing; anti-sticking, polymers and other materials- and substrate condition and properties.

5.3.1. Design role and condition

The imprint equipment needs to have the possibility to heat the substrate in order to soften or melt the resist on the substrate. There is also a need for homogeneous heating with high accuracy control since the melted polymer must have the same floating rate on the entire surface, otherwise the polymer flow will vary and thus imprint resolution will decrease dramatically.

The heating system should thus be designed to achieve a fast imprint process combined with homogeneous heating over the entire resist area. The pattern penetrating into the resist is highly dependent on the temperature distribution. Heat variation on the substrate (resist) causes variation in the penetration into the resist, which results in different defects. One of the defects can be a loss of linewidth resolution in the subsequent pattern transfer step. Another defect corresponds to "resist plough effect". The "resist plough effect" occurs due to the differential flow of melted resist when it is transported in a narrow channel. This kind of flows creates a wave shaped resist profile that readily can be observed in the non-patterned areas close to the patterned area.

The applied pressure should be homogenous on the entire substrate surface in order to transfer the patterns from the stamp into the resist. The attached surfaces need to be highly parallel to each other to achieve a homogeneous pressure transfer. The parallelism condition is comparable to the pattern height in the stamp. As an experimental rule of thumb the parallelism should be controlled at a level of 10% of the pattern height. For instance, when using a stamp with patterns having a height of 100 nm, not more than 10 nm error in parallelism over the imprinted area can be tolerated. If this condition is fulfilled then the need for further optimisation in the pattern transfer step is minimised.

5.3.2. *Issues regarding polymer heating*

The temperature flow can be calculated both at thermal equilibrium and at thermal nonequilibrium. Here we will present a model[35] for this calculation that makes it much easier to calculate the junction or local temperature than using conventional diffusion equations. In our application, the model will be used for calculating, the interesting temperatures. To be able to present the model both thermal resistance and thermal capacitance need to be explained and defined. Thermal equilibrium is calculated using only thermal resistance but thermal nonequilibrium is calculated using both thermal resistance and thermal capacitance.

5.3.2.1. Thermal resistance. Every material has a certain thermal resistance, in other words they conduct heat in a very characteristic way depending on material parameters. Assuming one-dimensional heat flow (see Figure 3), Fourier's heat equation has the form:

$$P = \sigma_{th} A \frac{dT}{dx} \tag{1}$$

where P = applied power from heat source
σ_{th} = thermal conductivity
A = cross-sectional area of the thermal path.

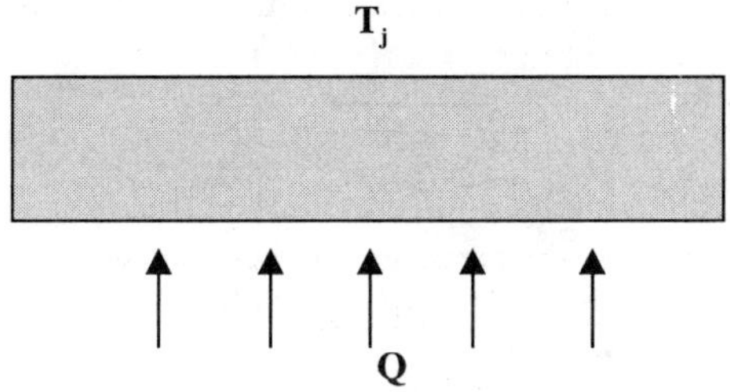

FIGURE 5.3. Cross section of the heat flow path in a simplified model.

If the thermal flow rate is constant then:

$$P\frac{x}{\sigma_{th}A} = \Delta T \tag{2}$$

where the term x/σ_{th} A = R_{th} is the thermal resistance. The following equation relates the temperature gradient ΔT to the applied power P and the thermal resistance R_{th}.

$$\Delta T = R_{th}P_{tot} \tag{3}$$

At thermal equilibrium the junction temperature T_j,can be calculated by:

$$.T_j = T_a + PR_{th} \tag{4}$$

Where T_a is the ambient temperature, P is applied power and R_{th} is thermal resistance.

5.3.2.2. Thermal capacitance. Before thermal equilibrium is reached, the temperature gradient ΔT = T_j - T_a will vary as shown in Figure 5:

This shows that there is a finite time required to reach steady-state thermal gradients. At time t_1 the cooling curve starts, it's the conjugate of the heating curve. The rate at which the temperature increases is identical to a RC-network. The thermal capacitance C_{th} is therefore needed in the calculations to determine at what rate equilibrium is reached. Thermal capacitance is given elementwise by:

$$C_{th} = V\rho_s c_s \tag{5}$$

where V is volume, ρ_s is density and c_s is specific heat.

5.3.2.3. Calculation of junction temperature at thermal nonequilibrium. The equivalent circuit shown in Figure 4 in the nonequilibrium situation is shown in Figure 6. At thermal nonequilibrium the junction temperature T_j, can be calculated by:

$$T_j = T_a + PZ_{th} \tag{6}$$

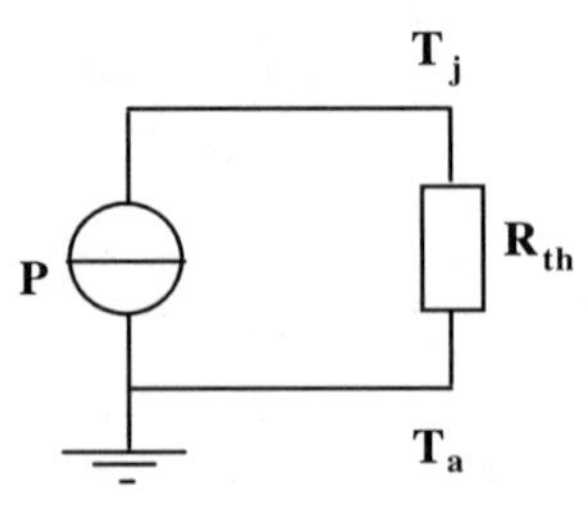

FIGURE 5.4. Equivalent circuit at thermal equilibrium of Figure 3.

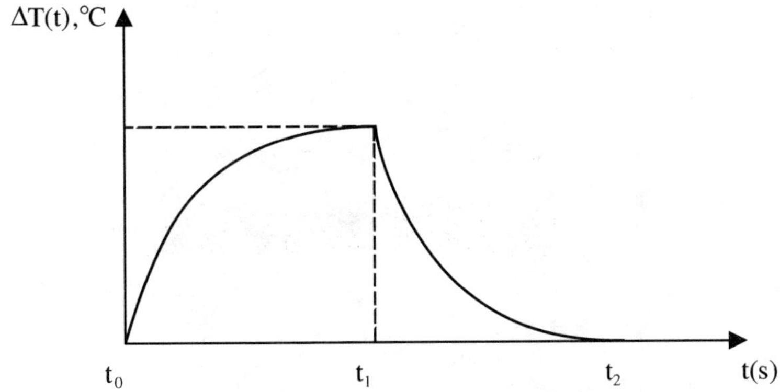

FIGURE 5.5. Transient junction temperature.

The thermal impedance can be written as:

$$Z_{th} = R_{th}(1 - e^{\frac{t}{\tau}}) \tag{7}$$

where $\tau = R_{th}C_{th}.$

5.3.2.4. Theoretically calculated temperature of the polymer film during imprint. A cross-section of the heater is illustrated in the Figure 7. The function of this configuration is that first of all due to the positive thermal resistance in the heat sheet, the produced heat is uniformly spread over the entire surface, and secondly is has a thin wall and thus a small thermal mass and therefore it can be heated and cooled quickly.[36,37]

There is a thin carbon layer, 0.0125 mm thick, through which a electrical current is sent. Above the carbon layer there is a one-millimetre thick aluminium plate with mono-layers of aluminium oxide on both sides. The aluminium oxide will serve as an electrical insulator. When the current pulse is sent through the carbon layer heat is generated that sequentially heat the polymer on a substrate e.g. PMMA on Silicon substrate.

Now we will consider the temperature variation in the polymer film, resulting from an applied power P_1. In Figure 7 a schematic picture is shown of the different material layers in the heater. The equivalent circuit is shown in Figure 8, where T_l denotes the local temperature (the temperature in the substrate/stamp), and T_{Poly} is the polymer temperature.

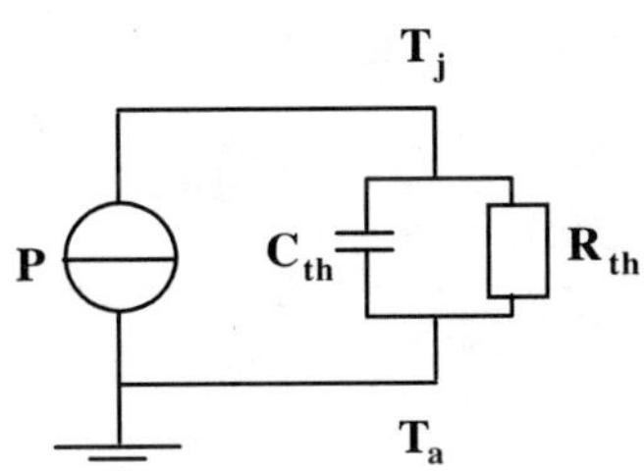

FIGURE 5.6. Equivalent circuit to Figure 4 in the nonequilibrium situation.

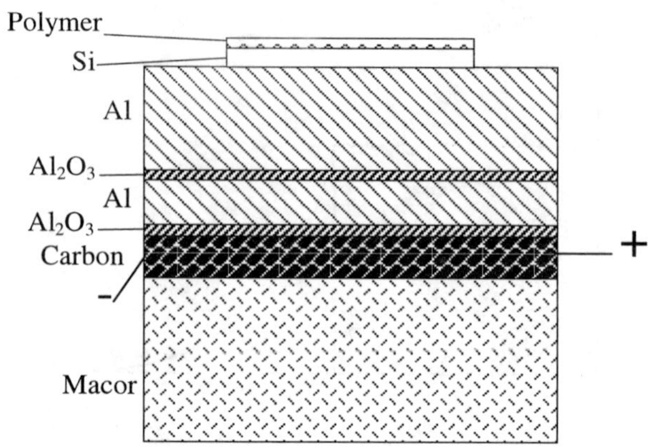

FIGURE 5.7. Schematic picture of the different materials layers in the heater setup.

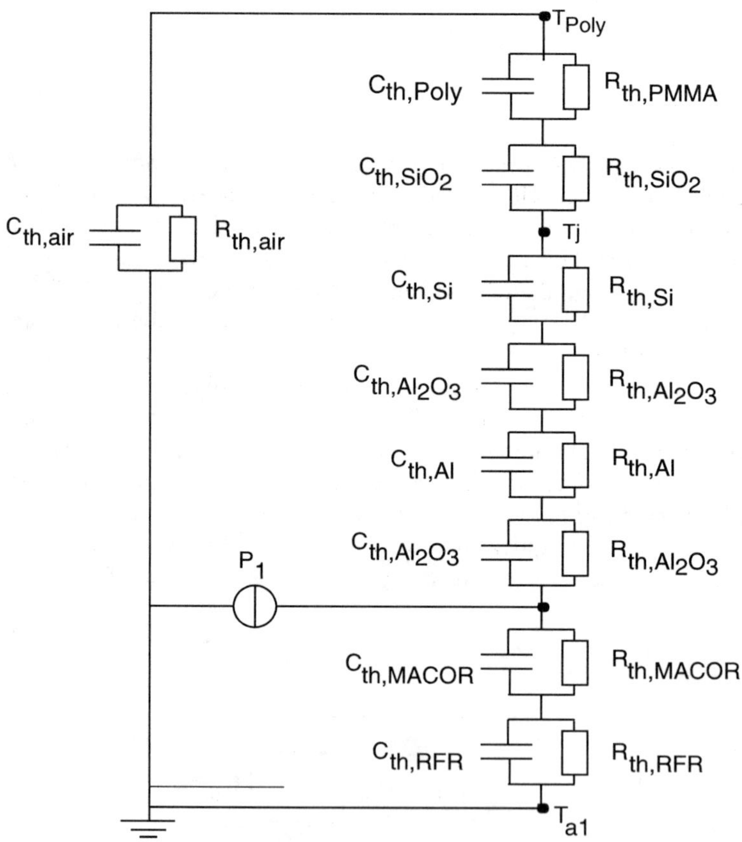

FIGURE 5.8. Equivalent circuit of the heater schematics shown in Figure 7. The term RFR in the figure means stainless steel.

At thermal nonequilibrium the junction temperature T_j can be calculated by:

$$T_j = T_a + PZ_{th} \tag{8}$$

The nonequilibrium model applied on the equivalent circuit results in the following time dependent equations for the local T_l which is the polymer temperature T_{Poly}:

$$T_l = T_{a1} - P_1 Z_1 + P_1 Z_2 \tag{9}$$

The power P_1 is applied in the carbon layer as a pulse with pulse length t_P. Immediately after that, the heat is transferred through the different material layers and reaches the surface. This heat transfer corresponds to the second part, namely $P_I Z_2$, The first part, $P_I Z_I$, is the heat flow from the carbon to the ambient, in other words, the heat loss.

$$T_{Poly} = T_l + P_2 Z_3 - P_3 Z_4 \tag{10}$$

The equation above describes the polymer temperature which, in this application, is the really interesting temperature. The term $P_2 Z_3$ in the expression corresponds to the heat that is transferred from the silicon layer to the polymer layer and the term $P_3 Z_4$ is the heat loss to the ambient air. Note that the applied power is now P_2, this is not an external power but the power that has reached the polymer from the applied power P_I within a certain time interval $\delta\ t$. P_2 can be represented by a local blackbody radiator within the certain time interval, which yields:

$$P_2 = \frac{\sigma \cdot T_l^4}{\delta\ \mathrm{t}} \tag{11}$$

σ = Stefan-Boltzmann constant. Of course:

$$P_2 = \xi \cdot P_1 \tag{12}$$

where $0<\xi<1$.

P_3 is the cooling power from the ambient air and it can be expressed in the same way as P_2:

$$P_3 = \frac{\sigma \cdot T_{a1}^4}{\delta\ \mathrm{t}} \tag{13}$$

The ambient temperature is signifcantly lower than the heating temperature, therefore:

$$P_3 = \zeta \cdot P_1 \tag{14}$$

where $\zeta<1$.

The theoretical model gives the polymer temperature T_{Poly}:

$$T_{Poly} = T_{a1} - P_1 Z_1 + P_1 Z_2 + P_2 Z_3 - P_3 Z_4 \tag{15}$$

The thermal impedance's Z_1, Z_2, Z_3 and Z_4 will be, according to Figure 8, as given below:

$$Z_1 = Z_{th,macor} + Z_{th,RFR} \tag{16}$$

$$Z_2 = Z_{th,Al_2O_3} + Z_{th,Al} + Z_{th,Al_2O_3} + Z_{th,Si} \tag{17}$$

$$Z_3 = Z_{th,SiO_2} + Z_{th,Poly} \tag{18}$$

$$Z_4 = Z_{th,air} \tag{19}$$

Using these expressions, the polymer temperature is:

$$\begin{aligned} T_{Poly} = {} & T_{a1} \\ & + P_1(Z_{th,Al_2O_3} + Z_{th,Al} + Z_{th,Al_2O_3} + Z_{th,Si} - (Z_{th,macor} + Z_{th,RFR})) \\ & + \xi \cdot P_1(Z_{th,SiO_2} + Z_{th,PMMA}) - \zeta \cdot P_1 Z_{th,air} \end{aligned} \tag{20}$$

where the thermal impedance is defined already and both ξ and ζ is like above.

It would be interesting to study how fast the different material layers reach thermal equilibrium when heat is transferred through them. Therefore the equilibrium time (τ_Θ), which is a measure of the time required to reach thermal equilibrium in a given thickness of material, is introduced [38-42]

$$\tau_\Theta = \frac{\pi}{4} \frac{\rho_s c_s}{\sigma_{th}} l^2 \tag{21}$$

where l is the thickness of the material layer, σ_{th} is thermal conductivity, ρ_s is density and c_s is specific heat.

The result of the calculation gives approximately 10 ms to create thermal equilibrium in all material layers when heat is transferred from the carbon layer to the polymer layer. This is the maximum time required to reach thermal equilibrium, because in our model we require that one layer has to be in thermal equilibrium before heat is transferred to the next. The result above should be compared with the time to get thermal equilibrium when heat is transferred through the Macor layer, which is approximately 26 second. This shows that most of the heat is transferred to the polymer layer (as it should be) and that the Macor material is a very good thermal insulator.

5.3.3. Balance and surface parallelism

The thickness variation on each stamp or substrate can be as large as a micrometer on a large wafer area (see Figure 9).

In the imprint process the polymer thickness is between 50 nm and 500 nm, which is less than the thickness variation of the underlying substrates. This thickness variation of the substrates appears as flat but with small long-range fluctuations.

FIGURE 5.9. A plate where the surface is flat but with small long-range fluctuations.

This means that merely keeping parallel stamp and substrate surfaces is not enough to make contact over the entire surface of the stamp and substrate. There is thus a need for an arrangement that reduce the sum of the errors when the entire surface of the stamp and the substrate is in contact. One solution is to apply a pressure in order to counteract (neutralize) the forces for small deformation of substrates and achieve a flat and parallel contact surface. One way to solve this problem is to apply an oil-balance system in combination with a division of the applied pressure into infinitesimal small area elements by using a rubber sheet. By using this method we minimize the height requirements of substrate surface and material.

5.4. NIL EQUIPMENT DESIGN

Two different types of NIL equipments have been built in the authors laboratory.

5.4.1. *Rigid press equipment*

The first prototype[43] was based on solid parts to press a substrate onto a stamp. The press consists of two solid substrate holders, part 3 and 4 in figure 10 below, where part 3 holds a stamp and it can move back and forth in the direction horizontal to the part 4. This movement is implemented using a hydraulic piston applying a regulated force between the stamp and the substrate. The heater that holds the substrate, part 4, has a thin wall and a small thermal mass and therefore can be heated and cooled quickly. The substrate holder can move freely in all direction for balance purposes. An oil chamber, part 5, where the pressure is also measured, makes the balance of substrate holder possible. In Figure 11 a photograph of the complete set-up is shown. The minimum force needed is less then five bar pressure in order to make the surfaces of the substrate and the stamp in parallel contact. In this construction we apply a constant force and therefore the actual pressure has to be calculated. The pressure valve is the force value divided by the contact area of stamp and substrate. Imprint pressure is dependent on the size of the stamp area,

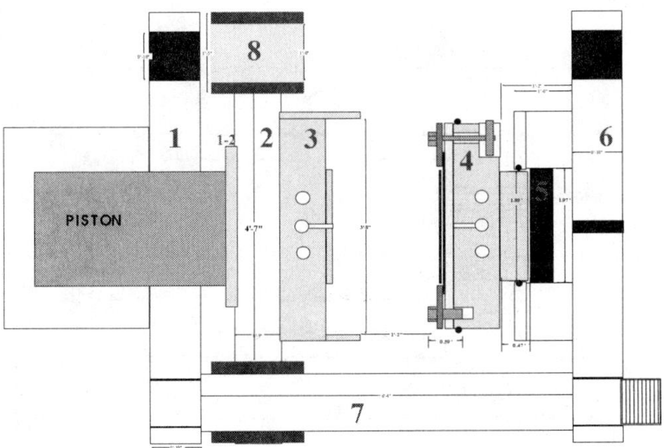

FIGURE 5.10. Schematic imprint equipment setup according to first approach.

FIGURE 5.11. Imprint equipment setup according to first approach with the computer controlled electronics.

(when stamp is smaller then substrate) according to:

$$P_{imprint} = \frac{P_{measured}}{A} \cdot \left(25^2 \cdot \pi\right) \tag{22}$$

where A is area of the stamp in square millimetres. From this it may be concluded that the imprint pressure can not be measured with a very high resolution. Therefore the weakness of this kind of equipment is the resolution of the applied pressure.

5.4.2. Soft pressing equipment

The second prototype was based on a solid substrate holder and a soft pressure chamber (see Figure 12 and 13). This equipment has been manufactured (ref.44) and it has shown perfect pressure homogeneity over a large area.

The unique feature of the press is that the pressure is applied to the stamp hydraulically or by gas without any rigid piston (soft press). For thin stamps (300-600 μm) this feature practically excludes effects of stamp/substrate thickness variation or waviness. For instance, with a substrate waviness of ~5 μm it is possible to achieve ~20 nm variation of resist thickness in the imprinted area.

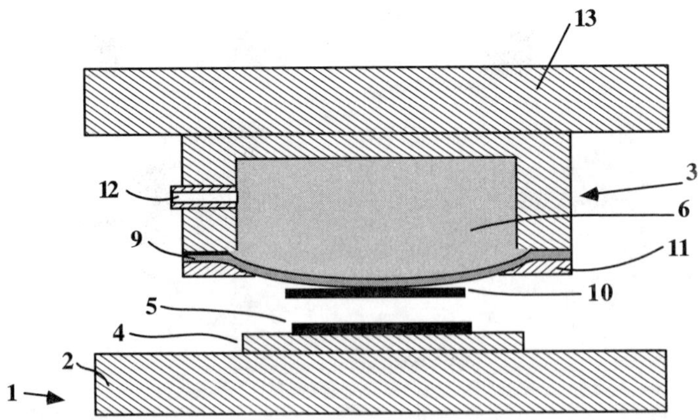

FIGURE 5.12. Schematic imprint equipment setup according to the second approach.

FIGURE 5.13. Photograph of imprint equipment manufactured according to the second approach.

5.4.2.1. Precise process control. A proprietary design of the heater embedded in the substrate chuck provides a uniform temperature field across the whole imprint area, which improves the process latitude and critical dimension uniformity. Due to the soft pressing the pressure over the whole imprint area is independent on the substrate size and shape and thus can be easily controlled.

Detail numbered (1) in Figure 12 shows the heater part that include no. (2) baseplate, no. (4) heater as it was described before, and no. (5) is substrate. The surface flatness and mechanical set up of this part it critical. The mechanical/thermal/physical properties such as thermal expansivity/conductivity, porosity and shear strengths of the different materials should be considered. This is due to the need of a mechanical stability during both high differential temperature and under high pressure conditions.

The purpose of upper part (3) is for parallelism and levelling of surfaces. Part no. (13), (12) and (11) make a cavity for oil or other medium like gas. The chamber (6) is limited in volume by a membrane (9). This membrane is flexible and can be deformed in a small range. This small flexibility of the system produces the necessary parallelism and flatness of a stamp (10). The pressure on the stamp/substrate is measured in the chamber (6) and this reading corresponds to the actual imprint pressure.

5.5. IMPRINT PROCESSING

5.5.1. Theoretical adaptation for imprint process

The relationship between the imprint parameters can be theoretically described using a simple two dimensional flow theory. The parameters like viscosity and glass transition temperature are material dependent. Other important parameters like time, pressure, pattern depth and width are related to each other.

In order to have a simple model the stamp is estimated to have a 50% coverage of the patterns evenly distributed over the entire surface. The pressure, P, required to press two plates of radius, r, together through a liquid is according the equation below[45]

$$P = \frac{3 \cdot \eta \cdot r^2}{4 \cdot h^3} \cdot \frac{\partial h}{\partial t} \tag{23}$$

where η is viscosity of the liquid, h is initial thickness of the liquid and $\partial h/\partial t$ is the rate of the displacement. This equation describes transport of a liquid out through the entire surface to the edge of the plates. The rate of a liquid displacement is proportional to $\partial h/\partial t$ but inversely proportional to the viscosity of the liquid.[46]

The equation should not be different by replacing the liquid with a polymer. The viscosity is strongly dependent on temperature. The polymer viscosity is lowered enough to flow using high temperature. If the temperature is constant then the viscosity of polymer is constant and it is directly proportional to the molecular weight for each polymer.

With the assumption that the displacement rate of a polymer (with fluid-like viscosity) is the same as the filling rate of the polymer in the cavities of a stamp, and if this displacement rate is constant within the time t, one can write:

$$\frac{\partial h}{\partial t} \approx \frac{h}{t} \tag{24}$$

using this assumption the pressure equation can be simplified to:

$$P \approx \frac{3 \cdot \eta \cdot r^2}{4 \cdot h^3} \cdot \frac{h}{t} = \frac{3 \cdot \eta(T) \cdot r^2}{4 \cdot h^2 \cdot t} \tag{25}$$

where $\eta(T)$ is viscosity of the polymer and it is inversely proportional to the temperature. The pressure equation gives a simplified picture of the time and the effective height change h_x, of the polymer according to:

$$P \approx \frac{\eta\,(T) \cdot r^2}{h_x^2 \cdot t} \tag{26}$$

The value of r is dependent on the distance the polymer flow and it can appear in two different cases.

Case one is that the polymer fills only the cavities of stamp. The polymer is thin and the amount of polymer is just enough to fill only the cavity of the stamp. The r, in the pressure equation is small and its value is equal to the largest feature size on the stamp.

In case two, the polymer flows out through the gap between the surfaces of the stamp surface and the substrate surface and some amount of polymer fill the stamp cavities. The polymer is thick. The r in the pressure equation is then equal to radius of the stamp.

When the temperature is constant the viscosity is expected to be constant for each polymer. And finally, the relation between the applied pressure, height of stamp-cavities, d, and the feature size of the stamp can be written as:

$$P_{\eta(T)} = \frac{k}{t} \cdot \left(\frac{r}{d}\right)^2 \tag{27}$$

$P_{\eta(T)}$ is the pressure that is required to fill the stamp-cavities of depth d and width of r within the time t. The filling medium is a polymer having low viscosity of $\eta(T)$ at temperature T.

This means that the fill time or pressure needs to be increased to fill a very small depth while a larger depth needs smaller pressure or less time to be filled. These aspects have been discussed in greater detail in Chapters 3 and 4.

5.5.2. *Stamp fabrication*

Stamps can be made by different methods. Here we describe three different types:

1. By metal lift-off process. The structure is produced by metal structures with a thickness chosen approximately equal to the thickness of the polymer layer(s). This kind of stamp is suitable for short and fast use.
2. Si or SiO_2 stamp made by plasma etching process using a metal lift-off layer as an etching mask. This kind of stamp is a long-term use stamp.
3. Electroplated metal (Nickel) stamp on Si-substrate, using V-shape patterned structures in PMMA and electroplating Nickel to build a Nickel-plate. This kind of stamp represents a long term and stable stamp. This process resembles the way stamps for compact discs are made. Using electroplating it would be possible to make low cost replications of a "high-cost" master which is a pre-requisite for a commercial use of NIL.

5.5.2.1. Si-Metal stamp. We have fabricated patterns of various nanosized features, 50 to 100-nm wide lines and holes with 50-nm diameter, respectively in thick (up to 350 nm) resist coated on silicon wafer. These are made on a Si/SiO_2 substrate using conventional electron-beam lithography and lift-off technique, where a 200 nm layer of Aluminium was evaporated on the substrate followed by lift-off cleaning (Figure 14,15). The 3-dimensional Aluminium pattern was used in the subsequent imprint step. Pattern depths varied between 170-nm and 200-nm.

5.5.2.2. Si-stamp. We have fabricated patterns of various nanosized features, 50 to 100 nm wide lines and holes with 50-nm diameter resist of thickness down to 50 nm coated on silicon wafer. These are made on a Si/SiO_2 substrate using conventional electron-beam lithography and lift-off technique, where a 20 nm layer of Chromium was evaporated on the substrate followed by lift-off to make an etch-mask (Figure 16,17). The Chromium pattern on the Si/SiO_2 substrate was transferred to a 3-dimensional SiO_2 pattern using CHF_3 RIE (Reactive Ion Etching). Pattern depths varied between 300 nm and 400 nm.

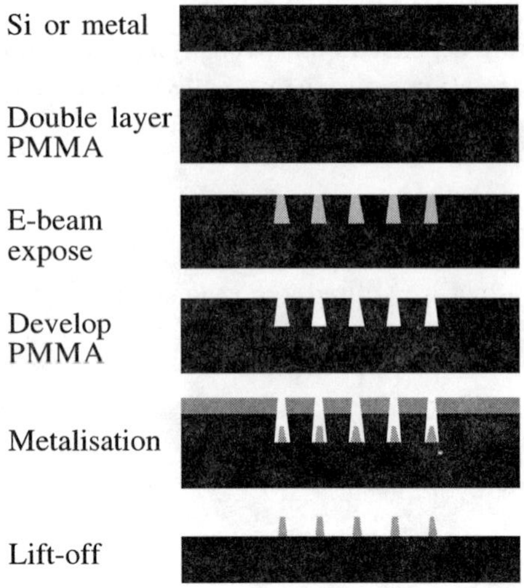

FIGURE 5.14. Stamp manufacturing method using a lift-off process.

5.5.2.3. Nickel stamp. As a starting material we used a 6-inch silicon wafer. After a cleaning procedure, the sample was spin-coated by a double-layer resist. To obtain a forward-leaning profile of the patterned structures, 150 nm high-molecular weight PMMA (950K) was spun on the substrate and then hard baked in 180°C for 60 minutes.

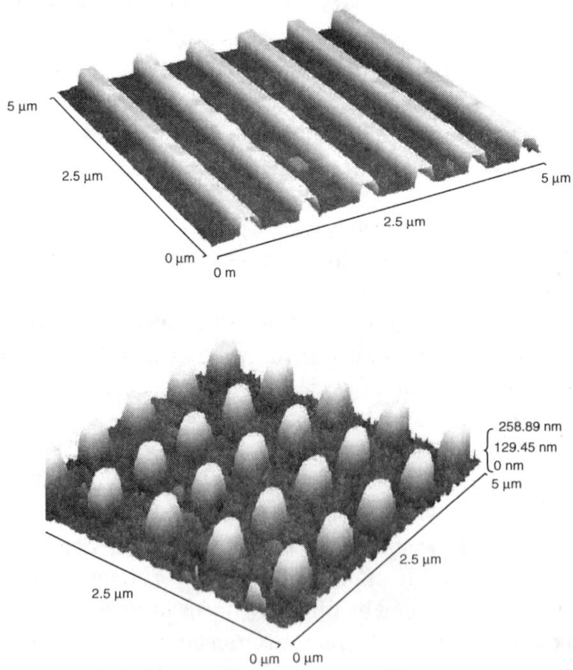

FIGURE 5.15. Aluminium structures on SiO2 The thickness of the metal is 200nm.

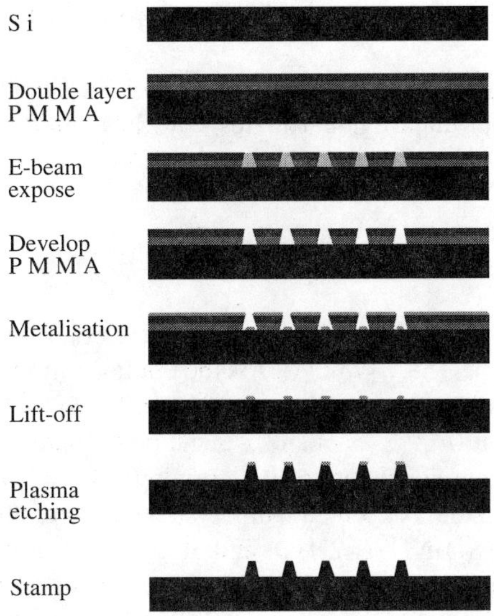

FIGURE 5.16. Stamp manufacturing method using etching process.

Thereafter, 150 nm low-molecular weight PMMA (200K) was spun on the first layer and baked at 180°C for another 60 minutes. The resulting thickness of the resist was approximately 300 nm, as measured by atomic force microscope (AFM).

Electron beam exposure was performed in a commercial scanning electron microscope (SEM), equipped with a pattern generator and a beam blanker. The accelerating voltage and the electron probe current were 35 kV and 100 pA respectively. The double layer PMMA was developed at 20°C in a 1:3 mixture of methylisobutylketone and isopropanol for 60 seconds. The test pattern consisted of 50 and 100 nm lines and 50 nm dots in 100 μm by 100 μm areas.

This electron-beam defined "master" was then was coated with a 20 to 50 nm thick silver, gold or nickel layer by thermal evaporation or sputter coater. The metal layer

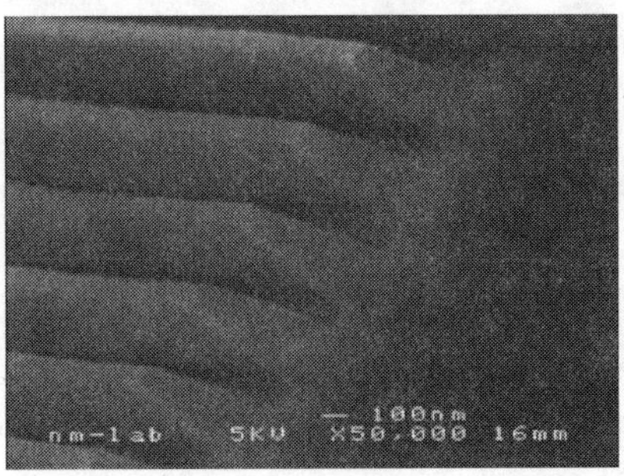

FIGURE 5.17. 100 nm lines and 300 nm spacing in SiO2 The height of the structures is 300nm (© AVS The Science & Technology Society 2000, reprinted from reference 43).

served as the cathode in the subsequent galvanic process where a nickel-father was made.

The Nickel electroplating took place in a Nickel Sulfamate type bath. This type of plating is known to produce Nickel plates with low internal stress.[47] The Nickel deposition rate on a flat surface is given by the equation:

$$Rate = \frac{a \cdot I}{d \cdot F \cdot n \cdot s}. \tag{28}$$

The rate is the thickness of the nickel grown per second. In the rate equation, a is the atomic weight of Nickel (58,7 g/mole), I is the plating current, d is the density of the Nickel (8,9 g/cm^3), F is Faraday's constant (96500 C/mole), n is the number electron required to plate each ion ($n_{Nickel} = 2$) and s is the plating area. The result of a Nickel electroplating is very much dependent of the growing rate and condition of the plating bath. In general a good result is expected with bath condition of 300 to 400 gram per litre of Nickel Sulfamate ($Ni(SO_3NH_2)_2$-$2H_2O$) and 30 gram per litre Boric acid in water to obtain pH of 4 in the bath. The stirring and the temperature of the plating bath also influence the plating result. The optimum stirring of 100 to 150 rpm and temperature of 50°C improve the quality of the finished nickel plate. The process starts with very low current in the range of mA/cm^2 and it increases gradually to several A/cm^2 in order to produce a thick layer of Nickel. Using this process, a 6-inch nickel stamp was fabricated successfully and thickness of 250μm thick Nickel-plate was achieved with very low stress.

Below (Figure 18) a schematic sequence shows the Nickel stamp fabrication process and in the AFM-picture (Figure 19) is displayed the structure made in Nickel using this process. A print using this stamp was shown in Figure 1.

5.5.3. Chemical surface modification (stamp release surface and Anti-adhesion layers)

In the imprint process a stamp is pressed into a resist and onto a polymer surface. High temperature and high pressure affect the contact between the stamp and the polymer surface. When the temperature decreases and the pressure is removed, there will be some residual bonding force between the polymer that remains in contact with the top of the stamp surface, which may result in the polymer becoming attached to the stamp surface. The extent of this attachment depends on the attraction of the polymer to the stamp versus the bonding energy of the polymer to the base substrate. If the stamp surface is coated with an anti-adhesion layer one may dramatically decrease the sticking of polymer onto the stamp surface.[48-50] There exist numerous ways how to attach different anti-adhesion layers onto various kinds of stamp surfaces.[51-55]In the author's laboratories we employ a flour- contaning silane utilizing well known silane chemistry to fabricate a monomolecular thin layer (reference 48) on Si/SiO2 stamps.[56]To have a monomolecular thin layer is particularly important when the structures have dimensions well below 100 nm, e.g. in the 10 nm domain. If the anti-adhesion layer would be too thick it could effectively change the dimensions on the stamp surface leading to uncontrolled feature sizes. In order to avoid reaction with water we work in the gas phase allowing the molecule Tridecaflouro (1,1,1,2)-tetrahydrooctyl-thriclorosilane[57] to be

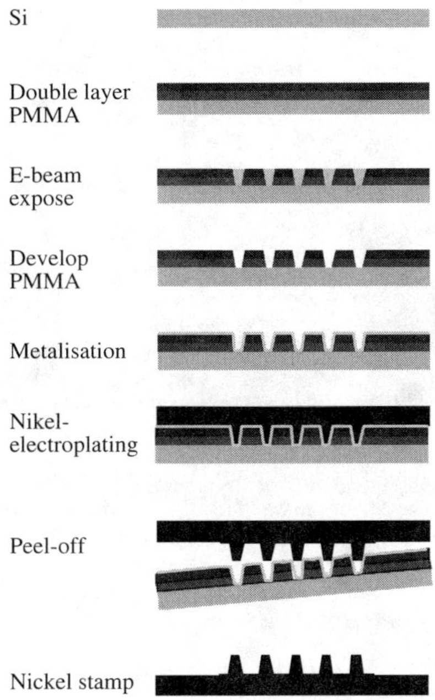

FIGURE 5.18. Schematics of a Nickel stamp fabrication.

covalently attached by spontaneous reaction with the hydroxylated Si/SiO_2 surface under elimination of hydrochloric acid. In this way we obtain a thin flouro-containing layer that works as an excellent anti-adhesion agent. For further details see reference 36. The thickness of the silane layer has been determined by X-ray Photoelectron Spectroscopy to be equal to 3.1 Å in good agreement with a monolayer thickness. Below we show typical imprint result (Figure 20a) using an anti-adhesion coated Si-stamp (Figure 20b). The resist combination used in this imprint was a 40 nm thick PMMA (200k) bottom layer and a 100 nm thick top layer of mr-I 8000 resist.[58]

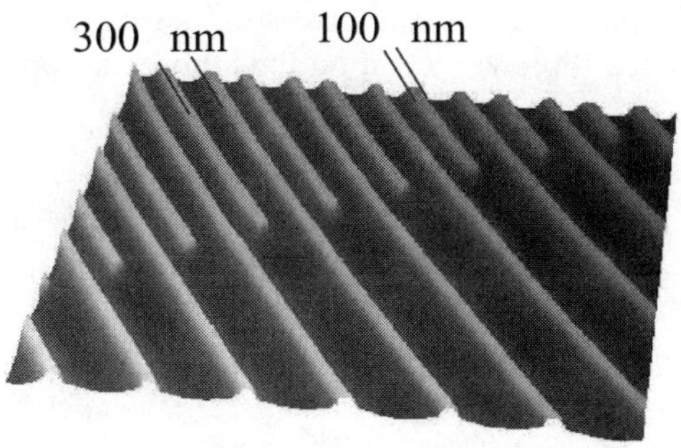

FIGURE 5.19. AFM of a Nickel stamp made according to description as above (reprinted from ref 15; ©AVS The Science & Technology Society 2000).

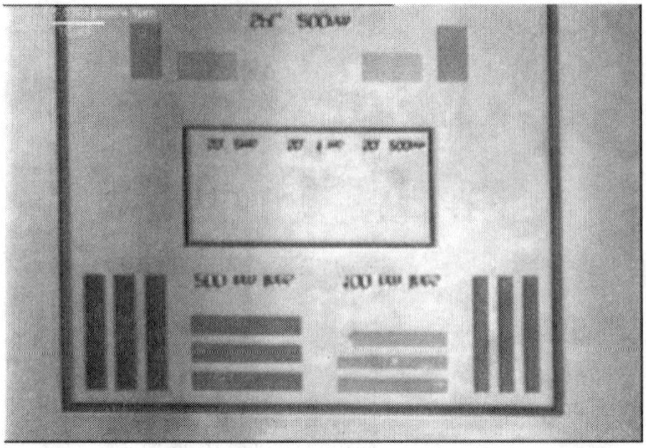

FIGURE 5.20. Optical micrograph of the complementary substrate area to the stamp shown in Fig.7a. No resist lift-off can be observed. The defects in the third structure (bottom from left) are due to a defect in the stamp pattern produced by electron-beam lithography (reprinted from reference 48).

5.6. DISCUSSION AND CONCLUSIONS

The ability to manufacture such small devices, down to a few nanometers, is crucial for applications in many different areas including microwave circuits, high-speed optics, as well as for realization of bio-electronic devices. For patterning at the nanoscale and to achieve overlap with current mass production techniques, there is a demand for a high resolution lithography capability combined with extreme reproducibility over large areas. Nanoimprint lithography is an ideal fabrication method since it allows multiple fabrication of identical nanostructures by replication of a patterned pre-defined stamp. In table 1 below we have summarized the alternatives for fabrication of nanostructures that exist today. As can be seen from the table, Nanoimprint Lithography is a good candidate for large area nanolithography. The drawback of the method is, at this stage of its development, the lack of a good multilevel technology. However, alignment at the micrometer range has already been shown. Furthermore, not all structures rely on the possibility to make multilevel processing, examples of this are nanobiosensors, MSM ultrafast photodiodes and nanopatterning of conducting organic structures.[59]

Nanoimprint has quite large similarities with fabrication of compact discs (CDs). The success of CD-fabrication has relied on the embossing process and the cutting of

TABLE 5.1. Comparison of various techniques for fabrication of nanometer sized structures. (Abbreviations as usual : Exp=Exposure, h=hour, min=minute, s=seconds, gen=generation, etc)

Technologies	UV-lith	E-beam	EUV	X-ray	Nanoimprint
Pattern gen.	Mask	Serial exp.	Mask reflection	Mask	Stamp/Mold
Exp. field	Wafer	1-10 mm	Wafer scan	Wafer	Wafer
Min. dimens.	250 nm	6 nm	20 nm	50-100 nm	8 nm
Throughput/h	10-50	0.1	20-60	10-40	30-100

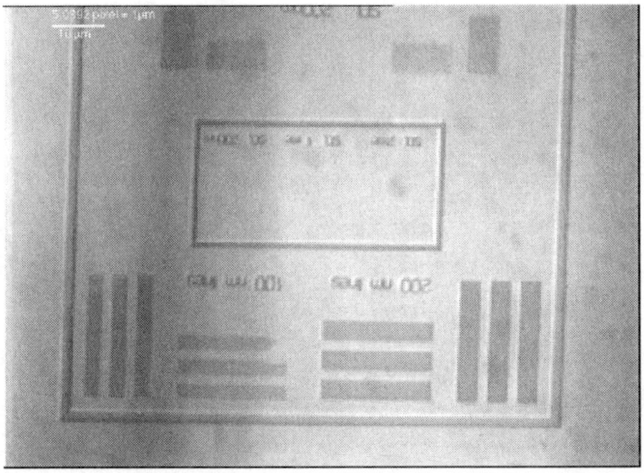

FIGURE 5.21. Optical micrograph of an F13-TCS coated imprint stamp containing several nm-structured test-patterns (not resolved) after twenty imprints. Immobilization was performed by the vapor phase deposition method *described* in this section. No resist residues can be observed on the stamp (reprinted from reference 48).

costs by using copies of the expensive master as stamps for the subsequent printing. The copying technique employed is galvanic replication. This method can also be used for NIL. As a matter of fact there exists already several studies of how far a standard CD production technology would go in resolution if one would use an electron-beam defined stamp. This could very well be an extremely interesting alternative to NIL when making e.g. nanostructured plastic components such as nanofluidic devices or devices for single use, such as various kinds of sensors. The smallest feature replicated into a plastic material using CD-technique, i.e. injection compressing technique known today is down to some few nanometers when using small stamps and printing a very small area only

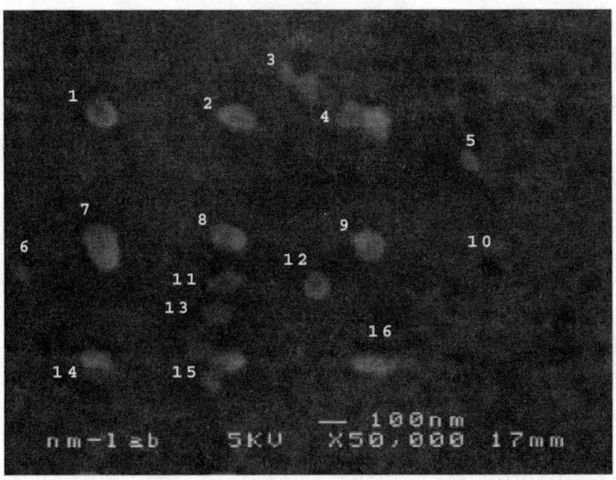

FIGURE 5.22. SEM micrographs of dots on a Ni stamp. The structures are numbered in order to compare easily the dot in the stamp versus the structures in the polycarbonate disc (from reference 15; ©AVS The Science & Technology Society 2000).

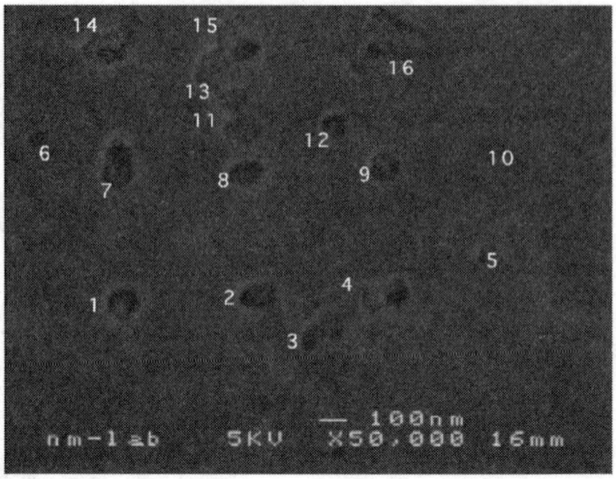

FIGURE 5.23. SEM picture shows structure on a polycarbonate disc produced by the stamp as shown in Figure 22 (reprinted from ref 15; ©AVS The Science & Technology Society 2000).

(≈ few cm^2).[60] Printing using injection compressing has also been done with electron-beam defined stamps with regular line features at the 100 nm scale and irregular features at the scale of some few nm's (due to artefacts when fabrication of the CD stamp using electroplating). These small features were shown to reproduce beautifully on the printed plastic (polycarbonate) discs (reference 15) revealing a resolution around 20 nm over the full 6 inch wafer. Hence, it would be possible to make all plastic nanofludic patterns in combination with structured and surface modified surfaces for various biotechnology applications at a very low cost.

There exist several different approaches and variants to the original NIL-process. For instance, work is directed towards investigation of low temperature processes[61,62] and recently imprint into the semiconductor surface direct without using a resist material has been shown.[63] These developments fall however beyond the scope of this review.

In summary we have shown successful 6 inch wafer scale NIL and there is no doubt that the method can be further up-scaled to meet future demands on the 8 and 12 inch wafer scale lithography with sub-30 nm features. We believe that large area NIL will become an enormously important nanotechnology method that will be employed in many applications outside traditional top-down microelectronic applications as well as inside the fields of microelectronics. We are convinced of that we have only seen the very beginning of a technology that may make a large difference both in the near and distant future.

ACKNOWLEDGEMENTS

The authors are grateful to all people at the divison of Solid State Physics and The Nanometer Consortium who contributed to and supported the work presented in this chapter. Especially we would like to put forward the following individuals that in various capacities have been deeply engaged in the wafer scale Nanoimprint project: Tord Stjernholm at the workshop for skilled work and knowledge that made it possible to realize the various kinds of NIL-equipments that have been used, Eva-Lena Sarwe and Ivan Maximov for help with etching, metallization, SEM investigations and lithography issues, Mariusz Grazcyk for help with various general as well as NIL processing issues and last but not least also the present students in the NIL group Marc Beck and Patrick Carlberg for their invaluable efforts and work with the development of the NIL technology. Finally, we would also like to thank all the members of the EU-IST MEL-ARI/NID projects NANOTECH and CHANIL for helpful discussions and support, i.e. all the people in the labs of Profs. Scheer and Sotomayor-Torres in University of Wuppertal, Prof Ahopelto at VTT, Prof Cardinaud at CNRS/Univ of Nantes, Prof Mayer at Univ of Duisburg and all the employees at MicroResistTechnologies GmbH, Berlin headed by Mrs Gabi Gruetzner. Finally we would like to acknowledge the financial support provided by the Swedish Strategic Research Council and the European Commission.

REFERENCES

1. S.Y. Chou, P. R. Krauss and P. J. Renstrom, Appl. Phys. Lett. **67**, 3114 (1995), Science **272**, 85 (1996).
2. S.Y. Chou, P.R. Krauss and P.J. Renstrom, J. Vac. Sci. techn. **B14**, 4129 (1996).
3. Stephen Y. Chou, Peter R. Krauss, Wei Zhang, Lingjie Guo and Lei Zhuang,J. Vac. Sci. Technol. **B15**, 2897 (1997).
4. S. Y. Chou and P. R. Krauss, Microelectron. Eng. **35**, 237 (1997).
5. P. R. Krauss and S. Y. Chou, Appl. Phys. Lett. **71**, 3174 (1997).
6. L. Kong, Q. Pan, B. Cui, M. Li and S. Y. Chou, Journ. Appl. Phys. **85**, 5492 (1999).
7. Z. Yu, S. J. Schablitsky and S. Y. Chou, Appl. Phys. Lett. **74**, 2381 (1999).
8. S. Brittain, K. Paul, X. Zhao and G. Whitesides, Physics World, p.31, (May 1998).
9. N. B. Larsen, H. Biebuyck, E. Delamarche and B. Michel, J. Am. Chem. Soc. **199**, 3017 (1997).
10. H. Tan, A. Gilbertson, S. Chou, J. Vac. Sci. Technol. **B 16**, 3926 (1998).
11. T. Bailey, B. J. Choi, M. Colburn, M. Meissl, S. Shaya, J. G. Ekerdt, S. V. Sreenivasan, and C. G. Willson, J. Vac. Sci. Technol. **B 18**, 3572 (2000).
12. X. Sun, L. Zhuang, W. Zhang, and S. Chou, J. Vac. Sci. Technol. **B 16**, 3922 (1998).
13. Two EU projects have been concerned about wafer scale NIL, the NANOTECH and the CHANIL projects, both belonging to the IST-MELARI/NID initiatives and coordinated by one of the authors of this article (LM).
14. B. Heidari, I. Maximov, E.-L. Sarwe, and L. Montelius, J. Vac. Sci. Technol. **B 17**, 2961 (1999).
15. B. Heidari, I. Maximov and L. Montelius, J. Vac. Sci. Technol. **B18**, 3557 (2000).
16. W. Zhang and S. Y. Chou, Appl. Phys. Lett. **79,** 845 (2001).
17. T. Haatainen, J. Ahopelto, G. Gruetzner, M. Fink and K. Pfeiffer, Proc. of SPIE, 3997 (2000).
18. Z. Yu, W. Wu, L. Chen and S. Y. Chou, J. Vac. Sci. Tecnol. **B19**, 2816 (2001).
19. K. Pfeiffer, M. Fink, G. Ahrens, G. Grützner, F. Reuther, J. Seekamp, S. Zankovych, C.M. Sotomayor Torres, I. Maximov, M. Beck, M. Graczyk, L. Montelius, H. Schulz, H.-C. Scheer, F. Steingrüber, Microelectronic Engineering **61-62**, 393 (2002).
20. I Maximov, E-L Sarwe, M Beck, K. Deppert, M Graczyk, M.H. Magnusson, and L Montelius, Microelectron. Eng.**61-62**, 449 (2001).
21. H. Schulz, D. Lebedev, H.-C. Scheer, K. Pfeiffer, G. Bleidiessel, G. Grützner and J. Ahopelto, J. Vac. Sci. Technol. **B 18**, p 3582, (2000).
22. H.-C. Scheer, H. Schulz, Microelectronic Engineering **56**, 311-332 (2001).
23. K.Pfeiffer, M.Fink, G.Grützner, G. Bleidiessel, H. Schulz, H.-C. Scheer, Microelectron Engineering, **57-58**, pp 381-387 (2001).
24. H.Schulz, H.-C. Scheer, T.Hoffmann, C.M.Sotomayor Torres, K.Pfeiffer, G. Bleidiessel, G. Grützner, CH. Cardinaud, F. Gaboriau, M.C. Peignon, J. Ahopelto, B. Heidari, J Vac Sci Technol B **18**, 1861-1865 (2000).
25. D. Lyebyedyev, H. Schulz, H.-C. Scheer, Mater. Sci. Eng. **C 15**, p 241 (2001).

26. Ch. Finder, C. Mayer, H. Schulz, H.-C. Scheer, M. Fink, K. Pfeiffer, in GMM-Fachbericht 36, S. 195, VDE Verlag GmbH Berlin, 2002.
27. Ch. Finder, M. Beck, J. Seekamp, K. Pfeiffer, P.Carlberg, I. Maximov, E.-L. Sarwe, S. Zankovic, C. Mayer, L.Montelius, C.M. Sotomayor Torres, *Optical Microscopy for quality control in nanoimprint lithography*, in preparation for Microelectronic Engineering.
28. C. Cardinaud, M.C. Peignon, P.Y. Tessier, Appl. Surf. Sci. **164** 72-83 2000 .
29. F. Gaboriau, M.C. Peignon, A. Barreau, G. Turban, C. Cardinaud, K. Pfeiffer, G. Bleidissel, G. Gruetzner, Microelectronic Engineering **53,** 501-505 (2000).
30. S. Zankovich, J.Seekamp, S. Romanov, C.M. Sotomayor Torres, I. Maximov, M. Beck, I Shorubalko, L. Montelius, D. Reuter, P. Schafmeister, A.Wiek, *Nanoimprint-Induced Effects on Electrical and Optical Properties of Quantum Well Structures, to appear in* Microelectron Engineering.
31. H.C.Scheeer, H.Schultz, H.Hoffmann and C.M.Sotomayor-Torres, J. Vac. Sci. Technol. **B16**, 3917 (1998).
32. Y. Hirai, M. Fujiwara, T. Okuno and Y. Tanaka, m. Endo, S.Irie, K.Nakagawa and M.Sasago, J. Vac. Sci. Technol. **B19**, 2811 (2001).
33. J. Bicerano, Prediction of Polymer Properties, 2d edition (Marcel Dekker, Inc., New York) 1996.
34. L.J. Fetters, D.J. Lohse, D. Richter, T.A. Witten and A. Zirkel, Macromolecules **27**, 4639, (1994).
35. Paolo Antognetti. *Power Integrated Circuits*. McGraw-Hill 1986.
36. Carl Fjelkner, "Design, Fabrication, and Characterisation of an integrated surface temperature sensing device" M Sc thesis at Solid State Physics, Lund University 1999.
37. Babak Heidari, Lena Hallén. *SOA-Kurvor för DMOS-Transistorer*.M Sc thesis at Solid State Physics, Lund University, Lund 1991.
38. William B. Burford III, H. Grey Verner. *Semiconduct*or Junction and Devices. McGraw-Hill Book Company New York 1965.
39. S. M. Sze. *Semiconductor Devices Physics and Technology*. John Wiley & Sons New York 1985.
40. Max J. O. Strutt. *Semiconductor Devices volume* one Semiconductors and Semiconductor Diodes. Academic Press New York 1966.
41. Paolo Antognetti. *Power Integrated Circuits*. McGraw-Hill 1986
42. W.E.Newell. Transient Thermal Analysis of Solid-State Power Devices - Making a Dreaded Process Easy. IEEE Power Electronic Specialists Conference. pp 234 (1975).
43. L. Montelius, B. Heidari, M. Graczyk, T. Ling, I. Maximov, and E-L. Sarwe, SPIE Proceedings Microtechnology, 28 Feb-2 March, 2000.
44. Obducat AB, Malmö, Sweden.
45. R.B. Bird, R.C. Armstrong, O. Hassager, Fluid Mechanics, Dynamics, of Polymeric Liquids, Vol. 1, (John Wiley) (1987).
46. P. Ruchhoeft, M.Colburn, B Choi, H. Nounu, S. Johanson, T. Bailey, S. Damle, M. Stewart, J. Ekerdt, J.C. Wolfe, C.G. Willson, J. Vac. Sci. Technol. **B 17**, 2965 (1999).
47. L.J. Durney, editor, Electroplating Engineering Handbook, 4th edition (Van Nostrand Reinhold, New York) 1984.
48. M Beck, M Grazcyk, I Maximov, E-L Sarwe, T G I Ling, M. Keil and L Montelius, 1st International IEEE-Nanotechnology conf, Hawaii Oct 28 (2001); Microelectron. Eng. **61-62**, 441 (2002).
49. H. Schulz, H.-C. Scheer, T. Hoffmann, C.M. Sotomayor Torres, K. Pfeiffer, G. Bleidiessel, G. Grützner, Ch. Cardinaud, F. Gaboriau, M.-C. Peignon, J. Ahopelto, B. Heidari, J. Vac. Sci. Technol. **B18**, 1861 (2000).
50. K. Pfeiffer, M. Fink, G. Bleidiessel, G. Gruetzner, H. Schulz, H.-C. Scheer, C.M. Sotomayor Torres; T. Hoffmann, Ch. Cardinaud, F. Gaboriau, Microelectr. Engineering **53**, 411 (2000).
51. V. J. Novotny, T. E. Karis, Appl. Phys. Lett. **71**, 7 (1997).
52. A. Mueller, T. Kowalewski, and K. Wooley, Macromolecules **31**, 3 (1998).
53. G. J. Vancso* ; Lattice Imaging of Self-Assembled Monolayers of Partially Fluorinated Disulfides and Thiols on Sputtered Gold by Atomic Force Microscopy, Langmuir; **13**, 3769 (1997).
54. John F. Rabolt ; Vapor Phase Self-Assembly of Fluorinated Monolayers on Silicon and Germanium Oxide, Langmuir; **13**, 1877 (1997).
55. Chidsey, C. E. D.; Bertozzi, C. R.; Putvinski, T. M.; Mujsce, A. M. J. Am. Chem. Soc. **112**, 4301 (1990).
56. I. Maximov, E.-L. Sarwe, M. Beck, K. Deppert, M. Grazcyk, M. H. Magnusson and L Montelius Microelectron, Eng. **61-62**, 449 (2002).
57. Can be purchased from e.g. www.abcr.de

58. MicroResistTechnologies in Berlin sell a variety of imprint resists dedicated for various NIL processes, see www.mrt.de/com
59. T Mäkelä, T. Haatainen, J. Ahopelto and H. Isotalo, J. Vac. Sci. Technol. **B19**, 487 (2001).
60. Nikolaj Gaadegard, Ph D Thesis defense, June 2002, Copenhagen University.
61. S. Matsui, Y Igaku, H. Ishigaki, J. Fujita, M. Ishida, Y. Ochiai, M. Komuro and H. Hiroshima, J. Vac. Sci. Technol **B19**, 2801 (2001).
62. H. Schulz, M. Wissen, N. Roos, H.-C. Scheer, K. Pfeiffer, G. Gruetzner, Low-Temperature Wafer-Scale. 'WARM' Embossing for Mix & Match with UV-Lithography, Proceedings of the SPIE **4688**, p 24 (2002).
63. D. Y. Khang and H. H. Lee, Appl. Phys. Lett. **76**, 870 (2000).

6

Step And Stamp Imprint Lithography

Jouni Ahopelto and Tomi Haatainen
VTT Centre for Microelectronics, Finland

6.1. INTRODUCTION

As the semiconductor industry continues to push to smaller device geometries, lithography becomes more and more a crucial process step. Traditionally, the semiconductor industry has relied on optical lithography and new generations of optical steppers are being developed for shorter wavelengths. Anyhow, in order to produce sub-50 nm structures researches have started to look for new solutions, such as X-ray lithography, fast electron beam (e-beam) lithography and nanoimprint lithography (NIL).

The reseach on NIL has been quite active during recent years. After the demonstration of sub-10 nm features it was presumed that NIL would accomplish a new generation lithography tool to be used in many applications, replacing optical lithography, e.g., in semiconductor production.[1,2] Large area nanoimprinting can also be much faster than conventional sequential e-beam litography and much cheaper than X-ray lithography. However, optical lithography systems currently used in semiconductor industry are capable to produce lines down to about 100 nm and below.[3] Recently, it was reported that an optical stepper with line widths down to 11 nm will be in the market in 2005 or 2006.[4] The system will be very expensive, but it can be expected that CMOS industry will rely on optical steppers also in the future. Moreover, much faster e-beam writing systems are under development and, with pattern recognition options, they provide relatively fast patterning tool also in wafer scale in mix and match mode. These systems will be probably fairly expensive but will be still competing with NIL.

Although it may be expected that within semiconductor industry it is difficult for NIL to compete with optical steppers, some niche applications requiring a relatively low-cost patterning method may take advantage of NIL. Fabrication of interdigitated metal fingers for sensor and or detector applications,[5] gratings,[6] filters,[7] polymer lasers[8] and

Alternative Lithography: Unleashing the Potentials of Nanotechnology
Edited by Sotomayor Torres, Kluwer Academic/Plenum Publishers, 2003

Fresnel lenses[9] in plastics may be such applications. It was also found that imprinting is a suitable method for patterning conducting polymers without any significant reduction in the conductivity.[10] There are promises in mass production of, e.g., polymer LEDs utilizing approaches used in paper production. Here a multiple layer polymer foil is imprinted in a roller type system, providing a way to a very low-cost mass production. Optical steppers are designed and optimized to handle wafers and are not capable to handle substrates with arbitrary shape. An advantage of imprinting lithography is that the substrate does not necessarily have to have the shape of a wafer. Another advantage is that the desired structure, say a grating, can be created directly without etching and resist removal steps, making the process relatively simple.

Two different approaches have been suggested to realize large area imprinting lithography: large area parallel imprinting and sequential imprinting. In parallel imprinting a large area stamp, for example a 100 mm or 150 mm patterned wafer, is used as the stamp to pattern a substrate of similar size in one imprinting step.[11,12] This approach is described in Chapter 5 in this book. In many casesa a hydraulic press is enough to perform parallel imprinting, and the first results were obtained with very simple systems. More sophisticated tools, dedicated solely for imprinting, have been developed later on.[11,13]

The sequential process, step and stamp imprinting lithography (SSIL), mimicks the operation of an optical stepper in which the substate is exposed chip by chip.[14] The stamp corresponds to the reticle of the stepper and is used to pattern one chip at a time. In this approach the requirements for the imprinting apparatus, in addition to heating and pressing capabilities, include the possibility to move the stamp with respective to the substrate and to align the stamp with the substrate. In our experiments we have used a commercial flip-chip bonder without any modifications. The system, although not developed for imprinting lithography, can be used to heat the stamp and substrate and to press them against each other with well controlled force. It also allows x-y movement and aligning of the stamp relative to the substrate. The system is suitable for demonstration of SSIL, but for effective use in production some improvements, e.g., in stamp attachment and stamp to substrate aligning, are required.

Besides wafers, SSIL can be used to pattern substrates with different shapes. The ability to align the imprinted pattern relative to the existing features of the substrate is a useful feature when patterning objects with arbitrary shape. These objects may be for example a light source and detector casted in a polymer block on which a grating or Fresnel lens is added by imprinting. SSIL is especially well suited for such patterning because only the stamp is heated above the glass transition temperature and the substrate temperature is kept lower. It may well be that in the future many applications utilize this advantage of imprint lithography.

The SSIL method is in principle a relative flexible process which can be used for large area imprinting, multi layer imprinting, mix and match with UV-lithography and to create patterns on other substrates than wafers.

Recently developed step and flash imprint lithography (SFIL) is based also on step and repeat sequence.[15] The method relies on photo-curable, low-viscosity organosilicon precursor used as resist. The grooves in a transparent stamp are filled by capillary action and the organosilicon is then hardened by UV exposure through the stamp. More details of the method are given in Chapter 7 of this book.

In the next sections we will first describe the SSIL process and then show, how it can be applied in creating matrices in silicon by etching or by lift-off and how it can be combined with conventional UV-lithography for mixing and matching.

6.2. STEP AND STAMP IMPRINTING LITHOGRAPHY

The step and stamp imprinting process is shown schematically in Figure 1. A stamp is pressed into a polymer layer to create the first imprint. The stamp is then lifted and moved to the next site to create the next imprint and so forth. The pattern on the stamp may consist of a complete device structure that can be created in a single layer lithography step or contain only a small segment to be incorporated into the device. The latter approach, mix and match imprinting with UV-lithography, is described in more detail in section 4 in this Chapter.

The apparatus which is used in SSIL must be capable to heat the stamp, press it against the substrate and to position and level the stamp relative to the substrate. In the development of the SSIL process a commercial flip-chip bonder, Karl Suss FC150, was used. The system was used in the experiments without any specific modifications. The relevant parts of the system regarding imprinting are the bonding arm, substrate holder and aligning/leveling optics. The bonding arm is used to level the stamp with the substrate, to heat the stamp and to press the stamp against the substrate. The substrate holder can rotate and move the sample in the x-y plane. The optics is used to align the stamp relative to the substrate and to level the stamp so that the surfaces are parallel when pressed together. The maximum force that can be applied with the standard bonding arm

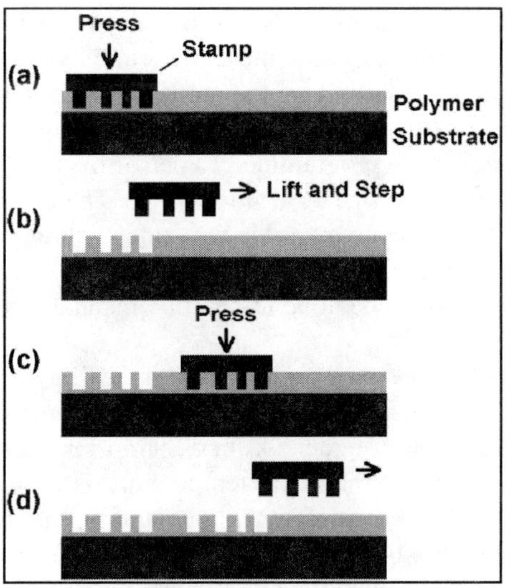

FIGURE 6.1. Schematics of the step and stamp imprinting process. (a) The stamp is pressed into polymer resist to define the desired pattern. (b) The stamp is lifted and moved to the next site. (c) The imprinting is repeated at the new location. (d) The sequence is repeated again.

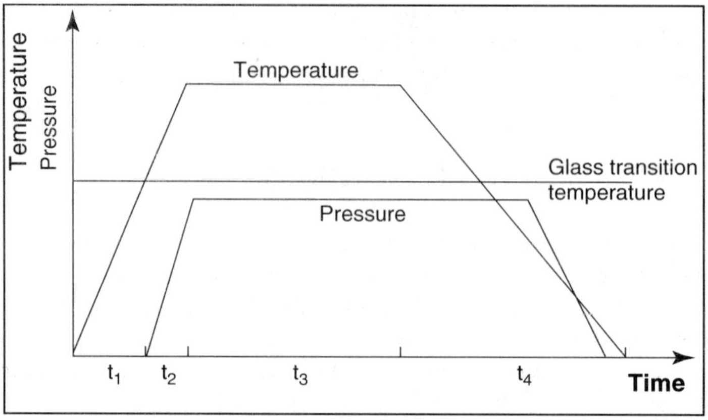

FIGURE 6.2. Temperature and pressure profiles during a step&stamp cycle. The temperature of the stamp is raised above the glass transition temperature of the polymer before the pressure is exerted onto the stamp. Before the stamp is lifted up, the stamp temperature is allowed to fall below the Tg.

is about 500 N, which in practice limits the maximum chip size to 5×5 mm^2. For larger chips this force is not high enough to displace the relative viscous polymer.

In the experiments the stamp was attached to a 40×40 mm^2 SiC plate with a high temperature silicone adhesive. The SiC plate is attached to the bonding arm by vacuum. The plate and the stamp are heated from the back side using an IR lamp and the temperature is measured by a thermocouple attached to a small piece of silicon wafer behind the plate. The substrate is held on a similar SiC plate by vacuum and is heated by a separate IR lamp. The temperature of the stamp and the substrate can be controlled separately. The maximum temperature both in the arm and the substrate is 400 °C. The temperatures given below are those measured by the thermocouples behind the SiC plates, and do not necessarily give the correct surface temperature of the stamp or the substrate.

The flip chip bonder can be programmed to perform an imprinting sequence, i.e., the temperature and force profiles can be set in advance. The movement in the x-y plane can also be programmed. The accuracy of the movement is high enough for imprinting of matrices but not high enough for mixing and matching with UV-lithography. The leveling and the aligning have to be done in practice manually while carrying out mix and match lithography.

Typical temperature and pressure profiles used in SSIL are shown in Figure 2. The temperature of the stamp is raised to the glass transition temperature of the polymer prior to pressing it against the polymer. A final stamp temperature typically around 80-100 °C higher than the glass transition temperature is required to reduce locally the viscosity of the polymer. In a typical step and stamp process the substrate is also heated to a temperature slightly below the glass transition temperature. It is important to keep the substrate temperature below the glass transition temperature while imprinting matrices because higher temperature could degrade the already imprinted patterns while imprinting the next patterns.

After the temperature of the stamp has reached the glass transition temperature, it is pressed into the polymer and the force and temperature are increased until the

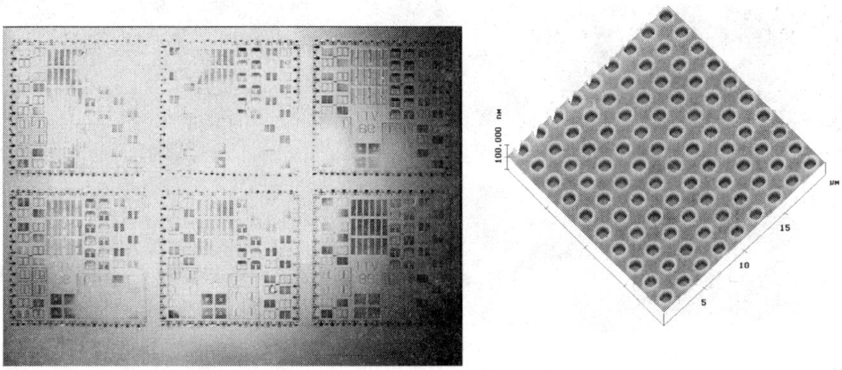

FIGURE 6.3. (a) Optical micrograph of repeated imprints into PPM polymer. (b) AFM image of an array of 1 mm holes. A detail of Figure 3(a).

preprogrammed values are reached. The pressure is applied and the temperature kept constant to let the polymer flow and adapt the pattern on the stamp. Then the stamp is let to cool down and is lifted from the substrate when the temperature of the stamp is below the glass transition temperature of the polymer.

Typical process parameters for mr-I 8000 resist[16] are a stamp temperature of 200-220 °C, a substrate temperature of 80 °C, a pressure of 10-20 MPa and imprinting time of 60 s. There is no forced cooling in the bonder and the cooling down to 20-30 °C below the glass transition temperature takes about 180 s. Ramping up the temperature and force is quite fast and can be done in less than 30 s. A complete step and stamp cycle takes less than 5 minutes with the FC150 bonder.

The silicon stamps used in the experiments were patterned using UV-lithography and dry etching. No anti-adhesion layer was used, only the careful removal of resist residue in solvents after dry etching was necessary. No adhesion was observed during imprinting of several 4×4 matrices into mr-I 8000 resist spun onto silicon wafers. When the resist residue was removed from the stamp in oxygen plasma after dry etching, severe adhesion was observed.

Optical micrographs and an AFM image of structures imprinted into PPM polymer (currently commercialized as mr-I 8000) using SSIL are shown in Figure 3.

The flip-chip bonder used in the experiments is not designed for imprint lithography, and some modifications are required to make it more suitable for SSIL. Some deficiencies that need improving are, for example, the stamp attachment to the bonding arm, which has to be modified, the force that the bonding arm can exert onto the stamp should be higher to allow the use of larger stamps, the x-y table movement from one chip site to another should be more accurate, and the cooling should be forced to reduce the cycle time.

6.3. PATTERN TRANSFER USING STEP AND STAMP IMPRINT LITHOGRAPHY

Lithography is usually used to transfer a desired pattern into the substrate. In case of imprint lithography, this may be direct molding of the substrate surface to create,

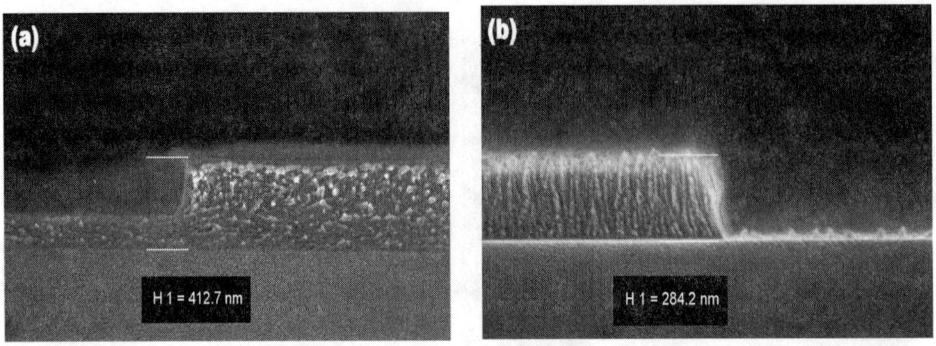

FIGURE 6.4. (a) Cross section of as-imprinted PPM polymer showing the residual layer in the bottom of the imprinted trench. (b) Cross section of similar pattern after oxygen ashing. The residual layer has been removed but the thickness of the resist has also been reduced.

for example, gratings in plastics. If the substrate to be patterned is of hard material, an etching process is needed. In optical lithography the windows of the exposed and developed pattern open down to the substrate surface and the resist can be used directly as etching mask. In imprint lithography a thin layer of polymer remains at the bottom of the window and an extra oxygen plasma treatment is required to remove the residual layer. The drawback here is that the total thickness of the polymer layer in areas which are not imprinted decreases during oxygen ashing and this decrease is somewhat uncontrollable. Moreover, the width of the imprinted trenches tends to increase slightly during ashing, as will be shown later. SEM images of as-imprinted and ashed polymer profiles are shown in Figure 4. In the resist profile an increase of the thickness of the resist at the edge of the imprinted trench can be seen. This is due to the stamp replacing the resist underneath the ridge. This shows also that the stamp is not in full contact with the resist surface.

In Figure 5 an AFM image is shown of trenches etched in silicon through imprinted mr-I 8000 polymer mask. The windows were opened in oxygen plasma at 125 mTorr using 150 W rf power. The ashing time was 30 s, resulting in a remaining thickness of the polymer of 140 nm in areas that were not imprinted. The pattern was transferred into silicon by reactive ion etching (RIE) in CF_4/CHF_3 plasma. The selectivity in etching between mr-I 8000 and Si was 1:2, i.e., the etch rate of mr-I 8000 is twice that of Si, which is too low for many applications.

To etch deeper trenches the polymer must be replaced with a hard etching mask. A typical etching mask is aluminum patterned by lift-off. In imprint lithography the oxygen ashing required to open the pattern windows thins down the remaining polymer layer. Consequently, the thickness of the metal layer evaporated has to be thinner to achieve the discontinuity in the metal film at the edge of the pattern window required for successful lift-off. This restriction can be avoided by using multilayer resists in imprinting. [11,17-19]

In many cases only single patterned metal layer, such as one for interdigitated metal finger electrodes, is needed. The fabrication of such a structure is relatively straightforward by imprint lithography and lift-off. As an example, a silicon stamp and the resulted interdigitated aluminum fingers fabricated by SSIL and lift-off are shown in Figure 6.[15]

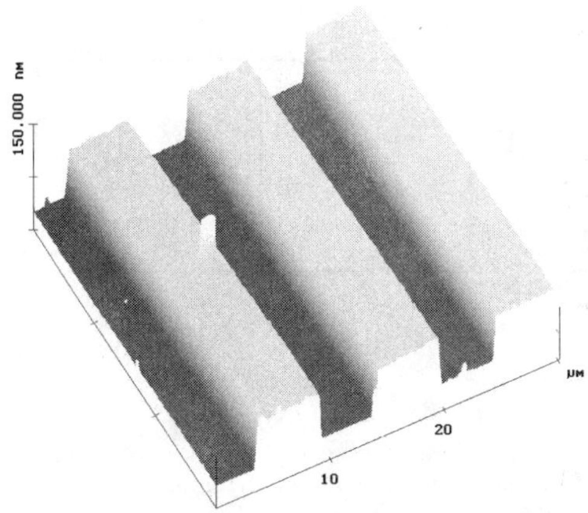

FIGURE 6.5. AFM image of trenches etched into silicon using mr-I 8000 resist as etching mask. The depth of the trenches is 65 nm.

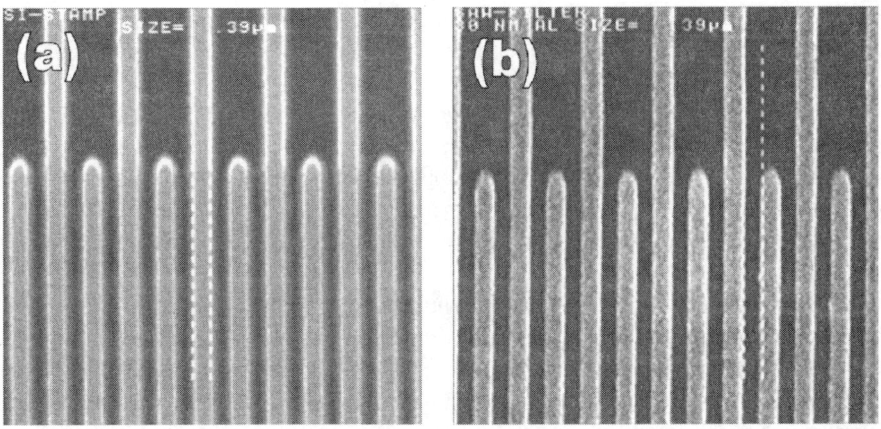

FIGURE 6.6. (a) SEM image of a silicon stamp and (b) interdigitated aluminum fingers defined by imprinting and lift-off. The pitch is 800 nm.

The silicon stamp was defined by UV-lithography and dry etching. The pitch is 800 nm and the depth of the trenches is 500 nm. The imprinted pattern was opened in oxygen plasma and the metal fingers were defined by lifting-off 30 nm of aluminum. In the next section the fabrication of single layer metal structures using the combination of SSIL and UV lithography will be described.

6.4. MIX AND MATCH WITH UV LITHOGRAPHY

During the imprinting experiments it was found that it is not very easy to imprint large and small features using the flip-chip bonder in one simultaneous step. This is due to

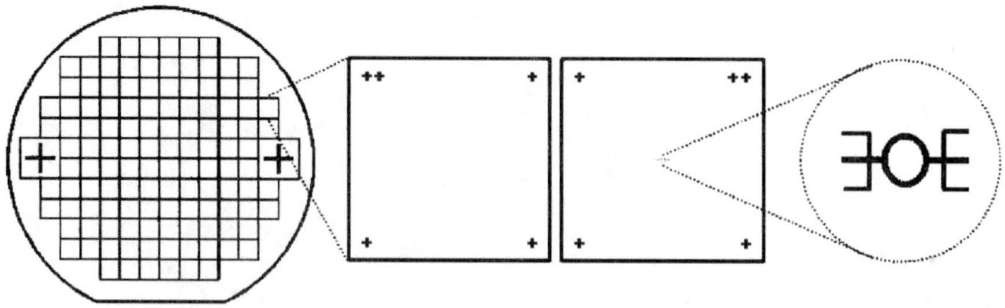

FIGURE 6.7. Alignment marks used in mix&match SSIL. On the wafer there are global marks for UV lithography. Each chip has local marks for imprint lithography. The stamp has corresponding marks as mirror image. In mix and match the imprinted pattern is aligned with the marks on a single chip and the global marks are used to define the leads and contact pads.

the different optimum imprinting parameters for large and small patterns. In many cases it is convenient to create the small features by imprinting and use standard lithography techniques for the large patterns, such as connecting leads and contact pads. SSIL can be combined with UV-lithography rather easily, because the flip-chip bonder is capable to align the stamp with the substrate.

A UV-lithography contact aligner utilizes global alignment marks located typically at the edges of the wafer. SSIL requires alignment marks on every chip. For mix and match lithography the wafers have global marks for contact aligner and local marks on every chip for imprinting. The stamps have the corresponding local alignment marks, only as mirror image, to align to the chip. The global and local marks are shown schematically in Figure 7.

We have tested the mix and match approach to by fabricate various metal structures by imprinting the small features and adding the leads and contact pads using UV-lithography. An example is shown in Figure 8. The stamp used was an Ahranov-Bohm (A-B) ring pattern dry etched into silicon. No anti-adhesion surface treatment was used. A-B rings with a diameter of 3 μm and width of 0.5 μm were made first by imprinting and lift-off. In mix and match the pattern typically covers only a very small fraction of the total surface area of the stamp. Consequently, the force used is much smaller than described in the previous section. In this case we used only 5 N force. Otherwise the process is similar to that described in section 3. The residual layer of polymer from the bottom of the pattern was removed in oxygen plasma and 30 nm of Al was sputtered and lifted-off. The leads and contact pads were created by UV-lithography and lifting-off 60 nm of Al.

In the FC150 bonder the optics block is between the stamp and substrate when the levelling and aligning are performed. At this stage the stamp is 175 mm above the substrate. During imprinting the stamp is brought into contact with the substrate and the alignment accuracy is limited by the accuracy of the movement of the bonding arm relative to the substrate holder. The typical alignment accuracy for the bonder is about 2 μm and for the contact aligner less than 1 μm. Compensating for some systematic errors in the bonder alignment procedure, an overall aligning accuracy of about 2 μm was obtained. This accuracy is not very good but it is high enough for contacting leads and

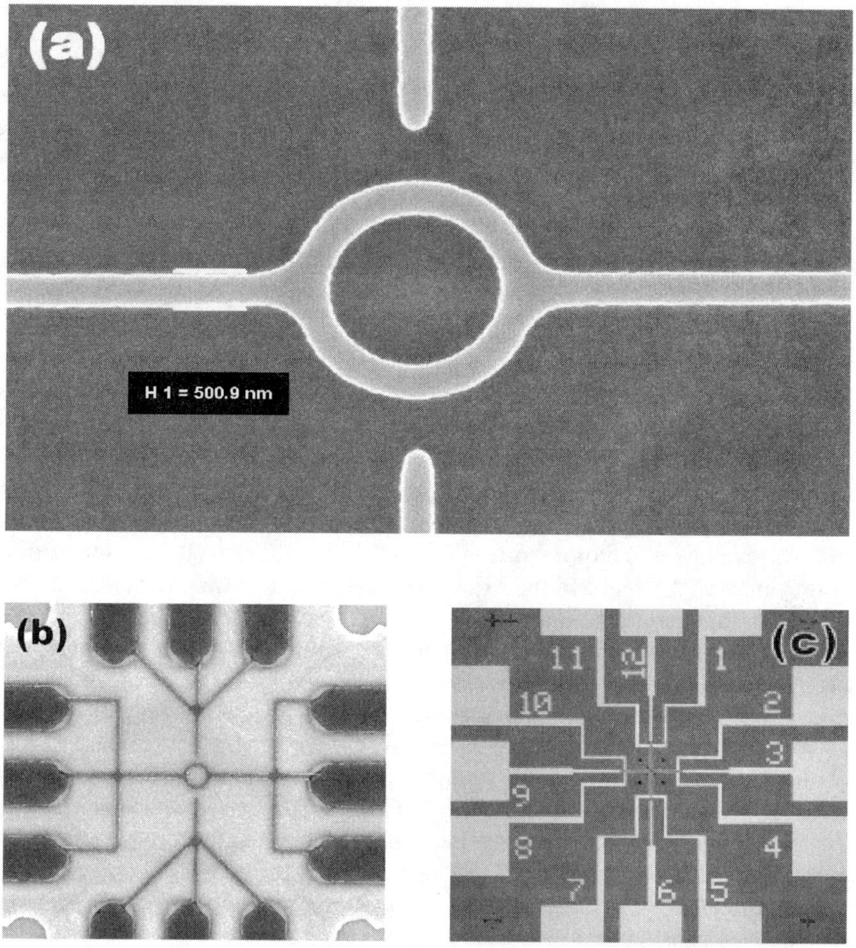

FIGURE 6.8. **(a)** SEM micrograph of an A-B ring pattern etched into silicon. **(b)** SEM micrograph of an A-B ring defined by imprinting. The ring was imprinted using the stamp shown in (a) and by lift-off. The wider leads were defined by UV lithography and lift-off. **(c)** Optical micrograph of the whole chip showing the contact pads. The pads and leads were defined by UV lithography and lift-off. The A-B ring shown in (b) is in the very center of the image.

pads to imprinted structures, and to demonstrate the feasibility of mixing and matching of SSIL and UV-lithography. The reproducibility of the width of the rings will be covered in the next section.

6.5. PATTERN REPRODUCTION

The fidelity of pattern transfer was checked by impriting several 4 by 4 matrices of 5 μm lines and spaces[21] and by repeating the mixing and matching of A-B rings described in the previous section.

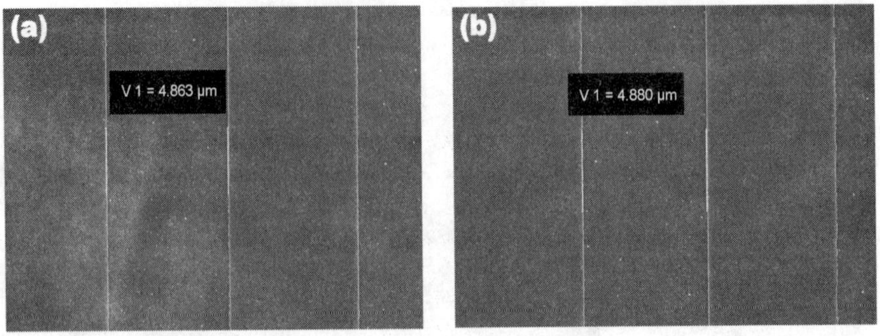

FIGURE 6.9. Top-view SEM micrographs of nominally 5 mm wide imprinted and dry etched trenches in silicon. (a) First of the 16 imprints. (b) Last of the 16 imprints.

The 3 × 3 mm^2 stamp used to create the 4 × 4 matrices consisted of 5 μm lines and spaces dry etched in silicon. The depth of the trenches was 300 nm. The width of the ridges was measured using scanning electron microscopy (SEM) at six random points. The average width of the ridges was 4822 nm. We estimate that the error in the measurement was ±30 nm. In the experiments, no anti-adhesion layer neither surface treatment was applied to the stamp.

The matrices were imprinted into 345 nm thick mr-I 8000 resist spun onto 100 mm Si wafers. During imprinting the substrate was held at 80 °C and the stamp at 220 °C. The release temperature was 20-30 °C below the glass transition temperature, i.e., 80-90 °C. The force during the 240 s imprinting was 60 N, corresponding to a pressure of 13 MPa, assuming that only the ridges in the stamp were pressed against the resist. The ramping time for the pressure was 60 s and the cooling time to reach the release temperature was 180 s. The temperature of the substrate increased unintentionally during each imprinting from 80 °C up to 105-110 °C.

The residual layer at the bottom of the windows was removed on oxygen plasma and the opened resist pattern was used both as etching mask and as lift-off mask. The selectivity of mr-I 8000 to silicon in a CF_4/CHF_3 plasma was found to be about 1:2. The selectivity is relatively low and the resist is not a suitable mask for etching deep trenches. For etching deeper trenches the resist was first used as a lift-off mask for aluminum. 30 nm of Al was sputtered and lifted-off using the mr-I 8000 mask. The aluminum mask was used to etch the trenches in CF_4/O_2 plasma. SEM images of the resulting 200 nm deep Si gratings are shown in Fig. 9. In Fig. 9(a) is shown the first grating and in Fig. 9(b) the last, 16th grating. The first and the last grating reproduces the stamp features well. No degradation of the previous patterns was found during the subsequent imprintings.

The width of the etched ridges was measured from each grating at about the same position. The results are given in Figure 10 (a). The average width is 4856 nm and the standard deviation 29 nm. We estimate that the accuracy in the measurement is ±30 nm. The average of the imprinted and etched ridge widths is about 34 nm larger than the average ridge width of the stamp. This difference can be attributed to a slight broadening of the imprinted trenches in the resist during the oxygen plasma treatment. Similar results were obtained also for other imprints of patterns in 4 × 4 matrices. The distribution of the width of the imprinted gratings is small and results in deviation less than 1 % from

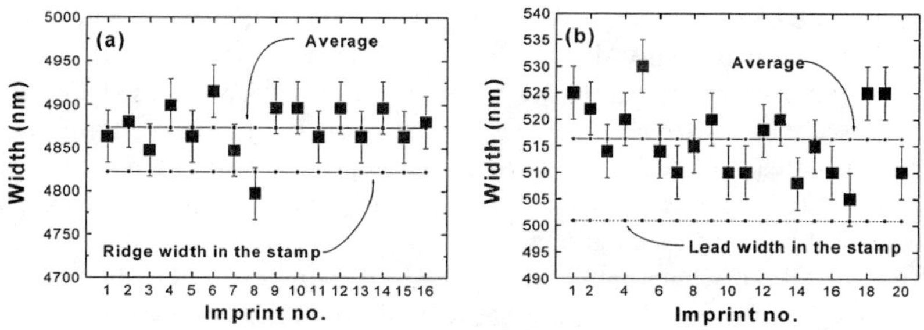

FIGURE 6.10. (a) Width of trenches etched in silicon measured from the 16 samples in the imprinted 4 by 4 matrix. The aluminum mask used in dry etching was formed by lift-off after opening the imprinted pattern in oxygen plasma. (b) Width of 20 A-B rings formed by imprinting and lift-off. The stamp and one of the rings are shown in Figure 8.

the original ridge width in the stamp. This does not necessarily reflect other than the distribution of the original ridge widths in the stamp. Nevertheless, what is important here is the fidelity of the pattern reproduction after repeated imprinting, lift-off and dry etching. It is important to note that no adhesion was observed during repeated step and stamp although no anti-adhesion surface treatment was used.

Similar results were obtained when several A-B rings were imprinted during the mix and match experiments. The width of the leads in the stamp was 501 nm (see Figure 8(a)). The width distribution of 20 consecutive A-B rings after imprinting and lift-off are shown in Figure 10(b). The average width is 516 nm and the standard deviation 7 nm. Again, a slight increase in the width occurred, probably resulting from oxygen ashing.

The somewhat uncontrolled broadening of the trenches during oxygen ashing can become a problem when very small patterns are imprinted and the patterns are opened in plasma. For transferring the imprinted pattern the resist residue removal is generally required in NIL. To minimize the the broadening of the features, the thickness of the residual layer must be mininized. This probably means that the optimized process parameters scale with the size of the features to be imprinted.

Despite of the slight broadening of the trenches during the ashing, the size distribution from imprint to imprint is very small, of the order of 1 % of the original feature size. Although the number of samples in the experiments was not very large, the overall results suggest that the repeated reproduction of patterns is relatively well controlled in SSIL.

6.6. CONCLUSIONS

Currently the emphasis in the field of imprinting lithography is mainly in developing techniques for large area parallel imprinting. We have shown that also sequential methods, mimicking the operation of an optical stepper, can be used to pattern large areas. To benefit fully from SSIL, more sophisticated imprinting tools, dedicated to step and stamp, must be developed. The leveling and alignment can be automated and the total

SSIL cycle time can be reduced by making the sample cooling efficient. Although a sequential process is naturally slowlier than a parallel process, with SSIL one can gain flexibility in imprinting in a similar way when an optical stepper is used, i.e., each chip can be exposed with a different reticle. Here, the stamp can be changed from imprint to imprint, which means that adjacent chips can be imprinted with different stamps.

A special feature in SSIL is that only the stamp, and the area under the stamp, is heated above the glass transition temperature of the resist. This is of course essential for the SSIL to be a viable patterning method. The quality of the already imprinted features must not degrade during the sequentially performed imprinting. This means that patterned substrates or objects which can not withstand high temperatures can be patterned by SSIL.

The capability to align the stamp to already existing features on the substrate makes it possible to use SSIL together with UV lithography for mix and match. We have demonstrated the combination of SSIL and UV lithography by fabricating A-B rings and other nanoscale metal structures. Here the critical issue is the accuracy in the alignment between the features created by SSIL and UV lithography. Thus, the mix and match requires two sets of alignment marks, one designed for UV and the other for imprint lithography. The overall accuracy achieved with the flip-chip bonder used in the experiments was of the order of 1.5 m, which is not very high but high enough to couple contact pads and leads to the imprinted devices.

Although it seems that NIL can not compete with optical lithography in, e.g., integrated circuit production, niche applications, such as patterning of functionalized polymers and applications requiring low-cost patterning techniques will form the application areas for NIL. Various imprinting techniques form an ensemble of relatively flexible lithography tools for the needs differing from those developed for major semiconductor industry. Step & stamp imprinting lithography is one variant of the imprinting methods, providing means both for large area imprinting and for mix and match with existing lithography methods.

ACKNOWLEDGEMENTS

This work has been partially funded by the European Union through projects NANOTECH (#28785) and CHANIL (#IST-1999-13415). M. Markkanen and J. Pekkala are gratefully acknowledged for technical assistance.

REFERENCES

1. S. Y. Chou, P. R. Krauss and P. J. Renstrom, Appl. Phys. Lett. **67**, 20 (1995).
2. S. Y. Chou, P. R. Krauss, W. Zhang, L. Guo and L. Zhuang, J. Vac. Sci. Technol.**B 15**, 2897 (1997).
3. T. Ito and S. Okazaki, Nature **406**, 1027 (2000).
4. See, e.g., Scientific American, p. 32 April 2001.
5. Z. Yu, S. J. Schablitsky and S. Y. Chou, Appl. Phys. Lett. **74**, 2381 (1999).
6. J. Seekamp, S. Zankovych, A.H. Helfer, P. Maury, C.M. Sotomayor Torres, G. Böttger, C. Liguda, M. Eich, B. Heidari, L. Montelius and J. Ahopelto, Nanotechnology, **13**, 1-6 (2002).
7. Puscasu, G. Boreman, R. C. Tiberio and D. Spencer, J. Vac. Sci. Technol. **B 18**, 3578 (2000).

8. M. Berggren, A. Dodabalapur, R. E. Slusher, A. Timko and O. Nalamasu, Appl. Phys. Lett. **72**, 410 (1998).
9. M. Li, J. Wang, L. Zhuang and S. Y. Chou, Appl. Phys. Lett.**76**, 673 (2000).
10. T. Mäkela, T. Haatainen, J. Ahopelto and H. Isolato, J Vac. Sci. Technol. **B 19**, 487 (2001)
11. B. Heidari, I. Maximov and L. Montelius, J. Vac. Sci. Technol. **B 18**, 3557 (2000).
12. W. Zhang and S. Y. Chou, Appl. Phys. Lett. **79**, 845 (2001).
13. H. Schultz et al., in Abstracts of The 45th International Conference on Electron, Ion and Photon Beam Technology and Nanofabrication, Washington, DC, May 29 - June 1, 2001, p. 248.
14. M. Colburn, S. Johnson, M. Stewart, S. Damle, B. J. Choi, T. Bailey, M. Wedlake, T. Michaelson, S.V. Sreenivasan, J. Ekerdt, and C.G. Willson, Proc. SPIE Vol. **3676**, 379-389 (1999).
15. T. Haatainen, J. Ahopelto, G. Gruetzner, M. Fink and K. Pfeiffer, Proc. SPIE **3997**, 874, (2001).
16. H. Schulz, H.-C. Scheer, T. Hoffmann, C. M. Sotomayor Torres, K. Pfeiffer, G. Bleidiessel, G. Gruetzner, Ch. Cardinaud, F. Gaboriau, M.-C. Peignon, J. Ahopelto and B. Heidari, J. Vac. Sci. Technol. B 18, 1861 (2000). The PPM polymer is now commercialized by micro resist technology GmbH, http://www.microresist.de.
17. Xiaoyun Sun, Lei Zhuang, Wei Zhang, and Stephen Y. Chou, J. Vac. Sci. Technol. **B 16**, 3922 (1998).
18. Hiroyuki Nakamura, Akiyoshi Baba and Tanemasa Asano, Jpn. J. Appl. Phys. Part 1 39, 7080 (2000).
19. B. Faircloth, H. Rohrs, R. Tiberio, R. Ruoff and Robert R. Krchnavek, J. Vac. Sci. Technol. **B 18**, 1866 (2000).
20. T. Haatainen and J. Ahopelto, Physica Scripta, in print.
21. T. Haatainen and J. Ahopelto, to be published.

7

Step and Flash Imprint Lithography

A low-pressure, Room-Temperature Nanoimprint Pattering Process

T.C. Bailey, M. Colburn, B.J. Choi, A. Grot, J.G. Ekerdt, S.V. Sreenivasan, C.G. Willson

Texas Materials Institute, The University of Texas at Austin, USA

7.1. INTRODUCTION

Does optical projection lithography have a limit? Is this limit related to the process physics and chemistry or the cost of ownership of lithography tools? A combination of improvements in optics, reduction in wavelength, and the introduction of more complex photomasks and processes has enabled printing of features smaller than 100 nm. These improvements came at a cost, however, and this cost has followed an exponentially increasing trend (1). The Semiconductor Industry Association International Technology Roadmap for Semiconductors[1] has identified alternative next generation lithography (NGL) imaging techniques based on X-ray and extreme ultraviolet (EUV) ionizing radiation, and electron beam lithography in both projection and direct-write modes. Each has its advantages and disadvantages and all are expensive. Many research groups are exploring alternative forms of imprint lithography, an inexpensive patterning method capable of sub-100 nm resolution on silicon substrates.[2-5] Photolithographic resolution follows the well-known relationship:[6]

$$R = \frac{k\lambda}{NA} \tag{1}$$

where k is dependent on the optical system and resist material, λ is the wavelength of the light, and NA is the numerical aperture of the lens; resolution is ultimately limited by λ. Imprint lithography has several important advantages over conventional optical lithography and NGLs; it is non-optical by design, and the resolution appears to be limited only by the resolution of structures that can be generated in the template or mold. In fact, sub-100 nm resolution is well documented for imprint lithography.[2,3] Imprint templates are typically fabricated using imaging tools such as electron beam writers that provide high resolution but lack the throughput required for mass production. The imprint process has a much shorter cycle time than direct write e-beam, therefore imprint lithography takes advantage of the resolution offered by e-beam technology for generating templates without compromising throughput during wafer patterning. The imprint equipment involves precision X-Y stages, overlay error measurement, and wafer handling, as with other lithography techniques, but imprint lithography realizes significant cost savings due to the lack of requirement for lasers and projection optics. An imprint platform built for a certain technology generation could conceivably extend indefinitely with slight modifications to tolerances such as the overlay alignment error correction and stage movement accuracy and precision. The main advances required to improve resolution are in the area of the imprint template development.

There are many imprint lithography techniques, all variations on a common theme. The basic premise is that a template or mold with a prefabricated topography is pressed into a displaceable material. That material takes on the shape of the pattern defined in the template, and through some curing process, the shaped material is hardened into a solid. The process is by nature a contact patterning process that transfers patterns without scaling. There are common challenges to all of these imprint techniques, the foremost being the dependence of this technology on 1-X imprint master resolution, and the potential for defect production and propagation from the contact nature of the replication process.

7.2. PROCESS OVERVIEW

The Step & Flash Imprint Lithography (SFIL) process is a high throughput, low cost approach to generating relief patterns with line widths smaller than 100 nm.[3,7] SFIL uses no projection optics, and operates at room temperature. The process relies on photopolymerization chemistry and low-pressure mechanical processes to transfer patterns. The two key differences between SFIL and other imprint lithography techniques are: SFIL uses a low viscosity, photo-curable, organosilicon liquid, and SFIL uses a transparent, rigid template that allows for layer-to-layer alignment. High temperatures and imprint pressures are avoided by using a low viscosity photo-curable liquid. The transparent rigid imprint template allows flood exposure of the photopolymer to achieve cure, and enables classical optical techniques for layer-to-layer alignment.

The SFIL process flow is shown in Figure 1. An organic transfer layer is spin-coated on a silicon substrate. A low viscosity, photopolymerizable, organosilicon solution is then dispensed on the wafer in the area to be imprinted. A transparent template bearing patterned relief structures, which has been surface treated to facilitate release after the photopolymerization step, is aligned over the coated silicon substrate. The template is lowered into contact with the substrate, thereby displacing the etch barrier that fills the imprint field, trapping the photopolymerizable liquid in the template relief. Irradiation with UV light through the backside of the template cures the etch barrier into a crosslinked polymer film. The template is then separated from the substrate leaving an

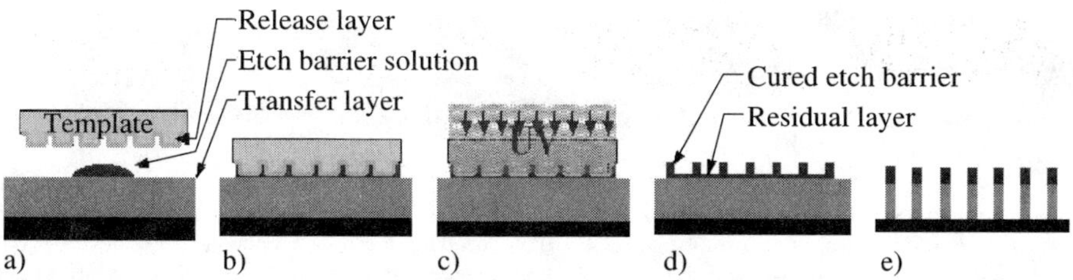

FIGURE 7.1. SFIL process flow 7. The process employs a template/substrate alignment scheme to bring a rigid template and substrate into parallelism (a), trapping the etch barrier (b). The gap is closed until the force that ensures a thin base layer is reached. The imprint is then illuminated through the backside of the template (c) to cure the etch barrier. The template is withdrawn (d), leaving low-aspect ratio, high resolution features in the etch barrier. The residual etch barrier (base layer) is etched away with a short halogen plasma etch, after which the pattern is transferred into the transfer layer with an anisotropic oxygen reactive ion etch (e), creating high-aspect ratio, high resolution polymer features.

organosilicon relief image on the surface of the coated substrate that is a replica of the template pattern. A halogen etch is used to break through the undisplaced etch barrier material, called the "residual layer," exposing the underlying transfer layer. An anisotropic oxygen reactive ion etch (RIE) is used to transfer the image through the transfer layer, thereby amplifying the aspect ratio of the imprinted image. The imprinting process is conducted at room temperature, and since the template is transparent, all of the alignment schemes that have been used successfully in mask aligners, photolithography steppers and scanners can be implemented without difficulty. The process is simple in concept, but every step in the process presents interesting challenges in materials engineering and science.

7.3. TEMPLATE FABRICATION

The SFIL process is designed with the capability for reliable, high-resolution patterning in mind. Layer-to-layer overlay alignment has always been a primary consideration in the development of the materials and the processes. Numerous layer-to-layer alignments must be achieved in order to pattern a semiconductor substrate with functional devices; there are five metal layers in, for example, the interconnect levels of the Intel P856 process flow[8] and many more steps are required to define the underlying devices, each of which requires alignment to substrate patterns. We have demonstrated that alignment to patterns on the substrate is possible with SFIL.[9]

SFIL templates have been fabricated using the process shown in Figure 2, which is similar to conventional phase-shift reticle processing. Other template fabrication schemes have also been investigated.[10] In the standard chromium process, a 6-in by 6-in fused silica reticle blank coated with ~1000 Å of chromium is spin-coated with an electron beam resist, and baked to drive off excess casting solvent. In general, the resist thickness after this baking step is between 200-300 nm, however the exact application and bake conditions are dependent on the specific resist and target resist thickness. The plate is then exposed by electrons in a direct-write e-beam tool; electron beam lithography is

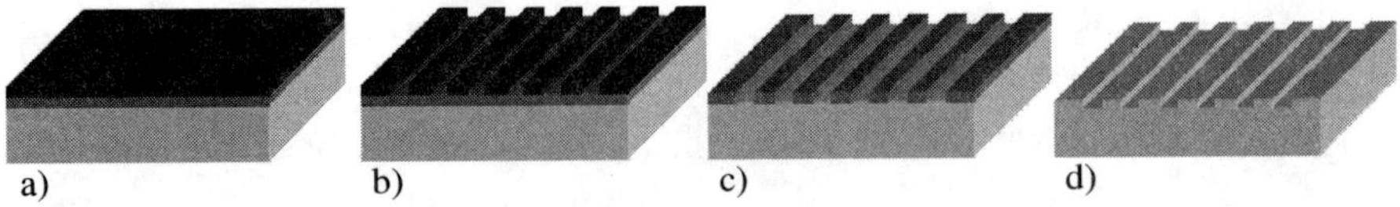

FIGURE 7.2. Template Fabrication Process. A 6-in square by 0.25-in thick fused silica plate is coated with a layer of Cr and a thin film of electron beam resist (a), exposed to an electron beam and developed (b). The resist is used as an etch mask to pattern the Cr, and is then removed (c). The Cr pattern is used as an etch mask to transfer the pattern into the silica substrate, and the Cr is removed (d), leaving a transparent, rigid template with a topographic pattern. Reprinted with permission from Resnick, et al., Proc. SPIE: Emerging Lithographic Technologies VI; v. 4688 (2002). Copyright 2002 SPIE.

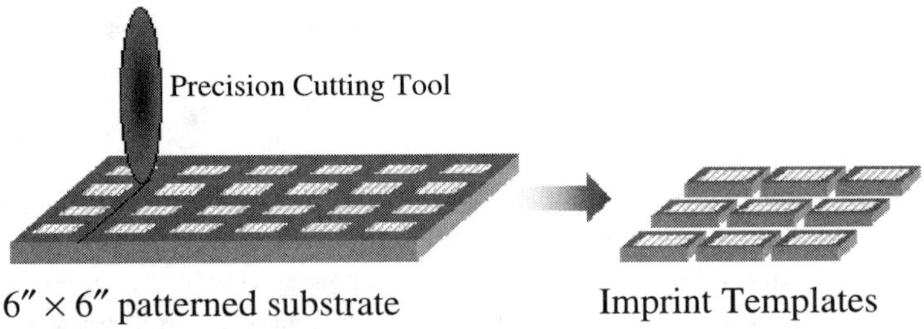

FIGURE 7.3. Template Generation. After patterning, the plate is cut into individual templates to fit in the SFIL stepper.

used because it is capable of writing very high-resolution patterns. The resist is then developed, leaving a resist pattern that exposes selected portions of the underlying Cr film. This resist pattern is used as an etch mask to pattern the Cr with Cl-based RIE. The resist is then removed, leaving a Cr pattern on the silica substrate. A fluorine-based RIE transfers the image into the fused silica substrate to a depth of 100 to 200 nm, depending on design constraints. The Cr layer is usually left on the templates until after the cutting process, which is described below; this facilitates template pattern recognition, and minimizes mistakes in cutting.

Each imprint template is 1-in by 1-in, which is larger than the largest die size predicted in the SIA Roadmap.[1] Several templates can be made on each 6-in substrate. The individual templates can be cut out using a variety of methods, such as a precision wheel saw, water knife, or cutting laser, as in Figure 3. Regardless of the cutting method used, the final template dimensions must meet the constraints of the imprint equipment. After cutting, the Cr is removed typically by immersion in a solution of ceric ammonium nitrate, and the surface is cleaned and treated as described below.

7.4. SURFACE TREATMENT

It is imperative, following exposure, that the polymerized etch barrier adhere to the underlying transfer layer, and release easily and completely from the template. The successful release of the imprint template from the imprinted polymer depends

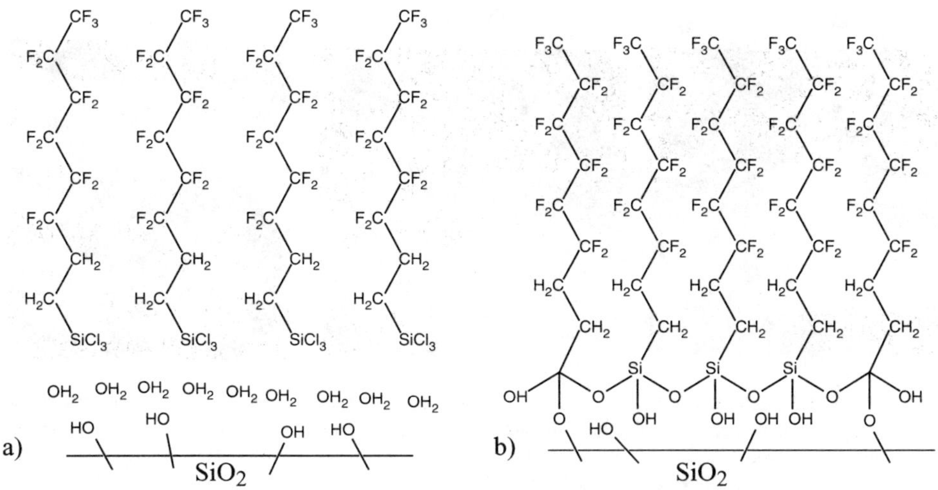

FIGURE 7.4. Self-assembling monolayer formation. A flourinated self-assembling monolayer is used to provide a low surface energy coating that ensures release of the cured etch barrier. Trichlorosilanes migrate to the template surface (a), where they react with surface water to form silanol intermediates. These intermediates undergo condensation reactions with hydroxyl groups on the quartz surface, as well as those on adjacent silanols, to form a networked siloxane monolayer (b). The orientation of the chains is away from the quartz, in a 3-D comb-like structure. Annealing further enhances the condensation, creating a highly networked, durable, low surface energy coating.lease provide a caption for this figure]

in part on the surface energies of the template and the imprinted polymer. The polymerized etch barriers we have worked with have surface energies in the neighborhood of 30 dynes/cm[3]. Lower surface energies on the template facilitate successful release. Hare, *et al.*[11]predicted that a surface composed of only -CF_3 groups would have the lowest critical surface tension of any system, ~6 dynes/cm. Nishino, *et al.* later verified that prediction, and demonstrated a surface energy of 6.7 dynes/cm for $C_{20}F_{42}$ adsorbed on pretreated glass.[12] Diffraction studies indicated that the $C_{20}F_{42}$ chains were aligned perpendicular to the glass substrate, which exposed -CF_3 end groups on the sample surface. As a comparison, TEFLON®, known for its release properties, has a surface energy of 18 dynes/cm.[13] A monolayer film with -CF_3 terminations clearly has the potential to be a superb release coating for the SFIL template. A fluoroalkyltrichlorosilane precursor is used to form such a monolayer on the template surface to promote this release.[3,14]

Liquid phase and vapor phase application methods have been studied. In both cases, The –OH groups on the SiO_2 surface react with the trichlorosilane, forming HCl as a byproduct [15]. The reaction of alkyltrichlorosilanes with SiO_2 proceeds very slowly in the absence of surface water, but quite rapidly in the presence of surface water.[16-19] Tripp and Hair studied the reaction of alkylchlorosilanes and fluoroalkylchlorosilanes with silica by monitoring the SiOH groups with infrared spectroscopy. They found that alkylchlorosilanes did not react with a completely dehydrated surface, and the fluorinated counterparts reacted slowly. In the presence of surface bound water, the precursors react readily to form networks derived from the formation of bonds between adjacent

FIGURE 7.5. *Water* contact angle on an untreated template (left), a newly treated template (center), and a treated template (right) that was used to imprint numerous times. The template retained its release functionality, and it can be seen that the water contact angle did not change significantly. Reprinted with permission from Bailey, *et al.*, J. Vac. Sci. Tech. B, 2000 **18**(6): p. 3572. Copyright 2000 AVS.

molecules. The water reacts with the self-assembled monolayer (SAM) precursor to form a silanol intermediate and acid in an irreversible reaction:[15]

$$\text{R-SiCl}_3 + 3\text{H}_2\text{O} \rightarrow \text{R-Si(OH)}_3 + 3\text{HCl} \quad (2)$$

The intermediate has three -OH groups, which can either bond to the quartz surface or to adjacent molecules through the loss of water. This network formation, pictured in

In one set of experiments the quartz templates were cleaned in an acetone ultrasonic bath, followed by O_2 RIE at 50 W, 10 sccm O_2, 20 mTorr for 10 min to remove organic contaminants on the surface. The clean templates were then treated with a release agent, tridecafluoro-1,1,2,2-tetrahydrooctyltrichlorosilane (Gelest), by vapor exposure at 1 atm total pressure (precursor plus N_2) for 90 min at room temperature, and annealed at 100 °C for 15 min. This surface treatment reaction has yielded surface energies in the neighborhood of 12 dynes/cm.[14]

The surface treatment must maintain its release characteristics through thousands of imprints in a manufacturing process. Preliminary results indicate that the current technique could provide films with the required durability. Figure 5 shows the water contact angle on an untreated template (left), a newly treated template (center), and a treated template (right) that was used for a period of two months and was cleaned vigorously. The template retained its release functionality, and it can be seen that the water contact angle did not change significantly. We have seen no evidence of catastrophic loss of release function, and work is currently underway to quantify film durability for a variety of surface treatment conditions.

7.5. ETCH BARRIER

The etch barrier material is subject to several design constraints. The etch barrier liquid must be readily displaced during the imprint step; it must photopolymerize rapidly

TABLE 7.1. Example components for the SFIL low viscosity photocurable imprint liquid.

Purpose	Component	Supplier	Trade name
High Si content	(3-acryloxypropyl) tris (trimethylsiloxy) silane	Gelest	SIA0210.0
	(3-acryloxypropyl) methylbis (trimethylsiloxy) silane	Gelest	SIA0194.0
Crosslinker	1,3-bis (3-methacryloxypropyl) tetramethyldisiloxane	Gelest	SIB1402.0
Photoinitiator	2-hydrozy 2-phenyl-propan-1-one	Ciba	Darocur 1173
	bis (2,4,6-trimethylbenzoyl)-phenylphosphineoxide	Ciba	Irgacure 819
	1-hydroxy-cyclohexyl-phenyl-ketone	Ciba	Irgacure 184
Low MW	acryloxytrimethylsiloxane	Gelest	SIA0320.0
	butyl acrylate	Aldrich	
	t-butyl acrylate	Aldrich	

during exposure; the polymer must release preferentially from the template while remaining adhered to the transfer layer; and it must exhibit sufficient etch selectivity during the O_2 RIE step to allow for high aspect ratios to be generated in the transfer layer.

7.5.1. *Chemistry*

The current etch barrier liquid is a multicomponent solution containing a photoinitiator, a monomer with high Si content to provide O_2 RIE etch resistance, a difunctional monomer to allow crosslinking, and a low molecular weight monomer to reduce solution viscosity. Some examples of typical monomers and components are listed in Table 1.

In separate studies of these acrylate monomers, the effect of fluence was examined using IR to follow the isolated C=C stretching absorbance peak at 1640 cm^{-1} and revealed nearly complete conversion with less than 30 mJ/cm^2.[3] The organic monomer, n-butyl acrylate, serves as a solvent and a mass-persistent component in the etch barrier formulation. This assists in minimizing shrinkage during polymerization. The silylated monomer provides etch resistance in the O_2 RIE "amplification" etch that transforms the low aspect ratio, high resolution relief structures into high resolution, high aspect ratio features. The crosslinker provides mechanical stability to the cured etch barrier relief structure and also serves to improve the cohesive strength of the etch barrier, which are both necessary for clean separation. For further information on UV curing of acrylate coatings can be found in works by Decker[21] and Kloosterboer.[22-24]

7.5.2. *Imprint Pressure*

The process for dispensing the etch barrier onto the wafer is critical. Several mechanisms of fluid delivery have been evaluated including capillary action, spin-coating, and compression filling. Capillary action involves orienting the template in close proximity to the wafer, dispensing an amount of etch barrier liquid at the template edge, and allowing capillary forces to draw the liquid into the void between the template and wafer. Given a fluid with viscosity similar to water (~1 cP) and a template-to-wafer gap of 100 nm, the time to fill a 1-in square template exceeds 1000 s, [25] which is not compatible with

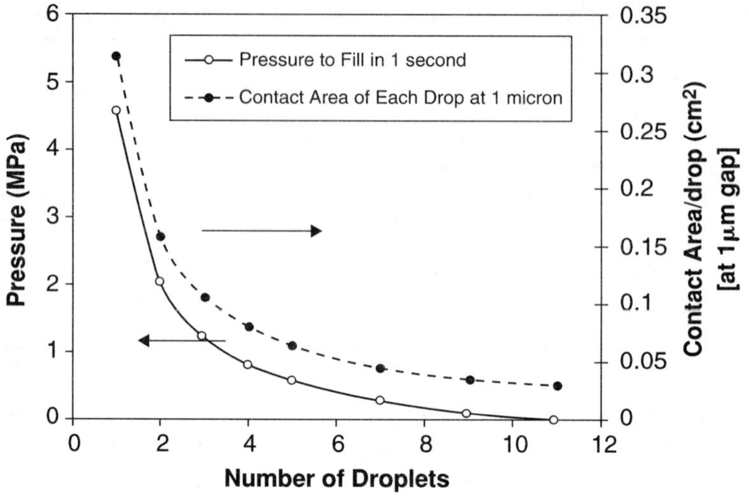

FIGURE 7.6. Imprint pressure required to achieve a 100 nm gap height in 1 second for a single drop in a multi-drop compression. The viscosity is 1 cP. Reprinted with permission from M.E. Colburn, Step and Flash Imprint Lithography: A Low-Pressure, Room-Temperature Nanoimprint Lithography. Department of Chemical Engineering; Ph.D. Thesis. 2001, Austin, TX: The University of Texas at Austin. Copyright 2000 University of Texas.

the typical throughput of 90 wafers per hour [1]. Spin-coating stable thin films requires a fluid with certain properties. A typical etch barrier solution would spin a stable film on substrates less than 0.1 mm in diameter,[25] which would not be useful in manufacturing. Compression filling involves dispensing a small amount of etch barrier solution on the wafer, and imprinting directly on top of it in order to displace the liquid. Compression filling is compatible with the SFIL process in its current form and is therefore the method of choice.

Figure 6 shows the imprint pressure required to achieve a 100 nm residual layer in 1 sec when the amount of liquid required to fill that volume is divided into a number of droplets. Imprinting on a single liquid droplet actually deforms (bows) the template to such an extent as to create severe nonuniformity in the residual layer, rendering the imprint unusable. In order to minimize this distortion, the volume of etch barrier is divided into many droplets, or spread it in a continuous pattern, over the active area to be patterned. Multi-drop and continuous pattern methodologies minimize the generation of localized forces that cause the imprint template to bend. Dividing the total liquid volume into n droplets reduces the imprint force required to displace the liquid by a factor of $1/n$, as shown in Figure 6.[25]

7.5.3. *Etch Selectivity*

The features are imprinted in etch barrier at a relatively low aspect ratio, typically less than 3:1. Transferring this pattern through the underlying transfer layer amplifies this aspect ratio, so long as the etch rate of transfer layer in O_2 RIE is greater than that

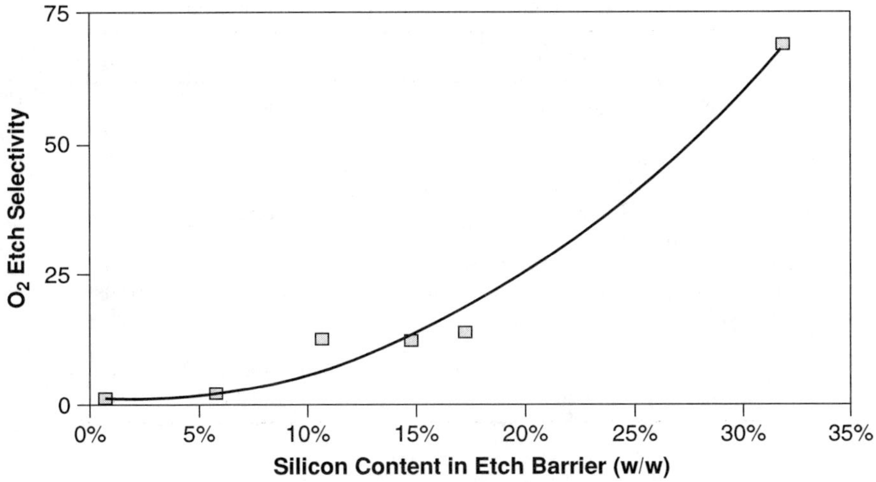

FIGURE 7.7. Effect of silicon content on O2 etch selectivity between the etch barrier and a polystyrene film. Etch barrier formulations containing >11% Si (w/w) yield the required 10:1 etch selectivity. Reprinted with permission from Colburn, et al., J. Vac. Sci. Tech. B, 2001. 19(6): p. 2162. Copyright 2001 AVS.

of the etch barrier. The target etch rate selectivity is 10:1. The etch barrier composition was modified to yield different amounts of Si incorporation, and the etch rate in O_2 RIE was measured for each film; this was normalized to the etch rate of the transfer layer (with no Si), and the results are shown in Figure 7. The O_2 etch selectivity starts at 1:1 for films with no Si, and increases to 60:1 for formulations with 30% (w/w) Si. At a Si weight percent of about 11% (w/w), the desired etch selectivity of 10:1 is achieved. We have used a variety of formulations in imprinting experiments, each with different physical properties, and different amounts of Si incorporation.[25] The original formulation, which was used to demonstrate the feasibility of the SFIL process, contained 21% (w/w) Si;[3,26] the formulation used in the UT-Agilent collaboration required a higher viscosity, and contained 30% (w/w) Si; [27] and a newer formulation contains 13.5% (w/w) Si.[9,28,29] The etch barrier chemistry is an important part of the process and an active area of research.

7.5.4. Volumetric Shrinkage

The SFIL process relies on *in situ* photopolymerization of a low viscosity, acrylate-based solution. Acrylate polymerization is known to be accompanied by volumetric shrinkage that is the result of chemical bond formation. Consequently, the size, shape, and placement of the replicated features may be affected. A series of measurements was performed on the potential etch barrier components listed in Table 2, which contain different pendant groups, in order to identify the volumetric shrinkage associated with the polymerization of each monomer.[25,30] HyperChem® simulations were utilized to determine the ratio of the pendant group volume to the total monomer volume, as shown in Figure 8.

TABLE 7.2. Volumetric shrinkage of SFIL components.

Monomer	Ave ΔV	St. Dev.
2-(acryloxyethoxy) trimethylsilane	9.10%	0.30%
(3-acryloxypropyl) dimethyl methoxysilane	7.80%	0.30%
(3-acryloxypropyl) methyl (bis- (trimethylsiloxy) silane	5.10%	0.20%
(3-acryloxypropyltristrimethylsiloxane) silane	4.60%	0.50%
Methacryloxyethoxy-Trimethylsilane	8.80%	0.80%
Acryloxytrimethylsilane	9.40%	0.30%
Methacryloxypropyl (tristrimethylsiloxane) silane	4.10%	0.10%
Monomethacryloxypropyl terminated dimethylsilane	2.20%	0.60%
Ethyl acrylate	16.40%	0.80%
Butyl acrylate	14.20%	0.30%
Hexyl acrylate	11.50%	0.40%
Lauryl acrylate	6.00%	0.40%

The volumetric shrinkage was measured using Archimedes' Principle, which relates the buoyant force acting on an immersed body to the force of gravity on the displaced fluid. Samples of known volume and mass were polymerized, tethered to a mass scale, and immersed in a fluid of known density. Assuming conservation of sample mass, the displaced volume was calculated from the known quantities, and was then used to calculate volumetric shrinkage.

Figure 9 shows the volumetric shrinkage of the components versus the pendant group volume percent. The linear trend for both organic and silylated monomers reveals that the shrinkage can be changed if necessary. An etch barrier solution consisting of 46% butyl acrylate, 46% SIA0210.0, 5% SIB1402.0, 1.5% Irgacure 184, and 1.5% Irgacure 819 shrinks 9.3% (v/v).[25,30]

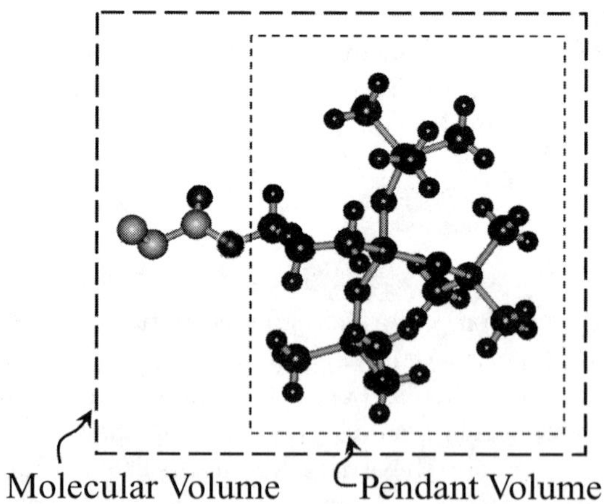

FIGURE 7.8. HyperChem® model of (3-acryloxypropyltristrimethylsiloxane)silane showing the size of the pendant group relative to the size of the acrylate segment.

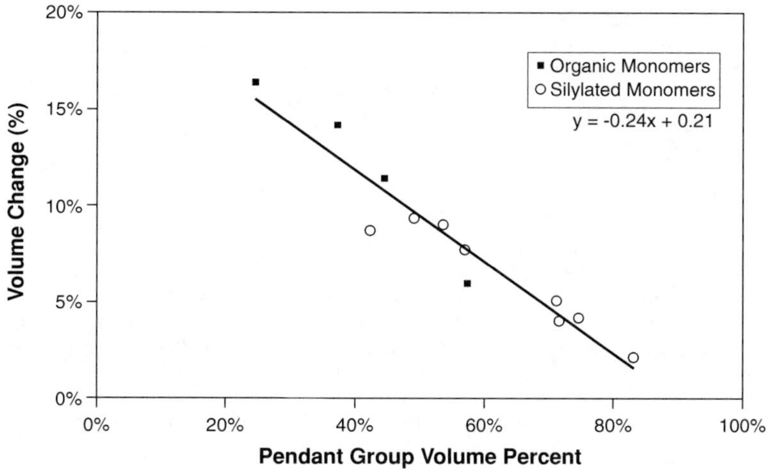

FIGURE 7.9. Volumetric shrinkage as a function of pendant group volume percentage for various SFIL acrylate components under consideration. Reprinted with permission from Colburn, et al., J. Vac. Sci. Tech. B., 2001. 19(6): p. 2685. Copyright 2001 AVS.

7.5.5. *FEM Analysis*

Finite element modeling (FEM) is a powerful tool for predicting effects such as distortion of the imprinted material due to volumetric shrinkage that arise from bulk material properties. The concern is that pattern dependent proximity effects are possible as the shrinkage within large features may impact the final placement of adjacent features.

The pattern in Figure 10 was used to evaluate pattern proximity effects.[25,30] All template features were 200 nm tall; the lines and boxes were separated by 100 nm; the

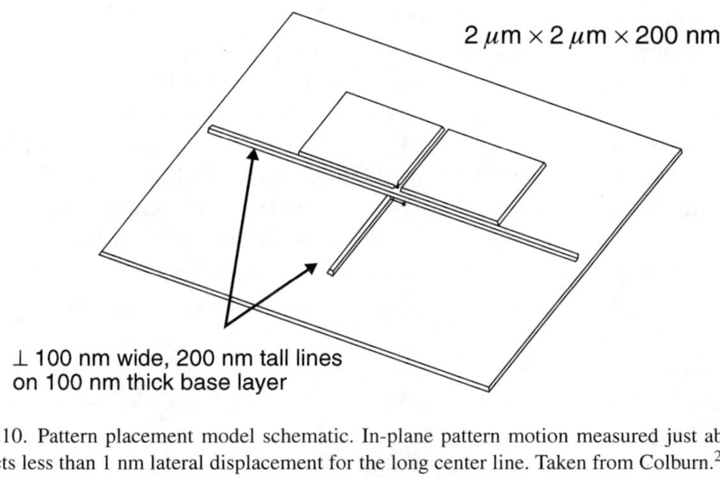

FIGURE 7.10. Pattern placement model schematic. In-plane pattern motion measured just above the base layer predicts less than 1 nm lateral displacement for the long center line. Taken from Colburn.[25]

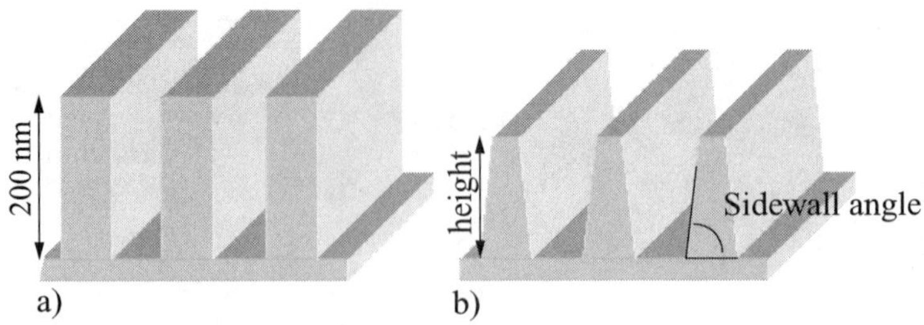

FIGURE 7.11. Grating model to investigate vertical shrinkage and sidewall angle distortion. Imprinted features as they would appear in the template prior to polymerization (a) are 200 nm in height. The feature height and sidewall angle (b) exaggerated for clarity, were calculated using the material property assumptions listed in the text showing a possible effect of polymerization.

boxes were 2 μm by 2 μm squares; the central line that runs along the boxes was 7.6 μm in length; the line that is perpendicular to the central line was 2.5 μm in length. The template was assumed to displace all but 100 nm of etch barrier solution, thus creating a 100 nm residual layer. The bottom of the etch barrier was fixed during polymerization to a rigid transfer layer, and the top of the etch barrier was unconstrained. The etch barrier was assigned a modulus of 1.6 MPa, representative of a low T_g polymer;[31] a Poisson's ratio of 0.3, since the Poisson's ratio for most polymers is between 0.25 and 0.44;[31] and a densification upon polymerization of 10% (v/v). The model assumed complete polymerization prior to densification; experiments on UV-cured acrylate systems by Philips Research Labs showed that under certain conditions the densification indeed lags behind the polymerization reaction.[22] The motion of the 7.6-μm centerline was analyzed at the base of the features to assess displacement motion in the plane of the residual layer. It is important to know whether proximity to small and large features would distort the long line. This displacement was predicted to be less than 1 nm from its original position. Therefore, no significant pattern proximity effect is anticipated for the SFIL process.

Volumetric shrinkage could manifest itself entirely in the z-direction, effectively lowering the aspect ratio of the polymer features. Shrinkage could also yield features with tapered sidewalls, which may result in an intolerable etch bias. The influence of densification and Poisson's ratio on the cross section of features, such as in Figure 11, has been investigated for features ranging from 100 nm to 10 μm in width. The polymerized feature height was taken to be at the top center of a feature in a dense array. The ratio of this height to the original height, 200 nm, is reported as the vertical shrinkage percent in Figure 12. Both the extent of densification and the Poisson's ratio affect the vertical shrinkage significantly. The maximum predicted vertical shrinkage is ~17% for a Poisson's ratio of 0.5 and 17% (v/v) densification (Figure 12). This represents the worst-case scenario for the SFIL replication process.

Significant deviation from the vertical in sidewall angle can cause a change in linewidth during etching of the residual layer. For a high volume manufacturing process, the sidewall angles should be no less than 80°. Analysis of the calculated sidewall angle data revealed a dependence on aspect ratio. Different features were modeled with height fixed at 200 nm and varying width, yielding different aspect ratios. Thus, large aspect

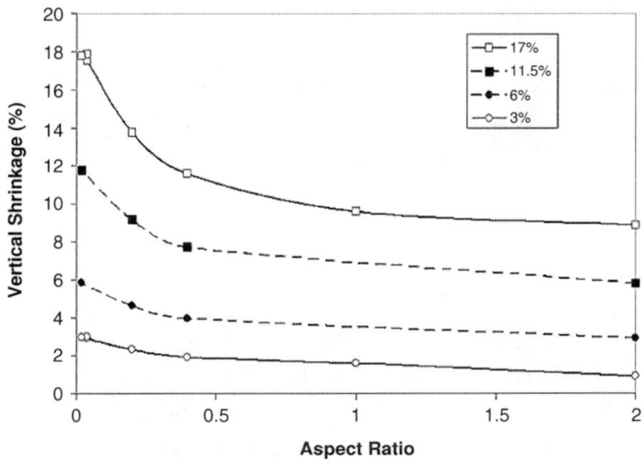

FIGURE 7.12. Predicted vertical Shrinkage of etch barrier features for varying volumetric densification percentages. For very low aspect ratio features the volumetric shrinkage is predicted to manifest itself entirely in the z-direction. Reprinted with permission from Colburn, et al., J. Vac. Sci. Tech. B., 2001. 19(6): p. 2685. Copyright 2001 AVS.

ratios refer to features with smaller linewidth, and conversely for smaller aspect ratios. Figure 13 shows that the sidewall angle is close to 90° for smaller features and that the sidewall angle approaches 80° for low aspect ratio features, using the worst-case model parameters 17% (v/v) densification and 0.5 Poisson's ratio. The line end angle was also studied as a function of aspect ratio; it follows a trend similar to that of sidewall

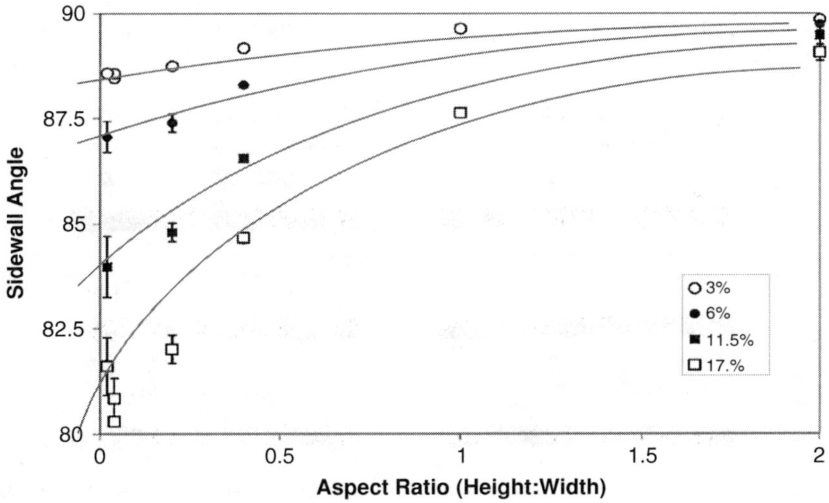

FIGURE 7.13. Predicted effect of densification on sidewall angle for varying densification percentages. High aspect ratio features experience minimal sidewall angle. Low aspect ratio features should exhibit 80 □sidewall angles in the worst case. Reprinted with permission from Colburn, et al., J. Voc. Sci. Tech. B., 2001. 19(6): p. 2685. copyright 2001 AVS.

angle.[25] As the aspect ratio becomes small, the sidewall angle and line end angle both decrease.

These models predict that densification of the etch barrier due to polymerization will not manifest itself as a problem in reference to pattern proximity effects, feature vertical shrinkage, or sidewall angle degradation.

7.6. RELIABILITY

As with any next generation lithography (NGL) being considered, the manufacturability of SFIL depends on the ability to transfer patterns from a master, in this case an imprint template, to the substrate without generating pattern defects. Since imprint lithography is a contact process, there is concern about defect generation and propagation. Fortunately the SFIL imprint process is self-cleaning for contamination on the imprint template.[14] The fundamental question to be answered is, does the imprint process create and/or propagate defects simply by its contact nature.

There are three extreme hypotheses illustrated in Figure 14. Figure 14a shows the case in which no defects are generated, and so no defects propagate; this is the perfect imprint process. Figure 14b shows the case in which some defects are generated during an imprint cycle, but the defects do not propagate through multiple imprints. They may exist for a few imprints, but they eventually disappear, regenerating the original defect-free imprint pattern. This case may be acceptable if the number of generated defects is very small and the number of steps required to clean the template is very low. The third case (Figure 14c) is that in which defects are generated, and most of these new defects propagate through multiple imprints, possibly even for the lifetime of the template, leading to catastrophic loss of pattern transfer fidelity. An initial investigation based on the effect of multiple imprints on the transfer fidelity using patterns comprised of features

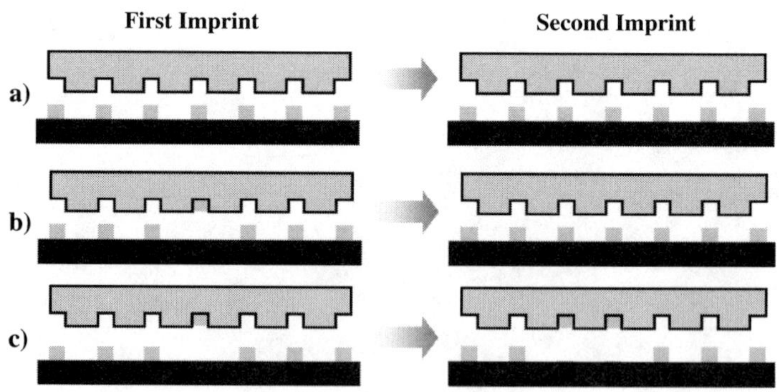

FIGURE 7.14. Three Hypotheses for defect generation and propagation. Perfect imprints (a) yield no defects and perfect pattern transfer. Defect generation without propagation (b) yields a random appearance of defects, which may be tolerable in very low numbers. Generation and propagation of defects (c) yields a case where the imprint template becomes dirtier over time, and would result in a geometric rise in defects with time. Reprinted with permission from Bailey, et al., J. Vac. Sci. Tech. B, 2001. 19(6): p. 2806. Copyright 2001 AVS.

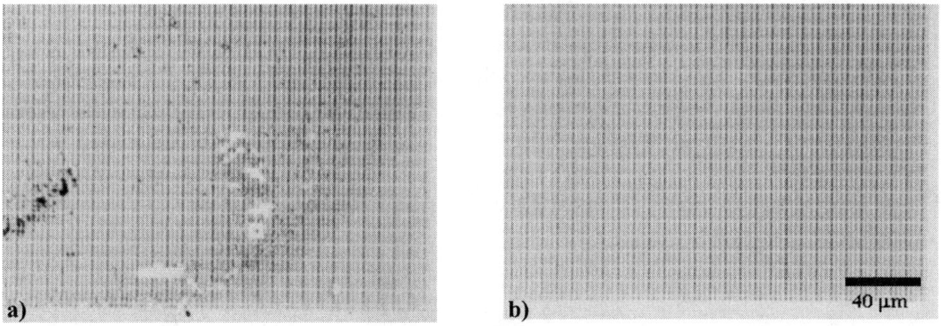

FIGURE 7.15. Images of an imprint template before (a) and after (b) two imprints. Template contamination is removed during imprinting. Reprinted with permission from Bailey, et al., J. Vac. Sci. Tech. B, 2001. 19(6): p. 2806. Copyright 2001 AVS.

with dimensions on the order of 1 micron is in progress. If there exists catastrophic defect generation processes, they should appear in patterns of such features.

An imprint template was inspected prior to installation in the SFIL stepper, and again after imprinting. After two imprints the contaminated template (Figure 15a) became visually clean (Figure 15b). The observation that the template becomes cleaner with successive imprinting allows immediate rejection of the third hypothesis, shown in Figure 14c. It appears that the best way to clean an imprint template is to use it.

Initial inspection of multiple imprints using a Leica INS2000 inspection microscope has been encouraging. Figure 16a shows a part of a field from the first imprint on a sample wafer, Figure 16b the fifth, and Figure16c the tenth imprint. The individual features in the field are 1 μm × 4 μm × 0.25 μm, and there are approximately 1100 bricks per image field. No defect generation is observed in the same field through the ten imprints.

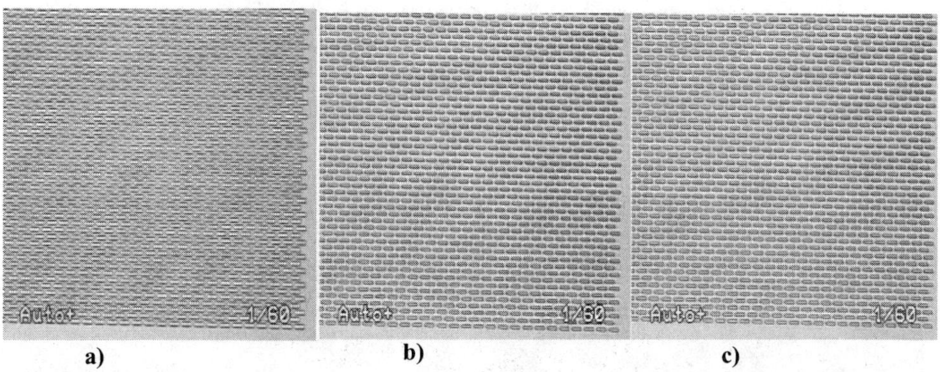

FIGURE 7.16. Visual inspection of multiple imprints. The same area of an imprint field was visually inspected through ten imprints. No generation of defects can be seen. The diagonal fringes are a Moiré effect manifested by the pattern regularity. Any observed difference in the pattern sharpness is due to manual focusing on the Leica INS2000, and not due to pattern transfer fidelity issues. Reprinted with permission from Bailey, et al., J. Vac. Sci. Tech. B, 2001. 19(6): p. 2806. Copyright 2001 AVS.

This field represents only a small fraction of the imprint area, of course, but the result shows that there are no catastrophic defect propagation processes operative.

Certain important characteristics of the imprint process can be gleaned from careful, manual analysis of the imprint images. These manual inspections show no pathological defect generation by, for example, release failure or feature fracture. No such behavior has been noted during inspection of many imprints on many wafers. Random defects have been observed, but all repeating defects have been associated with flaws in the template.

7.7. PATTERNING RESULTS

7.7.1. Patterning on Planar Substrates

Imprint lithography has been demonstrated to have astounding resolution[3,4,32,33] Figure 17 shows top-down and cross-sectional micrographs of 40 nm and 30 nm lines imprinted in etch barrier. It is believed that the resolution of this process is limited to the resolution of features on the imprint template, and not limited by imprint chemistry. SFIL has also produced features with aspect ratios greater than 10:1.[7]

7.7.2. Patterning over Topography

SFIL benefits from its ability to print high aspect ratio features even over relatively large topography.[7] An analog of SFIL was developed in collaboration with Agilent Technologies that allows patterning with a flexible imprint template using a roller press, and was used to demonstrate patterning over existing topography, as shown in Figure 18. A layer of PMMA was spin-coated on substrates bearing topography and baked to remove excess solvent and crosslink the film. The substrate was planarized by spin-coating a photopolymerizable organic film on the patterned substrate, imprinted with a blank (featureless) template and cured by UV exposure. This planarization layer consisted of 92% (w/w) cyclohexyl acrylate, 5% SIB1402.0 (Gelest), 1.5% Irgacure 819, and 1.5% Irgacure 184 (CIBA).[27] The standard SFIL process was then performed on the planar

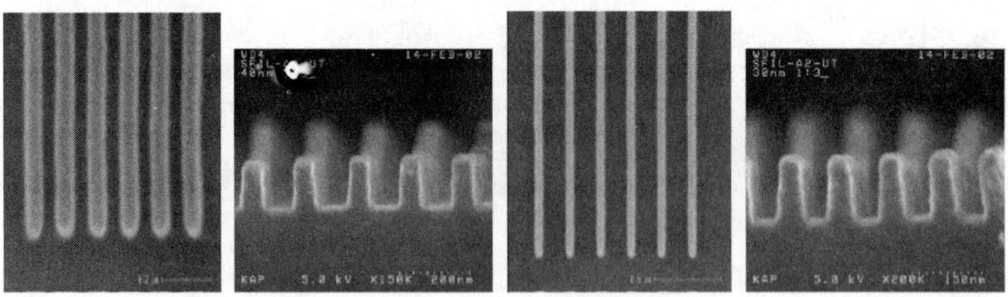

FIGURE 7.17. Top-down and cross-sectional SEMs of 40 nm (a and b, respectively) and 30 nm (c and d, respectively) semi-dense lines imprinted in etch barrier. Reprinted with permission of the author from D.J. Resnick, et al., Proc. Spie: Emerging Lithographic Technologies VI, v. 4688 (2002).

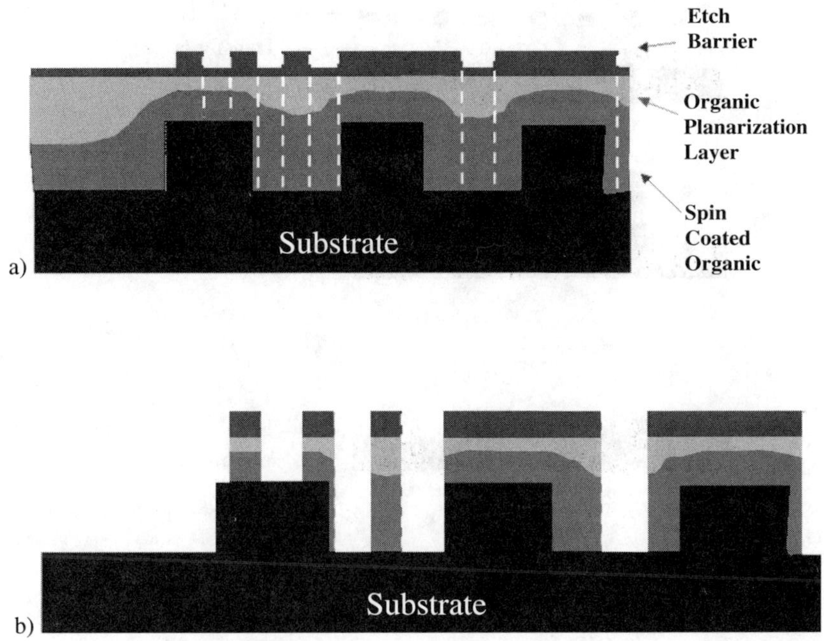

FIGURE 7.18. A schematic of the process developed under UT-Agilent collaboration, allowing SFIL to be used on substrates with existing topography. Cross-section prior to etch transfer (a), and after etch transfer (b), [7].

substrate. Finally, this pattern was etch transferred through the imprinted planarization layer and the organic transfer layer.

Figure 19 shows examples of the results obtained from the process. This work demonstrates that SFIL can in principle be used to pattern real device layers which are never perfectly planar.

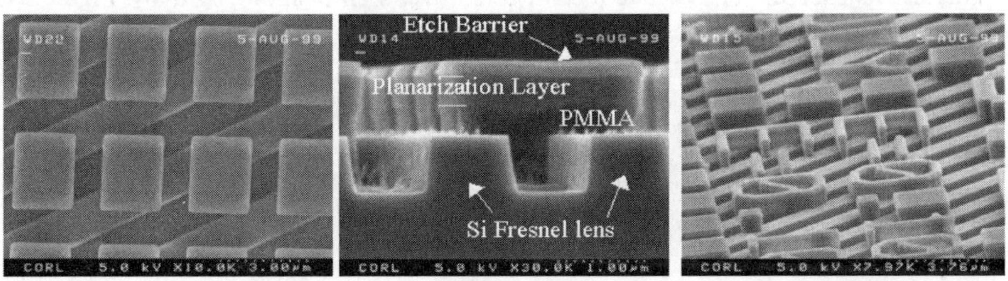

FIGURE 7.19. Top-down SEM (a) showing features printed over topography. The underlying lines are part of a 700 nm tall Fresnel lens. cross-section SEM (b) showing the silicon substrate, the spin-coated PMMA layer, the planarization layer, and the etch barrier. This particular sample was not etched completely, as evidenced by the residual "grass". features 250 nm and greater in with were transferred through 1.1 □ m PMMA and the planarization layer (c), to the bottom of the 700 nm tall Fresnell lens. Fig 20c reprinted with permission from Colburn, et al., J. Vac. Sci. Tech. B, 2001. 19(6): p. 2162. Copyright 2001 AVS.

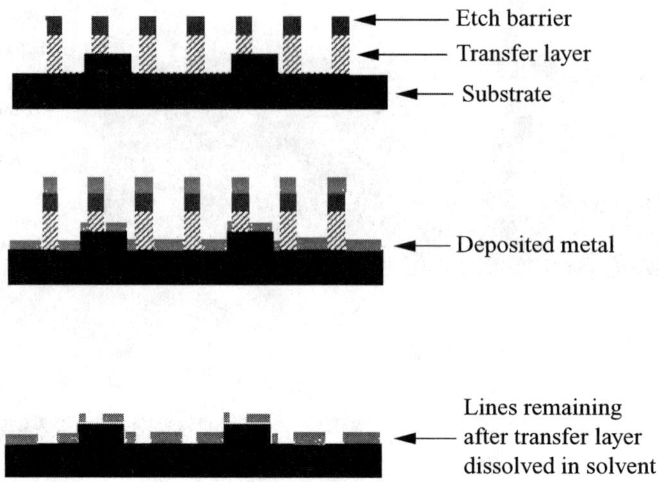

FIGURE 7.20. Additive metallization process by polymer lift-off. The sidewall profile of the transfer layer would be slightly undercut.

7.7.3. Metallization

Use of SFIL to make multilayer electronic devices requires compatibility with metallization schemes. Subtractive metallization, such as the techniques used to pattern blanket al.:Cu films, simply requires that the transfer layer possess etch selectivity to the metal film, and sufficient thickness. Additive metallization, such as damascene-like or lift-off processes, is also possible. Collaborations between the University of Texas at Austin and Agilent Technologies have yielded successful use of titanium lift-off to produce optical devices, and also demonstrate metallization over severe topography using SFIL. The transfer etch in Figure 1e was adjusted to produce a slight undercut of the transfer layer. Anisotropic metal deposition on the resulting structure yielded metal on the horizontal surfaces, and not on the feature sidewalls. This enabled dissolution of the polymer film stack to produce a metal pattern on the substrate, as shown in Figure 20. This process was used to demonstrate metallization over significant topography. Figure 21a shows 300 nm Ti lines patterned over 700 nm topography with no significant change in linewidth. Figure 21b shows a micropolarizer array comprised of 100 nm Ti lines/spaces illuminated through the backside using non-polarized light, and c shows the same array illuminated with polarized light.

SUMMARY

The goal of the SFIL development program is to enable patterning of sub-100 nm features at room temperature and with minimal applied pressure. We believe the use of low viscosity materials and photopolymerization chemistry will enable SFIL to achieve the throughput required for use in the microelectronics industry. Additionally, the rigid, transparent imprint template used in SFIL enables a precision in overlay alignment that is difficult to achieve in other imprint schemes.

FIGURE 7.21. Top-down SEM (a) showing Ti lift-off pattern over a patterned substrate with no apparent change in line width. Micropolarizer array comprised of 100 nm Ti lines/spaces under non-polarized (b) and polarized (c) illumination. The alternating sections of the checkerboard pattern allow perpendicularly polarized light transmission. Reprinted with permission from Colburn, et al., J. Vac. Sci. Tech. B, 2001. 19(6): p. 2162. Copyright 2001 AVS.

Three methods of fluid delivery were analyzed theoretically and tested experimentally. Droplet compression has proven to be the most effective method of fluid delivery using the rigid quartz template, while spin-coating was applied successfully to roller imprint equipment using a compliant imprint template.

The materials of the SFIL system have been tailored to ensure preferential release at the template-etch barrier interface. In a rudimentary study of imprint reliability, a field of ~1000 features were replicated faithfully and reliably over the course of 10 consecutive imprints without the appearance of visible defects.[34] Further, during a preliminary study of defect susceptibility,[14,34] it was shown that contamination on the imprint template was entrained in the etch barrier upon photopolymerization, effectively cleaning the imprint template.

The profile of replicated features was modeled using FEM techniques. Predictions based on the experimentally determined etch barrier material properties, including elastic modulus and volumetric shrinkage, indicated that no pattern motion or major geometric distortion will occur due to polymerization of the etch barrier. The etch selectivity of the etch barrier to transfer layer has been measured, and can be ensured by incorporating at least 11% (w/w) silicon into the etch barrier formulation. With these outstanding etch properties, an aspect ratio of 14:1 has been achieved in a 100 nm line/space array,[7,27] which was then used in a Ti lift-off process to produce a large area micropolarizer array. An advantage of SFIL chemistry was also demonstrated by the patterning of high aspect ratio features over 700 nm topography. Additionally, the resolution of the SFIL process appears to be dependent on the resolution of the features on the imprint template. We have successfully imprinted the smallest features patterned on SFIL imprint templates, which were smaller than 30 nm.

The design of SFIL materials is consistent with the requirements of high-volume manufacturing. SFIL has been developed to allow the process to meet the throughput, resolution, and die size requirements outlined in the SIA International Technology Roadmap for Semiconductors. We have shown SFIL to be a high-resolution technique capable of patterning a wide variety of substrates at room temperature, under low applied pressure, and in a reliable, cost-effective manner.

ACKNOWLEDGEMENTS

The authors thank Motorola Labs, International SEMATECH, Brewer Science, DPI-RTC, IBM, Ultratech Stepper, ETEC, 3M, NRL, and Compugraphics for generous gifts and technical consultation. Special thanks to Steve Johnson and Mario Meissl for their help in developing the SFIL process and equipment, and Britain Smith for help in the defectivity studies. We gratefully acknowledge the financial support of DARPA (MDA972-97-1-0010) and SRC (96-LC-460).

REFERENCES

1. S.I.A., *International Technology Roadmap for Semiconductors*, . 2001: available at http://public.itrs.net/Files/2001ITRS/Home.html.
2. Chou, S.Y., P.R. Krauss, and P.J. Renstrom, *Nanoimprint lithography.* J. Vac. Sci. Tech. B, 1996. **14**(6): p. 4129.
3. Colburn, M., *et al.*, *Step and Flash Imprint Lithography: A new approach to high resolution patterning.* Proc. SPIE: Emerging Lithographic Technologies III, 1999. **3676(I)**: p. 379.
4. Haisma, J., *et al.*, *Mold-assisted nanolithography: A process for reliable pattern replication.* J. Vac. Sci. Tech. B, 1996. **14**(6): p. 4124.
5. Xia, Y. and G.M. Whitesides, *Soft Lithography.* Angew. Chem. Int. Ed. Engl., 1998. **37**: p. 550.
6. Thompson, L.F., C.G. Willson, and M.J. Bowden, *Introduction to Microlithography.* 2nd ed, 1994, Washington, D.C.: ACS.
7. Colburn, M., *et al.*, *Step and Flash Imprint Lithography for sub-100nm Patterning.* Proc. SPIE: Emerging Lithographic Technologies IV, 2000. **3997**: p. 453.
8. Intel Corp., *http://www.intel.com/technology/itj/q31998/articles/art_1.htm.*
9. Choi, B.J., *et al.*, *Layer-to-Layer Alignment for Step and Flash Imprint Lithography.* Proc. SPIE: Emerging Lithographic Technologies V, 2001. **4343**: p. 436.
10. Bailey, T.C., *et al.*, *Template Fabrication Schemes for Step and Flash Imprint Lithography.* Microelectron. Eng., 2002: accepted for publication.
11. Hare, E.F., E.G. Shafrin, and W.A. Zisman, *Properties of Films of Adsorbed Fluorinated Acids.* J. Phys Chem., 1954. **58**: p. 236.
12. Nishino, T., *et al.*, *The Lowest Surface Free Energy Based on -CF_3 Alignment.* Langmuir, 1999. **15**: p. 4321.
13. Ulman, A., *An Introduction to Ultrathin Organic Films from Langmuir-Blodgett to Self-Assembly.* 1991, Boston: Academic Press. 442.
14. Bailey, T., *et al.*, *Step and Flash Imprint Lithography: Template surface treatment and defect analysis.* J. Vac. Sci. Tech. B, 2000. **18**(6): p. 3572.
15. Tripp, C.P. and M.L. Hair, *Direct Observation of the Surface Bonds between Self-Assembled Monolayers of Octadecyltrichlorosilane and Silica Surface: A Low-Frequency IR Study at the Solid/Liquid Interface.* Langmuir, 1995. **11**: p. 1215.
16. Zhao, X. and R. Kopelman, *Mechanism of Organosilane Self-Assembled Monolayer Formation on Silica Studied by Second-Harmonic Generation.* J. Phys. Chem., 1996. **100**: p. 11014.
17. Silberzan, P., *et al.*, *Silination of Silica Surfaces. A New Method of Constructing Pure or Mixed Monolayers.* Langmuir, 1991. **7**: p. 1647.
18. Le Grange, J.D. and J.L. Markham, *Effects of Surface Hydration on the Deposition of Silane Monolayers on Silica.* Langmuir, 1993. **9**: p. 1749.

19. Tripp, C.P., R.P.N. Veregin, and M.L. Hair, *Effect of Fluoroalkyl Substituents on the Reaction of Alkylchlorosilanes with Silica Surfaces.* Langmuir, 1993. **9**: p. 3518.
20. Tripp, C.P. and M.L. Hair, *Reaction of Methylsilanols with Hydrated Silica Surfaces: The Hydrolysis of Trichloro-, Dichloro, and Monochloromethylsilanes and the Effects of Curing.* Langmuir, 1995. **11**: p. 149.
21. Decker, C., *Photoinitiated Crosslinking Polymerization.* Prog. Polym. Sci., 1996. **21**: p. 593.
22. Kloosterboer, J.G. and G.J.M. Lippits, *Replication of video disks using photopolymerizatoin: process design and study of network formation.* J. Imaging Science, 1986. **30**(4): p. 177.
23. Kloosterboer, J.G. and G.F.C.M. Lijten, *Thermal and mechanical analysis of a photopolymerization process.* Polymer, 1987. **28**(7): p. 1149.
24. Kloosterboer, J.G., G.F.C.M. Lijten, and C.P.G. Zegers, *Formation of densely crosslinked polymer glasses by photopolymerization.* Polym. Mater. Sci. Eng., 1989. **60**: p. 122.
25. Colburn, M.E., *Step and Flash Imprint Lithography: A Low-Pressure, Room-Temperature Nanoimprint Lithography*. Department of Chemical Engineering; Ph.D. Thesis. 2001, Austin, TX: The University of Texas at Austin.
26. Ruchoeft, P., *et al.*, *Patterning curved surfaces: Template generation by ion beam proximity lithography and relief transfer by Step and Flash Imprint Lithography.* J. Vac. Sci. Tech. B, 1999. **17**(6): p. 2965.
27. Colburn, M., *et al.*, *Patterning non-flat substrates with a low pressure, room temperature imprint lithography process.* J. Vac. Sci. Tech. B, 2001. **19**(6): p. 2162.
28. Colburn, M., *et al.*, *Development and advantages of step and flash imprint lithography*. Solid State Technology, 2001. **46**(7): p. 67.
29. Bailey, T., *et al.*, *Step and Flash Imprint Lithography: A Technology Review.* Future Electron Devices (Tokyo), 2000. **11**(4): p. 54.
30. Colburn, M., *et al.*, *Characterization and Modeling of Volumetric and Mechanical Properties for Step and Flash Imprint Lithography Photopolymers.* J. Vac. Sci. Tech. B., 2001. **19**(6): p. 2685.
31. van Krevelen, D.W., *Properties of Polymers*. 3rd ed. 1990, New York: Elsevier.
32. Chou, S.Y., *et al.*, *Sub-10 nm imprint lithography and applications.* J. Vac. Sci. Tech. B, 1997. **15**(6): p. 2897.
33. Resnick, D.J., *et al.*, *High Resolution Templates for Step and Flash Imprint Lithography.* Proc. SPIE: Emerging Lithographic Technologies VI, 2002. **4688**.
34. Bailey, T., *et al.*, *Step and Flash Imprint Lithography: Defect Analysis.* J. Vac. Sci. Tech. B, 2001. **19**(6): p. 2806.

8

Using PDMS as a thermocurable resist for a mold assisted imprint process

L. Malaquin and C. Vieu
LAAS - CNRS 7, 31077 Toulouse, France

8.1. INTRODUCTION

Due to the recent major advances in the field of nanodevices during the last decade, the demand for patterning methods combining sub-micron resolution, low cost and high throughput is becoming a prime necessity. Nowadays, this demand arises not only from the microelectronics community where miniaturization has been a driving strategy since a long time, but originates from many fields: optics, photonics, nanoelectronics, nanomagnetism, nano-electro-mechanical systems seen as a small size version of MEMs, organic electronics and biochips. The emergence of a nanopatterning technique for mass production can transform these opportunities opened by the advances of basic research into a real invasion of nanodevices and nanosystems in our environment. Most of the prototypes of these innovative nanodevices, developed in many laboratories, are fabricated using Electron Beam Lithography (EBL). This mature technique combines high resolution (<10 nm),[1] dimensional control, alignment capability and high reproducibility and turned out to be the technique of choice for nanopatterning. However, despite the progress on electron beam resist sensitivity and low voltage EBL, this technique due to its sequential writing mode, is too slow for sustaining direct writing for mass production. Manufacturing nanostructures at low cost is the technological challenge of the forthcoming years.

The semiconductor industry in quest of the Next Generation Lithography (NGL) tool for patterning transistor gates of 50 nm and below, elected the Extreme UV (EUV) photolithography at a wavelength of 13 nm as the most serious candidate for industrial maturity on time. Although it is compatible with required throughput for semiconducting device manufacturing, EUV lithography cannot be considered as a low cost technique for large scale production of nanostructures for a large fan of application. Nano-Imprint lithography (NIL), introduced by S. Chou in 1995,[2] is an attractive technique for filling the gap where conventional lithographies fail. This low cost, high resolution technique simply involves the hot embossing of a thin resist layer with a re-usable master etched in a hard material. Sub-10 nm resolution[3] was readily demonstrated but some limitations have been identified mainly due to the high viscosity of the polymer melt.[4] The improvement of the polymer flow during embossing follows very simple rules. The higher the temperature beyond the glass transition temperature of the polymer: Tg, the longer the imprinting time and the higher the imprinting pressure, the higher is the quality of the imprinted features in the resist layer. These tendencies unfortunately are not really compatible with an industrial high throughput technique where the temperature/pressure/imprinting time budget needs to be minimized. Many efforts have been dedicated to optimize a low temperature, low pressure process enabling to replace the conventional process of hot embossing in a thermoplastic resist layer like PolyMethylMethAcrylate (PMMA) which requires typical temperature and pressure parameters in the range 150°C-240°C, 40-100 bars and imprinting times from 5 minutes to 1 hour. One possible way to solve this problem consists in imprinting into a monomer or a prepolymer fluid precursor exhibiting a very low viscosity and then cure the film inside the mould cavities by thermal heating at a moderate temperature or using UV illumination. This alternative method is called Mold-Assisted Lithography (MAL)[5]when UV curing is used. This latter approach enables to use a transparent mold like silica because required pressures are small, this greatly simplifies the alignment procedure between layers. Following M. Tormen and al.,[6] we have investigated a thermocurable process using a very well-known material: polydimethylsiloxane (PDMS). This elastomer is widely used when implementing the Micro-contact printing (μCP) technique.[7] In addition, this material can be easily synthesized and engineered in order to minimize the viscosity of the precursor fluid and to obtain specific mechanical properties after cross-linking. In this work we describe the experimental procedure we have established and we present some results of imprinting and pattern transfer by ion etching, lift-off and electroplating. We demonstrate that the same process is applicable over a large range of dimensions, it can be used to fabricate very high structures of resist (100 μm thick) and is also capable of nanometric resolution. Finally, we show the compatibility of PDMS material with classical microfabrication processes and we give some treatments for the engineering of this material. Applications of this new process range from microsystems when a thick resist layer is structured to biochip applications because PDMS is a material of choice for micro and nanofluidics, and also for optical components due to its transparency in the UV-visible spectral range. This imprinting process can also be used to create hard stamps of PDMS for achieving very high resolution μCP. The general purpose of this chapter is to show that using a thermocurable resist like PDMS, low pressure and low temperature imprinting processes can be achieved.

8.2. PDMS MATERIAL

Polydimethysiloxane elastomers have now become a very well known variety of material because of their usefull properties in the field of micro and nano engineering through many emergent applications such as micro contact printing, microfluidic, As mentioned by G.M. Whitesides,[7] this material have "a unique combination of properties resulting from the presence of an inorganic backbone and organic methyl groups attached to silicon". Apart from their electrical properties, the chemical characteristics of these polymeric chains confere very attractive capabilities in terms of adhesion, grafting, molding, stability and transparency. Moreover, the simplicity and the reproducibility of the reaction which transforms PDMS precursors chains into an elastomer appears to be very convenient with emergent nano-patterning techniques requirements, i.e., ease of use, low cost and adaptability to mass production.

8.2.1. PDMS chemistry

We first present a chemical description of the precursors needed for material fabrication and their influence on the bulk properties of the cross-linked material. Most of the applications described in the literature refer to a PDMS material purchased from Dow Corning under the reference Sylgard^{TM} 184. This commercial formulation is delivered as a two-parts kit. The first one contains vinyl terminated polydimethylsiloxane chains of about 60 monomers long ($n = 60$). The second part contains a ready to use mixture of methylhydrosiloxane dimethylsiloxane copolymers of about 10 monomers long ($m = 10$) and of a platinum based catalyst known as Karshtedt catalyst: $Pt^0 1.5((CH_2 = CH(Me)_2Si)_2O)$. A semi-developed chemical representation of these compounds is given on Figure 1.

PDMS precursor, i.e. the mixing of these three constituents, is cured by an organometallic catalyzed cross-linking reaction. The Karstedt catalyst enhances a polyaddition reaction of the SiH bond across the vinyl groups also called hydrosilylation, resulting in the creation of Si-CH2-CH2-Si linkages. A schematic description of the reaction that illustrates the conversion of the liquid prepolymer mixture into a solid elastomer is given in Figure 2.

The stability of each constituent, the low cost of initial products and the simplicity of the cross-linking reaction with no gas release appear to be very convenient for imprinting purposes.

8.2.2. General characteristics of PDMS chains

PDMS polymeric chains belong to the family of silicon oils. They are now widely used in many industrial applications as lubricants and impermeability agents because of

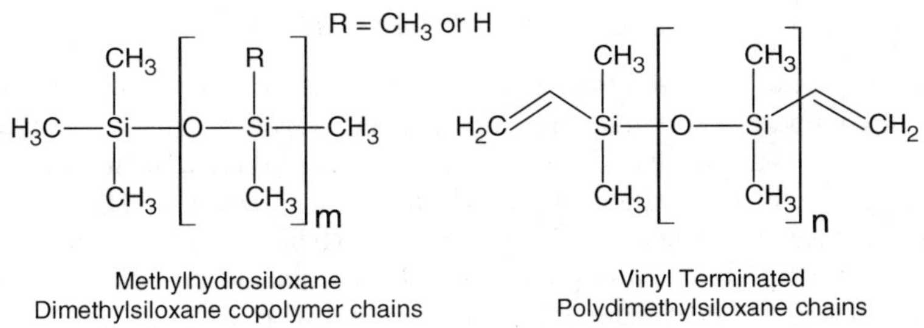

FIGURE 8.1. Polymeric chains used as precursors for PDMS material synthesis

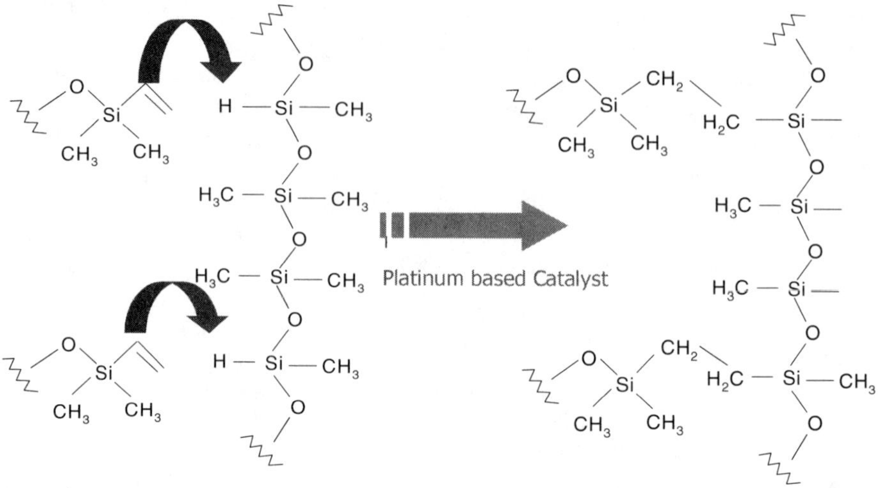

FIGURE 8.2. Cross-linking reaction of PDMS precursors

their many interesting capabilities like adhesion, and stability. These strong industrial interests give rise to pre-polymers mixtures that are commercially available in large quantities, with various conformations and functionalizations and with a large choice of chains length (i.e. polymerization rate), their properties appear to match with our process requirements.

As shown in[8] the chemical structure composed by a siloxane framework linked to methyl groups, confers to PDMS chains specific properties, such as non-polar behaviour, hydrophobicity, thermal stability which make them also usable as reference liquids in contact angle measurements and adhesion experiments. PDMS chains are very flexible and form a liquid material at room temperature. Their glass transition temperature is very low (-130°C). They are not volatile under standard conditions of temperature and pressure and present a very low surface tension (~21mN/m) which is quasi independent of molecular weight.

Viscosity of PDMS chains (η) is highly dependant of their molecular weight : η is multiplied by 6000 when increasing molecular weight from 400g/mol to 200000g/mol. Moreover, it slightly decreases with temperature with a linear dependence factor between η and T of about 10^{-2} K^{-1}.

8.2.3. *Physicochemical properties of cross-linked PDMS*

Concerning the chemical stability of the cross-linked material, most common applications of commercial formulations in microelectronics are dedicated to protection against moisture, environmental attacks, thermal shocks. Added to its thermal stability when in contact with standard atmosphere (from -50°C to 200°C), PDMS elastomers present an excellent overall stability. PDMS surface exhibits a very low surface tension (~21mN/m)[7]which prevents from polluting molecules adhesion and contribute to its preserving. However, PDMS material has some affinity with most of non-polar organic

TABLE 8.1. Cured PDMS (Sylgard 184) electrical and optical properties

Property	Value
Optical properties	
Refraction index	~1.49
Electrical properties	
Dielectric strength	21 kV/mm
Permittivity (100Hz&100kHz)	2.75
Resistivity	50 Ohm.cm

TABLE 8.2. Cured PDMS (Sygard 184) mechanical properties

Physical properties	Value
Durometer Hardness	50
Tensile strength	7.1 MPa
Elongation at break	140%
Tear strenght	2.6 kN/m

solvents such as toluene and hexane. The absorption of these solvents causes a reversible but problematic swelling phenomenon of the bulk material. As we will discuss later, cross-linked PDMS found many applications for microfluidics and lab-on-a-chip applications, due to its weak interactions with various bio-molecules ensuring no denaturation of the biological solutions. The very smooth surface of the cross-linked material is also favorable for the fabrication of pristine micro-channels. Another crucial characteristic of this material for imprinting purposes is the very low and reproducible shrinkage ~1% under cross-linking. This point turns out to be a real advantage when comparing to other possible materials for implementing mold assisted lithography.

PDMS material is also an optically transparent elastomer, which has been widely used for encapsulation applications in microelectronics because of its insulating properties. Table 1 gives some electrical and optical specifications of commercially available formulation Sylgard 184 as given by Dow Corning.

Main applications in this domain are dedicated to encapsulation of amplifiers, coils, circuit boards, and transformers. Concerning mechanical properties of Sylgard 184 elastomer, some commercially available data are given in the Table 2. These data show the typical elastomeric behavior of the cross-linked material. PDMS exhibits low durometer hardness, a low tensile strength, but an elongation at break of 140%.

These mechanical properties induce some limitations when implementing the micro-contact printing technique where the softness of PDMS causes possible sagging effects[9] but could be an advantage for demolding in a context of NIL.

8.2.4. PDMS as a material for Micro fabrication

As previously mentioned, the chemical composition of the material and the type of curing reaction present many advantages for an imprint process. First, the two low viscosity components do not need the use of any solvent during the process. Due to their low vapor pressure, an easy degassing step of the prepolymer mixture before molding under vacuum (down to 0.1 torr) can be performed to remove possible bubbles. In addition no harmful phenomenon of evaporation is taking place during the heating step.

On the other hand, the hydrosililation reaction ensures no production of any by-product and the stability of each component versus pressure and temperature allows simplicity of use. The reaction can only be slowed down or inhibited by amines, amides, azides, urethanes and organo-tin or sulphur containing compounds.

For applications in microfabrication it is very important that the material after the lithography step exhibits strong resistance towards temperature, etching solutions and ion etching. Thermal stability of cross-linked PDMS is better than many other conventional resists used in lithography. We have not observed any modifications of the PDMS structures even after heating up to 250°C. As we will show later, this material is a suitable mask resisting to many diluted acid solutions, to common electrolytes used for electroplating and exhibits interesting ion etching resistance. It is also worth noting that in contrast, a specific acid solution is available for stripping PDMS leaving a clean silicon surface. This material is stable against many solvents even if non-polar organic solvents can induce swelling.

8.2.5. *Engineering of PDMS*

The possible adjustment of the viscosity of the pre-polymer mixture is one of the most important engineering tools with respect to imprinting applications of PDMS. Indeed, the modification of polymers chains allows to control the viscosity of the pre-polymer and consequently to decrease the pressure during the embossing step as well as to decrease the thickness of the residual layer left on the surface of the substrate. This parameter has also a crucial influence on many physical properties of the material such as its mechanical resistance and can be adjusted in order to increase its capabilities for imprinting. It is important to note that these modifications do not induce any change in the wettability of the mixture on the mold which is a prerequisite condition for an imprinting process of high quality. Moreover, the control of the proportion of free reticulation sites across the methylhydrosiloxane dimethylsiloxane copolymers chains is also useful for a good mechanical properties control. Finally, by dosing the amount of catalyst in the pre-polymer mixture, the kinetics and the temperature of the reaction can be adjusted towards process requirements. Fastest processes used in our experiments were obtained with 0.1% of platinum based catalyst which corresponds to a complete curing in a few minutes at room temperature. This means that imprinting time and temperature can be significantly reduced compared to thermoplastic resists such as PolyMethylMetacrylate

TABLE 8.3. Compositions of PDMS precursors used in our experiments

Name	n	m	H/CH_3 (%)	Viscosity before curing (Pa.s)	Vinyl/Copolymer Chains ratio (%)	Catalyst amount (%)
Sylgard 184	60	10	50	5.5	~100%	-
ReformulatedPDMS	10	7	50	10.10^{-3}	110%	0.01%

FIGURE 8.3. SEM image of a $1cm^2$ patterned Si mold (micron size structures)

PMMA. Table 3 gives some details on the two main formulations of polymers chains used in our experiments.

8.3. TECHNOLOGICAL IMPLEMENTATION

A detailed description of the process is given in the following sections where we highlight the advantages of using curable materials and more particularly PDMS.

8.3.1. Masters fabrication

Two different techniques were used in order to pattern silicon masters. The first one involves proximity optical lithography on a Si [100] wafer for the fabrication of micrometric and millimetric structures. Using a negative or positive resist , we were able to obtain after transfer by reactive ion etching, structures covering a large range of dimensions from 2 μm to 500 μm. These molds were sawn in pieces of $1cm^2$. Figure 3 shows an image of a typical micrometric master.

For the master fabrication, transfer after lithography was achieved in a deep RIE machine using a high density plasma. Two different etching processes were implemented in order to transfer structures. The first one is a BOSCH process, which is generally employed to fabricate deep structures (etching rate of silicon: 2 $\mu m.min^{-1}$). Some examples of features are shown in Figure 4. As we can see on both positive (i.e. meaning that only a small part of the surface is in relief) and negative masters, the profile of the structures presents some regular defects corresponding to the alternative passivation and etching steps during the etching process.

Deeper masters (100 μm deep) featuring micronic structures were used for some specific experiments. The specific scalloped edge profiles obtained after the deep etching of silicon turned out to be quite deleterious for nanometric features. We thus designed a specific anisotropic etching process giving vertical and smooth edge profiles. Therefore, for structures height lower than 3μm, we have used a slowest ($200nm.min^{-1}$) but anisotropic homemade process (fluorine-based with SF6) on the same equipment, etching continuously the silicon wafer with no passivating cycles. Some examples of transferredfeatures with this process are given in Figure 5.

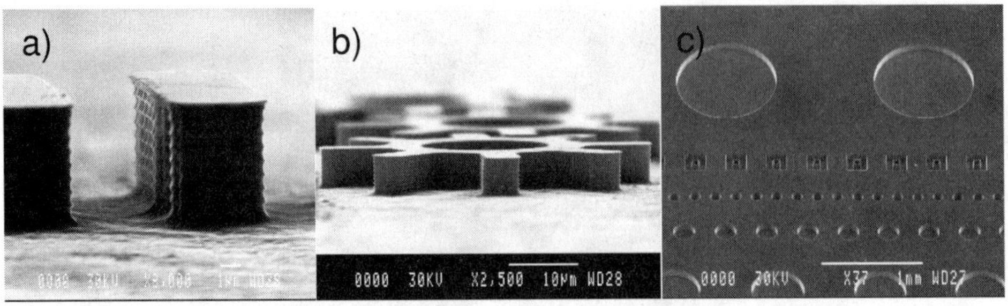

FIGURE 8.4. SEM images of positive (a,b : 6 µm deep) and negative (c :100µm deep) molds.

FIGURE 8.5. SEM images of negative nanometric molds obtained by HREBL and RIE (left: 100nm deep, right: 300nm deep)

The fabrication of molds featuring nanometric structures was performed by High Resolution Electron Beam Lithography (HREBL) using a 200kV TEM/STEM transmission microscope. Nanometric structures were produced on 4 × 4mm^2 Si[100] samples coated with PMMA. The resolution of the technique is about 5nm,[10] which allowed us to fabricate nanostructures such as those shown in Figure 5 with a high reproducibility.[11] The master fabrication was achieved using the anisotropic RIE of Silicon using the PMMA resist features as a mask. Depending on the PMMA thickness, the height of the transferred features ranged from 100 nm to 300nm.

8.3.2. Surface treatments

Two surface treatments are needed in our process in order to provide first a good adhesion between the cross-linked polymer on the substrate and secondly to allow demolding without affecting the wettability of the mold.

In most of our experiments we have used Si[100] or Silica substrates. Several well-documented processes are available to modify adhesion properties between these substrates and cross-linked PDMS. Even if some processes of covalent bonding are now available such as (PS)-PDMS diblock copolymers, vinyl-chlorosilane treatments or

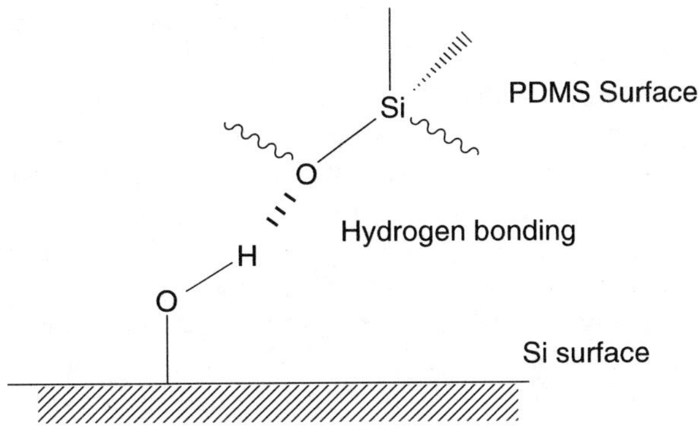

FIGURE 8.6. Schematic description of Hydrogen bonding between PDMS chains and a oxidized Si Substrate.

oligomer bonding,[12,13] we have decided to use more simple and stable processes able to enhance adsorption of the polymer. Indeed, the capability of siloxane groups to form hydrogen interaction with hydroxyl terminated Silicon oxidized surface is a good way to provide adhesion of PDMS on the substrate surface (Figure 6). The initial oxidized surface can be obtained either using a Si sample with its native oxide cleaned by UVOCS treatment or by using a conventional process based on a buffer HF treatment and then oxidation by dipping samples in a H_2SO_4-H_2O_2 treatment.[14]

Some experiments of molding have also been performed on metallic surfaces. Even if PDMS does not adhere on gold and Pt, it is possible to work on materials such as Ti, Ge and Ag with nevertheless a noticeable degradation of adhesion properties when compared to Si or SiO_2.

Concerning the anti-adhesive surface treatment of the master, processes are now well established.[7,15] In most cases they are based on alkyl chlorosilanes or fluorinated chlorosilanes treatments. Numerous studies have shown that these components are able to react on oxidized silicon surfaces and create a SAM (Self-Assembled Monolayer). The specific components we have used in our experiments are given in Figure 7.

The Si-Cl bonds of the OTS or HFTS molecules are broken by the hydroxyls groups to form Si-O-Si bonds.

Practically, Si masters are first cleaned in a 10% buffer HF solution followed by

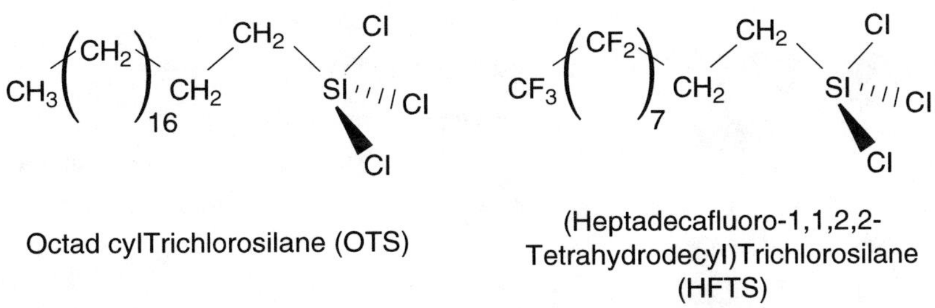

FIGURE 8.7. Chemical representations of SAMs precursors for oxidized silicon surface treatment

an oxidizing bath with a 50% H_2SO_4-H_2O_2 solution. After desiccation at 200°C for 30 min, samples are dipped in a 0.5 mmol.L^{-1} of OTS in Toluene. Samples are then rinsed with toluene and dried under Nitrogen blowing. Because chlorosilanes molecules are very sensitive to water, masters must be perfectly dried using hot plating or baking. Moreover, reaction was made under a constant Nitrogen stream to avoid any contact with atmosphere. The anti-adhesive properties of the master are still efficient after more than 10 imprinting processes at $T = 120°C$ but we have not carried out systematic studies on the aging of the master for much more imprinting steps.

The determination of the adhesion energy of adsorbed PDMS material is now well documented.[8] As said previously PDMS chains are widely used as reference liquids in adhesion experiments. The surface tension of silanisated surfaces is very low (~24 to 28mN/m according to ref 8) and depends on the coverage of the SAM layer. Our own measurements using contact angle measurements with three liquids exhibiting different type of affinities with the surface[16] (water, ethylene glycol and diiodomethane) give 24mN/m in good agreement with,[8] indicating a monolayer of high density. Despite the very low surface tension of the surface after silanisation, the specific chemical composition of PDMS gives an even lower surface tension, enabling suitable wetting of the master.

8.3.3. Imprinting process

In typical experiments, detailed in Figure 8, the precursor mixture containing the siloxane chains and the Pt catalyst is first pumped under primary vacuum for 10 min to allow the removal of any air bubble trapped during mixing. The precursor is then directly poured onto both substrate and master. We deposit a controlled volume of 10 μl on each part. The master is then put on the substrate and the sandwich is introduced in the press.

Imprinting processes were carried out on a commercial hydraulic press ESSYL equipped with 4 cm diameter plates. Initially destined to PMMA imprinting, this press was designed to work at very high pressures (300bar ± 3bars) and temperature until 300°C. Because of the drastic reduction of temperature and pressure possible with PDMS,

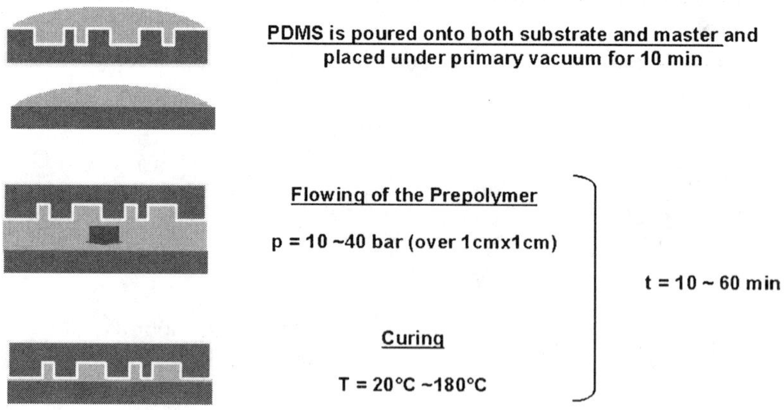

FIGURE 8.8. Schematic description of our imprint process

the press was modified in order to obtain more stability for low pressures (10 bars $\pm$ 0.1bar) and temperatures (200°C $\pm$ 1°C).

Accounting for the diameter of the press, we have limited the size of our masters to 1cm per 1cm except for EBL fabricated masters, which are limited by the size of the specimen holder (4mmx4mm) of our e-beam system. However, substrates were sawn into pieces of 4cm^2 to avoid flowing of the excess of pre-polymer under them.

8.3.3.1. Process Temperature. Using the commercial formulation of PDMS delivered by Dow Corning and known as Sylgard^{TM}184, the crosslinking reaction takes roughly 1h at 100°C or 15 min at 150°C. However, as explained before, by using a reformulated PDMS or by simply increasing the catalyst amount in the prepolymer, it turned out to be possible to reduce this time down to 5min at 80°C with a catalyst concentration of 0.01%. Imprinting time and temperature can thus be significantly reduced compared to thermoplastic resists such as PolyMethylMetacrylate PMMA. Even faster curing reactions can be obtained but are not compatible with manipulation times. Removals of the masters were accomplished at room temperature.

8.3.3.2. Pressure. Precursor mixture is very fluid: typical viscosity is around 4-5 Pa.s for the commercial kit and can lowered at 10mPa.s upon reformulating. This must be compared to the viscosity of PMMA (M_w = 950k) heated well above its glass transition temperature, which is around 400Pa.s at 230°C.[4]

As a consequence, typical pressure used in our experiments with 1cm^2 masters was considerably reduced compared to standard thermoplastics imprinting. The pressure was fixed at 10 bars using our 4 cm diameter press. Concerning $4 \times 4mm^2$ masters, pressure was limited at 3 bars i.e. 0.4kN. A few test experiments have been carried out at lower applied pressures with good success concerning the thickness of the residual layer (< 300 nm) and its homogeneity, however, we have not fully investigated what is the lower limit compatible with imprinting quality and uniformity.

8.3.4. Transfer procedures

8.3.4.1. Residual layer. The presence of a residual layer in the squeezed regions is a critical problem with respect to the transfer of imprinted features. Its removal generally involves an etching step, which must be as short as possible to preserve dimensions and quality of imprinted structures. Consequently, the thickness of this layer must be as thin as possible and very homogeneous all over the substrate. The topographical contrast obtained in cured commercial PDMS layers after imprinting has been characterized using a mechanical profilemeter. Typically, for 1cm^2 masters exhibiting etched cavities of 2200nm, we obtained imprinted depth of 2150nm. Accounting for the shrinkage of the material during curing and possible relaxation of the material upon pressure release, we think that the pre-polymer fluid fills completely the cavities of the master. With the commercial Sylgard184 we obtain routinely homogeneous residual layers of typical thickness 50nm. It appears that this thickness was quasi constant in a large range of applied pressures on the polymer (2 - 10 bars over 1cm^2) and independent of the feature size and shape.

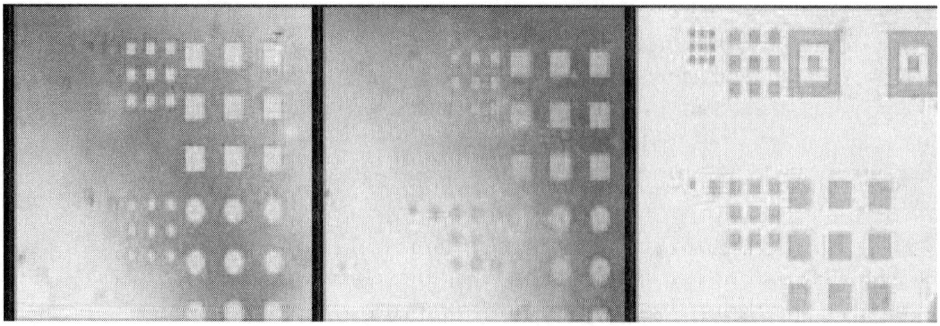

FIGURE 8.9. Optical microscopy images of the influence of an acid treatment on an imprinted surface (after 0, 2 and 10 min). after 10 min of treatment, the whole surface is cleaned from any residual layer.

Changing the viscosity of the pre-polymer mixture by reducing the length of the chains (n and m) allows us to decrease the thickness of the residual layer down to 20nm in the same range of pressure. This value is the minimum that we have obtained keeping the applied pressure below 10 bars.

In order to complete the imprinting for transferring the structures in the substrate, the removal of the residual layer of cross-linked polymer at the bottom of the features must be performed without affecting their shape and dimensions.

Due to its very low thickness, the residual layer can be easily stripped off by reactive ion etching using short etching times. At the micron length scale, we use an anisotropic process (home made fluorine based process) on a High-density plasma system and the removal of the residual layer does not induce any degradation of the edges of the features.

Alternatively, we found that it was possible to etch away this very thin residual layer by wet etching in an acid solution with no noticeable damage at the micronic scale. The removal of a 50nm thick resist can be achieved in 5minutes. The Figure 9 shows the evolution of the residual thickness after an inhomogeneous imprinting experiment during wet removal. The evolution of the residual layer is observed by optical microscopy. Brown color corresponds to a 50 nm thick residual layer.

With respect to the very small thickness of the residual layer, we think that the opportunity to complete the imprinting through a rapid wet etching can be regarded as a strong potential of our process, rendering possible the economy of any dry etching process before transferring.

We have not tested this wet removal step for nanometric features where possible loss of control of the dimension could be observed.

8.3.4.2. Reactive Ion Etching. Etching resistance of the PDMS layer has been investigated using the same conventional silicon etching processes as those used for master fabrication. Using the BOSCH deep etch process described before, we found a PDMS : Si ratio close to 1/6. In comparison PMMA turned out to be etched five times more rapidly. For the anisotropic home made process (fluorine-based with SF6) the selectivity between PDMS and Si was poorer (1/2). These values are independent on the formulation of the pre-polymer mixture.

8.3.4.3. Lift Off. After imprinting and removal of the residual layer either by wet or dry process, a 200nm thick Ti-Au layer is evaporated (10nm/min). The substrate was then dipped into a concentrated acid solution with ultrasonic agitation. In typical experiments successful lift-off operates within less than a 5 min.

8.4. RESULTS

8.4.1. Quality of imprinting

In this section we will first present results of imprinting using PDMS that demonstrate the quality of the replicated structures and their fidelity towards the master. We will also give a general overview of successful molded structures at the micro and nano scale-length and try to emphasize the interest of reformulating the starting material for nanometric features. Versatility and suitability versus structures shape and master conformation (positive or negative) will be discussed in the following sections.

Figure 10 shows some imprinted structures using a commercial PDMS material and a 2.2μm deep negative mold patterned with micrometric structures. SEM images of both masters and corresponding replicated features are given.

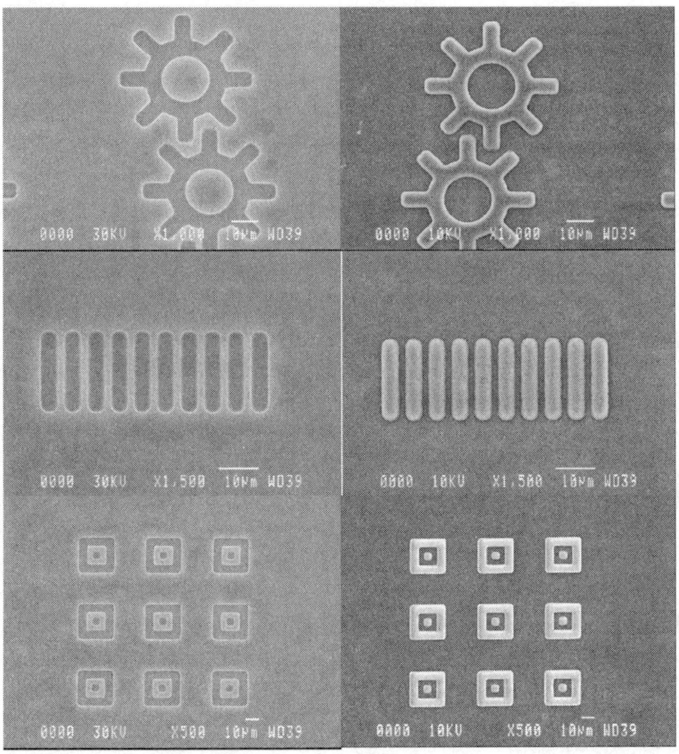

FIGURE 8.10. SEM images of a 2.2μm deep negative mold (left column) and corresponding molded structures in sylgard 184 pdms (right column).

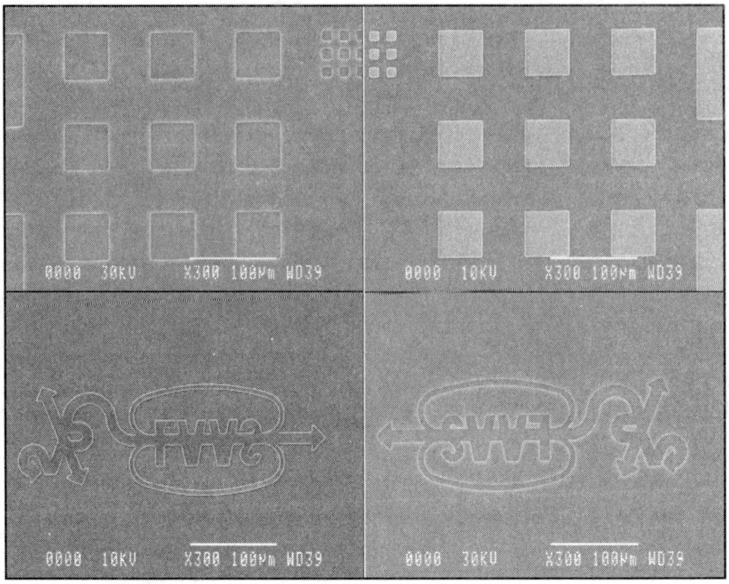

FIGURE 8.11. SEM images of a 2.2μm deep negative mold (left column) and corresponding molded structures in Sylgard 184 PDMS (right column).

As can be seen on the pictures, molded structures do not present any defects and exhibit shape and edge details, which reproduce perfectly the features of the master. Careful measurements indicate that the dimensions of the molded structures match with a very good accuracy (<1%) those of the master. Between the molded structures, the imprinted surface appears to be very clean and uniform all over the imprinted area. This confirms that the flowing process during molding is very easy due to the low viscosity of the pre-polymer solution.

Figure 11 shows some other examples of larger and more complex molded structures in order to illustrate briefly PDMS versatility. At the micronic length scale, we have not observed any limitation in the geometry, shape and dimensions of the patterns that are replicable. Due to the unique combination of low viscosity and wetting properties, we are able to mold simultaneously large (500μm) and small features (2μm) on the same level (see Figure 11).

Optical and SEM observations confirmed a good uniformity all over the printed area with a quasi-constant residual layer thickness of 60nm and no dispersion of the height of the imprinted features (2150 nm). Another important point is the state of the master surface after demolding. SEM inspections and many reuse of the same master attest for the good anti-adhesive properties of the treated mold. Pictures shown in Figures 10 and 11 were taken after the imprinting process and do not reveal any polymer residues or fragments. Some masters were reused 5 times successfully without any cleaning or treatment steps.

Using a master of only a few microns deep, it was not possible to evidence strong lateral shrinkage or deformation of the molded structures even for very large patterns. However, by increasing the depth of the master and consequently the thickness of the

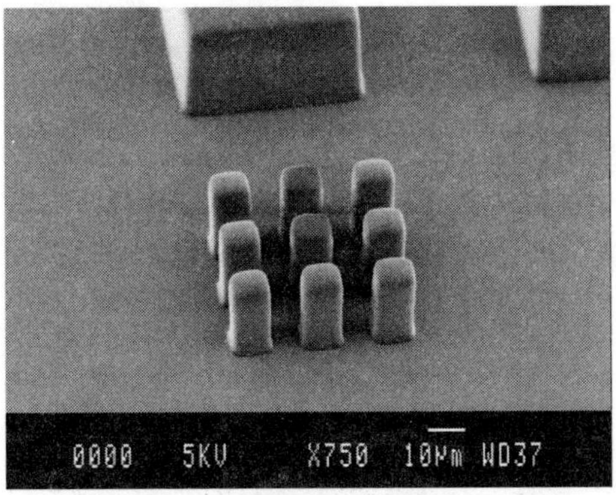

FIGURE 8.12. SEM image of 30μm high molded structures with Sylgard 184 PDMS. Note the rounded shape of the pillars.

imprinted PDMS layer some distorsions of the molded features become visible. When using the commercial formulation of the pre-polymer, thick molded structures appear rounded with smooth edges, see for example Figure 12 and the gears of Figure 17. These distorsions of the patterns are considerably attenuated when molding with a reformulated PDMS pre-polymer of lower molecular weight. However, such a reformulation also induces brittleness of the cross-linked material which can be detrimental in the case of deep masters and high aspect ratio patterns (see section 4.2.2). Consequently, reformulated pre-polymer mixtures were only devoted to the imprinting of nanostructures.

The good results obtained at the micronic scale for many different sizes and geometries of the imprinted structures confirm the suitability of PDMS material for high quality imprinting. However, some limitations turned out to appear when decreasing the feature sizes down to the nanometric scale. Indeed, at this length scale the rounding of the edges of the structures is no longer acceptable and prevents dimensional control. Better results were obtained using reformulated PDMS mixtures which attenuate this side effect. Periodic arrays of 200nm wide line with periods down to 400nm were successful replicated on a 200 nm deep Si wafer (Figure 13).

The maximum aspect ratio which can be obtained, is highly dependant on the sizes and shapes of the features. In the case of micronic features we have obtained aspect ratios as large as 1/15 as we will see later, but because of the elastomeric behavior of PDMS this ratio decreases in the case of small or isolated structures. For nanostructure replication best results were obtained for aspect ratios close to 1/1.

8.4.2. Versatility with respect to geometry

8.4.2.1. Master conformation (positive/negative). We have investigated both positive, i.e. elevated structures over a large recessed surface, and negative masters in order to analyze pre-polymer flow during the molding step. The size and the surface to emboss have a drastic influence in the embossing of high viscosity polymers such as melted thermoplastics[17] and renders more difficult the replication of negative masters. In our case, the low viscosity of PDMS pre-polymer guarantees an efficient flowing of the

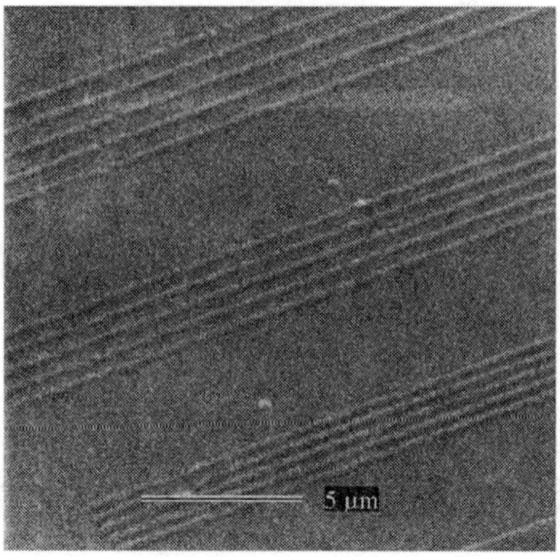

FIGURE 8.13. SEM image of arrays of 200nm wide lines (periods of 400, 600, 800nm) obtained with reformulated PDMS and a 200 nm deep master. The image shows the molded lines in PDMS explaining the weak contrast.

polymer inside the cavities of the mold enabling equivalent results for both master conformations even for short imprinting times. Figure 11 shows SEM images of molded structures obtained with the commercial Sylgard PDMS formulation and a negative master. Imprinting was performed with 2.2μm deep structures, at 100°C for 30min with a 1kN force applied on $1cm^2$ masters. The surface occupied by features was 5% for this negative master. The same kind of results demonstrating both positive and negative master replication can be observed also in Figure 18 after transferring through the Si substrate by ion etching.

Quality of imprinting, uniformity, thickness of the residual layer were not affected by the conformation of the master and equivalent results were systematically obtained for positive and negative molds. This advantage is the direct consequence of the capability of the low viscosity PDMS pre-polymer to flow easily through the patterns of the master. This point is confirmed by the fact that we have not observed any change of the residual layer after imprinting (50 nm for commercial PDMS, 20 nm for reformulated solution) over a large range of applied pressure (2-10 bars).

8.4.2.2. Sizes and aspect ratios of patterns. The elastomeric property of PDMS is a real advantage for demolding. However, its deformation capability is also a major drawback when replicating narrow features of high aspect ratio which tends to collapse when they are not attached to wider sections. This case is illustrated in Figure 14 where we can see that the softness of PDMS impedes the mechanical stability of high aspect ratio isolated features. This limitation has been well characterized in the context of microcontact printing where the maximum aspect ratio of the features of stamps made of PDMS Sylgard 184 was around 2.[9] In our case, the presence of a hard Si substrate appears to improve the mechanical stability and the highest aspect ratios obtained with the commercial PDMS were as large as 1/10. In addition to collapse events, the demolding of isolated and high aspect ratio structures can also results in a tear of PDMS despite surface treatments.

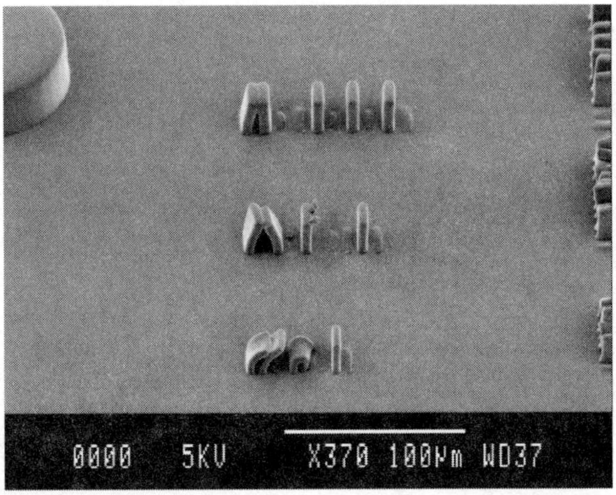

FIGURE 8.14. Collapse and demolding effects observed for isolated high aspect ratio structures molded in Sylgard 184 PDMS (SEM image).

As mentioned before, the reformulation of PDMS by reducing the length of the chains of the pre-polymer mixture improves the definition of the edges of the molded structures but increases brittleness. This effect is illustrated in Figure 15 for an extreme case of nanostructure replication where 30nm wide structures etched over 300nm (aspect ratio 10) in a negative master were imprinted in a reformulated PDMS. In addition to master conformation, dimension and depth, which are not very convenient for an imprint process, the reduction of molecular weight of pre-polymer chains leads to a too brittle material and causes a partial destruction of molded structures during the demolding step. This limitation imposes a convenient aspect ratio close to 1/1 when replicating nanometric structures below 200nm.

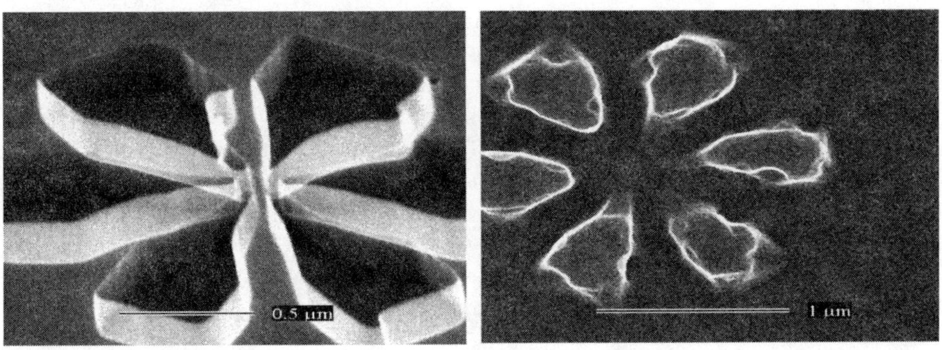

FIGURE 8.15. Illustration of tearing of high aspect ratio nanostructures using reformulated PDMS. The left image shows the negative master of high aspect ratio (30 nm wide structures, 300nm deep) and the right image the molded structures. Structures are present but fine details and edges were partially destroyed during demolding due to the excessive brittleness of the material.

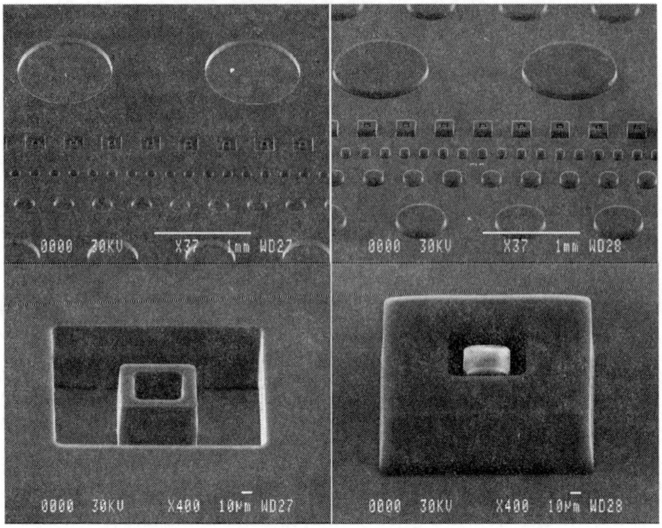

FIGURE 8.16. 100 μm high PDMS molded structures (right column) and the corresponding negative master (left column). Results obtained with the Sylgard commercial kit.

8.4.2.3. Thick resist process. All PDMS capabilities presented before in terms of viscosity, wetting properties and the specific curing process involving heating seem to be well adapted for the replication of micronic structures of very high aspect ratio. Our imprinting method can be consequently used as a thick resist process, with all the applications for microsystem fabrication. Figure 16 shows some typical images of a 100μm deep silicon master with features dimensions going from 20μm up to 1mm, that is used for such a purpose. The corresponding SEM images showing the replicated structures confirm that molding 100 μm high features is possible and do not require any advanced and specific optimizations. Even with the 2 cm^2 negative master used for these experiments, we obtained replicated structures which reproduce perfectly the master. In these series of experiments dedicated to thick film molding, the residual layer was uniform all over the imprinted surface with a thickness about 100nm using an applied force of 1kN. We have never encountered any problem of demolding even for the closed structures as those presented in Figure 16. The commercial formulation of PDMS (Sylgard) is suitable for this kind of realizations. The molding of thick PDMS layers in features of very high aspect ratio (up to 15) turned out to be much more straightforward than the replication of sub-100 nm nanostructures that requires reformulation of the pre-polymer mixture and a strong decrease on the aspect ratio of the features (down to 1).

Some other tests (Figure 17) have been successfully carried out with smallest features (2μm to 500μm) with a negative 30μm deep master.

8.4.3. Transfer techniques after molding

8.4.3.1. Reactive Ion etching. We have investigated the possibility to etch the Silicon substrate using the molded PDMS structures as a mask during RIE. Figure 18 shows

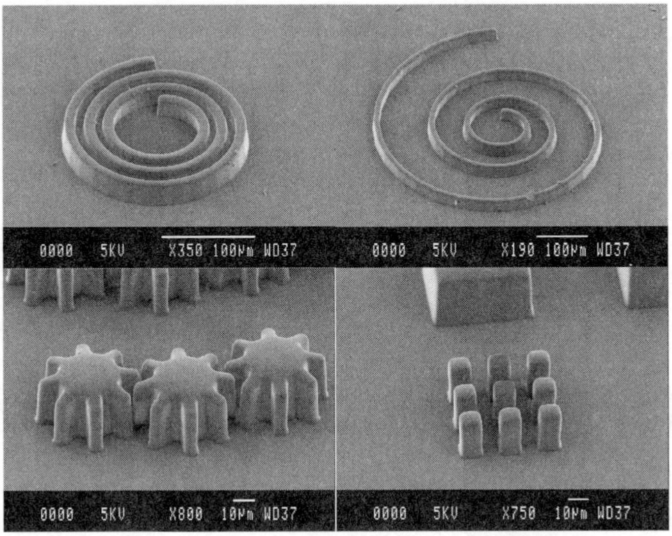

FIGURE 8.17. 30μm deep molded structures. Results obtained with the Sylgard commercial kit.

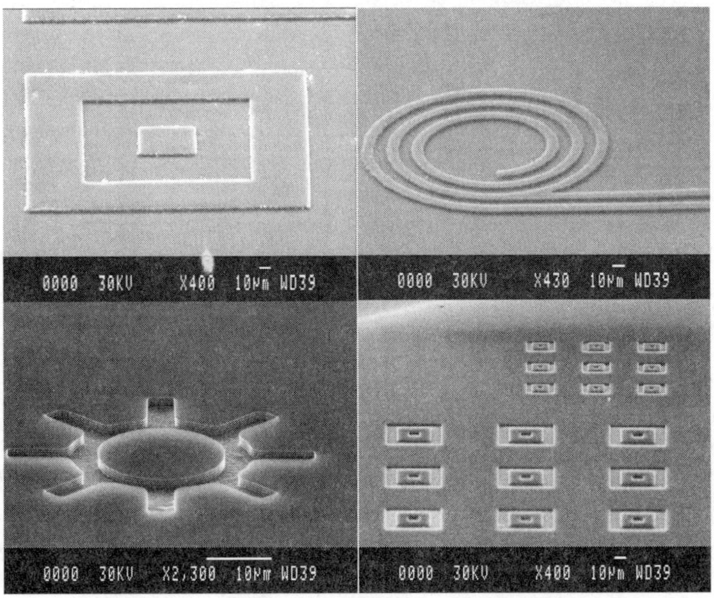

FIGURE 8.18. SEM images of 3μm deep structures etched in Si that are obtained by PDMS imprinting and RIE (BOSCH process). The two first images (top line) correspond to a negative master and the two last images (bottom line) correspond to a positive master.

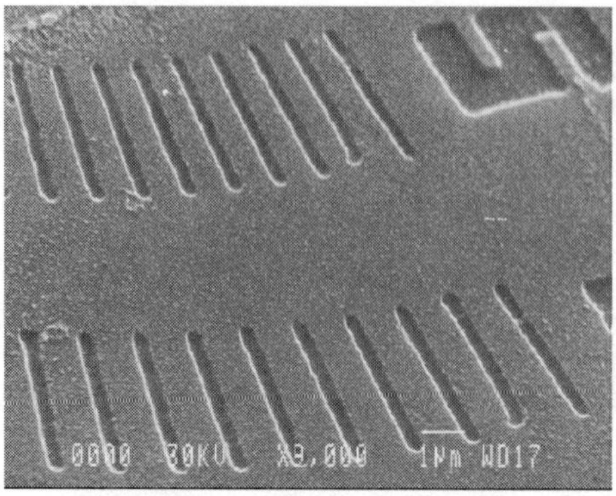

FIGURE 8.19. SEM image of 200 nm wide nanostructures etched in a silicon substrate after imprinting using a reformulated PDMS and pattern transfer through anisotropic RIE.

SEM images of transferred for both positive and negative masters. After the removal of the residual layer with RIE process, the etching of the Si substrate [100] was performed with a BOSCH process and stopped after an etch depth of 3μm.

These results reveal the versatility of the process towards the master's topography. Patterning the substrate by reactive ion etching does not seem to be dependent on master conformation (i.e. negative or positive molds). The removal of the residual layer does not affect the surface of the substrate. The results obtained for a negative master where the top surface after etching is perfectly smooth and flat, plane attest for the clean removal of the residual layer which confirms its homogenity. Indeed, the process used for removing the residual layer is not selective with respect to the Si substrate, consequently any uneven thickness or defect in the PDMS residual layer would have been transferred and amplified in the substrate through deep etching.

Only a few experiments have been made in order to transfer nanometric structures molded in PDMS. Figure 19 shows an array of 200 nm wide lines molded with a positive 100 nm deep master using a reformulated PDMS mixture and transported in the Si substrate using RIE. The residual layer thickness of PDMS after imprinting was estimated at 20nm ± 5nm.Transfer was achieved with the anisotropic RIE process detailed before. Because of the reduced etching selectivity obtained in this case, we have limited the etched depth at 100nm.

These few results confirm the potentiality of reformulated PDMS to reproduce high resolution features and its capability to be employed as a mask for substrate patterning. Further experiments will be dedicated to determine the exact limit of resolution of this process and to show the possible simultaneous replication of these nanometric features with larger structures in a one level process.

8.4.3.2. Lift Off. Another possible way to transfer molded features after the removal of the residual layer is the conventional Lift Off technique. Figure 20 gives SEM images of both master and replicated structures after removal of the residual layer by RIE and Lift-Off of a 200nm thick Ti/Au layer.

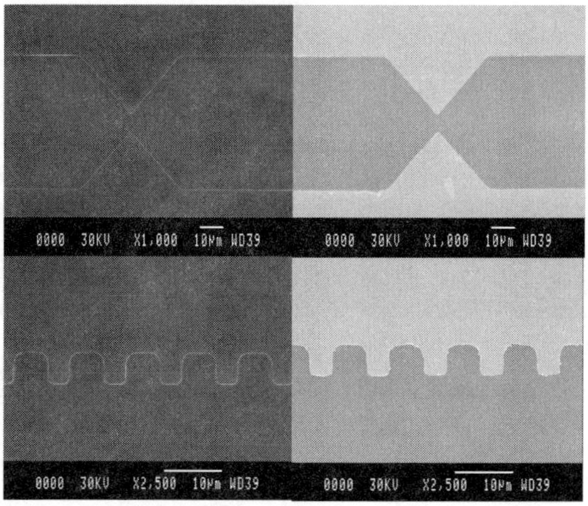

FIGURE 8.20. SEM images of lifted structures (Ti/Au : 50/150nm) (right side) after PDMS molding. The left images show the corresponding features of the master for comparison.

These pictures show the high fidelity of the replicated structures with respect to the master. The quality of the features and the defintion of their edges associated with the easy lift off process demonstrate the compatibility of PDMS with such a transfer technique. PDMS is not affected by the metal evaporation step, and provide a very clean surface after its removal in the acid bath used for lift-off. Figure 21 demonstrates that this lift-off technique can be successfully implemented for many different feature geometries and shapes.

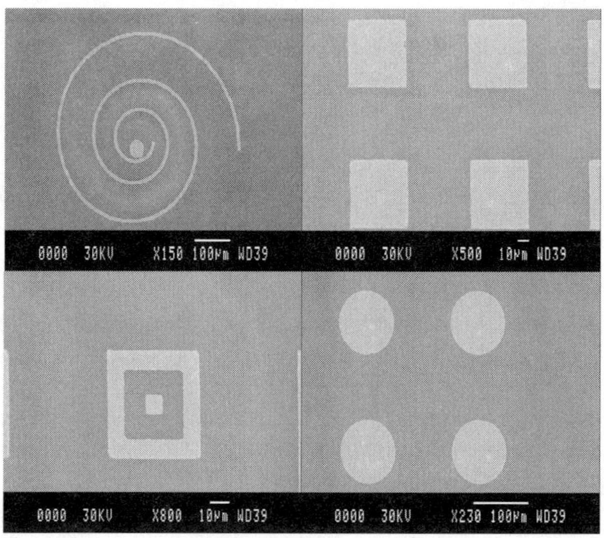

FIGURE 8.21. SEM images of lifted structures (Ti/Au : 50/150nm) after PDMS molding.

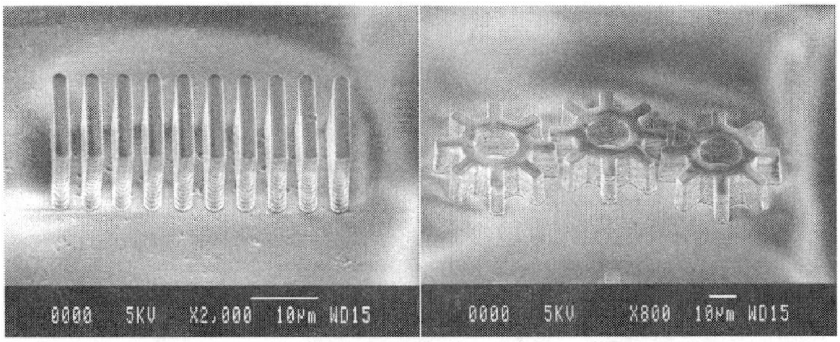

FIGURE 8.22. Microstructures molded in PDMS Sylgard 184 over a 1 µm thick BCB polymer layer with a 10 µm deep negative mold Note the scalloped profile of the edges which reflects intimately the edge definition of the master.

8.4.3.3. Multilayer processing. Some experiments have been carried out for the transfer of PDMS molded structures in polymers of optical interest such as BCB (BenzocycloButène).[8] These experiments have first shown the necessity of a surface treatment on the BCB layer to ensure a good adhesion of PDMS film. A combination of a plasma oxidation step followed by a wet oxidation treatment seems to be efficient to provide hydrogen bonding for PDMS adsorption. Figure 22 shows some SEM images of molded PDMS structures on a Si substrate coated with a 1 µm thick BCB layer. The negative master used in this experiment exhibited 10 µm deep micronic features. The contrast around the structures is only due to charging effect during SEM inspection. Process capabilities turned out to be equivalent to the similar tests made directly on a Si substrate in terms of aspect ratio and resolution. A close examination of the edges of the features molded in PDMS shown in Figure 22, enables to observe the scalloped profile imposed by the Bosch etching process used for fabricating the cavities of the master. This detail shows that the resolution of the process is better than 300 nm. Our soft imprinting process can thus be achieved on multilayered substrates and specifically on thin polymeric films. For this kind of application, specific treatments of the surface of the polymer film are generally required for promoting adhesion of the PDMS material during molding.

8.4.3.4. Electroplating. At this state of our study only a few experiments have been dedicated to the use of PDMS patterns obtained by imprinting as a mold for electroplating. We focused on the fabrication of micrometric gold structures by this technique. These experiments have shown the chemical resistance of PDMS versus electroplating cyanide solution, especially versus combined effects of acidity (pH~5) and temperature. The good insulating properties of the PDMS layer are also suitable for preventing growth under the structures. The main difficulty remains the adhesion of cross-linked PDMS on the seed metallic layers used in our first experiments. Indeed, as mentioned before, the imprint process must be performed on materials which present an oxidized surface in order to provide hydrogen bonding with PDMS. As a consequence a simple seed layer made of a thin Au film cannot be used. Following[15] for Ni electroplating after hot embossing in PMMA we think that a Cr/Ge layer can alleviate the adhesion problems and will be the purpose of our further experiments.

8.4.3.5. Post processing of PDMS for specific applications. In order to extend the field of possible applications of our process, we have investigated physicochemical treatments of the PDMS surface. In particular, we have devoted some of our experiments to the study of the plasma O_2 and UVOCS induced oxidation of the surface of the cross-linked material. Our goal was to demonstrate the possible conversion of the surface of this hydrophobic elastomer into an hydrophilic one and thus to increase its versatility towards specific applications involving PDMS as an interface. The influence of a O_2 plasma on cross-linked PDMS has been widely studied in the field of the μContact printing technique to provide hydrophilic capabilities and to functionalized PDMS surfaces,[19] and also to improve PDMS dielectric and protection capabilities. [20]

Another alternative procedure inducing surface oxidation of the PDMS is UV-Ozone treatment. This technique is generally used as a cleaning procedure for solid substrates (Si, SiO_2, AsGa, ...) in optoelectronics and microelectronics. It involves a surface treatment with one of the strongest oxidizing agent, ozone, produced from atmospheric oxygen by irradiation at 185nm and then dissociated into molecular and atomic species with a 254nm light irradiation. In the case of PDMS submitted to this treatment, previous works have shown the conversion of the surface into a silica like material which exhibits good hydrophilic properties.[22,23]

We have performed our experiments on a commercial UVOCS cleaning system. Si substrates were coated with 6 to 7μm thick Sylgard 184 PDMS layer obtained by spin coating and curing on a hot plate. Samples were then exposed to Ozone for various times. To allow a direct comparison a piece of a fresh Si wafer was put on the the sample surface in order to preserve a part of the PDMS surface from UV-Ozone treatment.

Results revealed first, a misunderstood swelling phenomenon of the volume of the cross-linked material exposed to ozone. A topographical step as deep as 300 nm is systematically observed attesting for the swelling of the exposed material.

Samples were then analyzed with both SIMS (Secondary Ion Mass Spectroscopy) and XPS (Xray Photoelectron Spectroscopy). These two techniques revealed a chemical modification of the material resulting in a silica like surface. XPS have demonstrated the conversion of the surface of PDMS of initial composition $Si_1O_2C_{1.9}$ to an oxidized material with a composition close to $Si_1O_2C_{0.33}$ after 60min of treatment.

Careful interpretation of SIMS spectra (not shown) have led to the conclusion of a complete modification of the polymer layer. Matrix effect have a drastic influence on SIMS spectra and cannot confirm an increase or decrease of atomic composition but revealed a modification of the whole 6$\sim$7 μm thick material.

Concerning physicochemical properties modifications, we have achieved contact angle measurements with water and a three liquid method detailed before (Section 3.2). As illustrated in Figure 23, measurements revealed good hydrophilic behavior after 30 min of UVOCS treatment and a modification of the surface energy.

A complete recovery of original PDMS properties can be obtained keeping samples in contact with standard atmosphere and was effectively observed after 10 days.

Finally, the modification of the PDMS material towards UV-Ozone treatment was confirmed by a chemical etching of the treated material in a diluted HF solution (5%). These experiments revealed a selective stripping between the original and the treated material resulting in the total conservation of the untreated material and a 100nm/min

FIGURE 8.23. Contact angle measurements on a cross-linked PDMS surface with water on a untreated surface (left image: $\alpha = 106°$) and after 30min of UVOCS treatment (right image: ($\alpha < 5°$).

etching rate of the modified polymer. This selective etching do not seem to be limited to the top surface layer of the polymer since etching was preformed over 2 μm deep and thus confirms our SIMS interpretation of a "bulk" modification of the exposed material.

This simple and low cost etching technique could be perfectly integrated in our mold assisted technique in order to remove the residual layer after imprinting and to improve process suitability. Moreover, the simple and rapid modification of PDMS interfacial properties should be useful in the field of micro and nano-fluidic applications since it is possible to engineer devices according to fluid characteristics.

8.5. CONCLUSIONS AND DOMAINS OF APPLICATION

In this chapter we have demonstrated the potential use of PDMS as a thermocurable resist for patterning micro and nanostructures using an imprinting technique. We discuss here the advantages of such a process in the perspective of applications. We distinguish applications of this process for micro/nanolithography where pattern transfer using the technologies described in the previous sections are necessary (lift-off, etching, electroplating) and other kind of applications where the micro/nanostructured PDMS thin layer can be used directly.

Our approach can be married into the family of Mold Assisted Lithography (MAL) even if thermo-curing rather than photo-curing is used. Developed in the sphere of influence of Nano-Imprint Lithography (NIL), our process differs deeply from conventional Hot Embossing Lithography (HEL) because the material flowing inside the cavities of the mold is a pre-polymer liquid of low viscosity while the polymer melt used in HEL exhibits very complex visco-elastic behaviour.[24] This specificity allows us to reduce considerably the pressure required for patterning down to a few bars (<10 bars). The temperature and the imprinting time are therefore dictated by the curing reaction taking place in the material once molded and does not depend in the flowing process of the embossed material. Since the curing reaction can be adjusted by simple chemistry through the catalyst concentration and the number of reaction sites along the pre-polymer chains, it turned out to be possible to perform micro and nano-patterning at moderate temperatures (<100°C) and reasonable imprinting times (<10 minutes). All these characteristics seem very interesting for potential applications of the technique as a micro/nano lithography tool. Indeed, low pressure and low temperature are suitable for patterning a large range of substrates that can be fragile or suffer from large differences in thermal expansion coefficients with the mold. Positioning accuracy and alignment control are also improved at moderate temperatures. Moreover, at low imprinting pressure the use of silica or hard polymers molds become possible allowing optical registration and sub-μm alignment between layers using diffraction or Moiré effects. The development of an industrial aligner capable of nanometric overlay accuracy for a x1 replication of a mold is not straightforward and the emergence of low temperature, low pressure processes using transparent molds is certainly a strong improvement eliminating many constraints. The MAL approach alleviates most of the problems of resist flow which are still a handicap for the hot embossing approach due to the complexity of visco-elastic mechanisms taking place during the squeeze flow of the polymer melt beyond Tg and possible hydrodynamic instabilities occurring when filling the cavities of the mold.[25] This enables us to pattern in the same run a large range of features covering many different length scales from the millimeter to the nanometer. This is another highlight of our process because in most applications for optics, optical components, biochips, nanometric and large micrometric features coexist in the same layer and it is required to replicate all of them keeping an imprinting time as low as possible. The compatibility of PDMS material with subsequent transfer technologies such as lift-off, etching and electroplating offers the possibility to use our process for replicating initial masters with inversion or not of the recessed regions (positive/negative). The interesting difference in etching speed between PDMS and Si when using a deep etch process offers the unique possibility to enhance the aspect ratio of the master during the replication. Possible replication of masters with high fidelity is an important issue when discussing the challenge of NIL for becoming an automated standard fabrication technique.

Thick resist processes are massively developed for microsystem fabrication. Conventional technologies use X-ray lithography for exposing resist layers several tens of μm thick or more recently UV lithography on SU8 photoresist. Our imprinting process on PDMS allows the fabrication of structures exhibiting aspect ratio up to 15. Micrometric structures on 100 μm thick PDMS films have been successfully obtained without any specific optimization of the process. We have not fully investigated the limitation of our process for higher aspect ratios but interesting opportunities are certainly opened.

As mentioned before, PDMS material is an interesting elastomer material for achieving Micro-Contact Printing (μCP) enabling the localized deposition of monolayer of molecules on a surface with sub-μm resolution. Some important limitations of the technique are caused by the poor stability and definition of the dimensions of the features of the PDMS stamp. Indeed thick PDMS replica obtained by molding a pre-polymer mixture over a hard master suffer from severe distorsions (thermal expansion, swelling by the ink, sagging, pairing between adjacent close features) .It has been shown recently[26] that the use of a hard PDMS stamps made of a thin PDMS layer (<1 μm) supported by a flat rigid substrate attenuates considerably these drawbacks and improves resolution, quality and registration accuracy. The process we have presented here is an elegant method to directly fabricate these hard stamps for μCP. Thin layers of PDMS, structured to nanometric dimensions with a very thin residual layer (<20 nm) can be readily obtained on various surfaces (silicon, silica, glass). In many aspects, the μCP approach is complementary to other nano-patterning techniques such as NIL and will be the candidate of choice typically when the material to be patterned can be degraded by heating, pressure, solvents ... or when the substrate itself is fragile or not planar. This technique widens considerably the range of processed materials well beyond silicon and other conventional materials for Microelectronics and Microsystems. We think that μCP can be relevant at the nanometric scale if PDMS hard stamps of high quality are used and our NIL process using PDMS as a thermocurable resist is probably the best procedure for fabricating those "hard" stamps.

PDMS is intrinsically an interesting material in itself for microfluidics and lab-on-a-chip applications,[27] in this context the structures directly obtained after imprinting can be used for microfluidic analysis chips. In this case, thick structures of microchannels and microchambers are required for moving, mixing or analyzing reactants or samples. The very smooth surface of PDMS material after imprinting and the possibility to generate deep and vertical microchannels are capital advantages for microfluidic structures. Moreover, the transparency of PDMS in a large spectral range enables the combination of microfluidic elements with fluorescence-based detection. Another major advantage of this material is the possibility to seal very easily the top surface of the PDMS layer onto flat surfaces with no recourse to elaborated bonding techniques. In more advanced works for fabrication of DNA electrophoresis chips,[28] deep microchannels and arrays of nanometric pillars must coexist in order to provide fluid transport and size dependent DNA molecule separation in the artificial gel formed by the nanostructures. The efficiency of our imprinting process for deep micrometric structures and nanometric ones should enable to simplify the fabrication of these chips combining both kind of features. For more complex chips containing for example electrokinetic fluid pumps,[29] the microchannels must be fabricated on top of integrated electrodes. More generally, the possibility to directly fabricate the fluidic structures on top of an intelligent chip where many electronic, mechanic and optical functions are integrated is highly desirable. This aspect is perfectly compatible with our imprinting process at low pressure and low temperature which can preserve the integrity of the pre-patterend elements of the substrate. This point is the main advantage of our imprinting process for microfluidic analysis chips compared to conventional molding over a master where the assembly of the PDMS replica containing the fluidic structures with the patterned chip containing for example the array of pumping electrodes is required.

PDMS material also exhibits interesting optical properties for various kind of applications. After the imprinting process the material is stable against temperature, it is transparent in the UV, visible an near-IR domains, its surface is smooth, its refractive index close to 1.43 is homogeneous and highly reproducible, it exhibits very low shrinkage during imprinting and finally various kind of optically active dopants or nanoclusters can be inserted in the pre-polymer solution. These specificities enable us to think that many new kind of structures like waveguides, gratings, 2D photonic band gap structures or micro-optical systems could be generated using our process. Once again the low temperature, low pressure imprinting process we have established should enable to integrate these passive optical structures on various kind of optically active materials or microsources like VCSELs.

ACKNOWLEDGEMENTS

The authors would like to acknowledge F. Carcenac, manager of the nanotechnology zone of the technological facility of our laboratory for advised councils and technical assistance, the TEAM group in charge of the clean room facility, E. Daran and B. Viallet for experiments on BCB and contact angle measurements, M. Mauzac from IMRCP for assistance on PDMS chemistry.

REFERENCES

1. C. Vieu, F. Carcenac, A. Pépin, Y. Chen, M. Mejias, A. Lebib, L. Manin-Ferlazzo, L. Couraud and H. Launois, Applied Surf. Sci. 164, 111 (2000)
2. S.Y. Chou, P.R. Krauss and P.J. Renstrom, Appl. Phys. Lett. 67, 3114 (1995)
3. S.Y. Chou, P.R. Krauss, W. Zhang, L. Guo and L. Zhuang, J. Vac. Sci. Technol. B15, 2897 (1997)
4. L.J. Heyderman, H. Schift, C. David, J. Gobrecht and T. Schweizer, Microelectronic Engineering 54, 229 (2000)
5. J. Haisma, M. Verheijen, K. van den Heuvel and J. van den Berg, J. Vac. Sci. Technol. B14, 4124 (1996)
6. M. Tormen, T. Borzenko, G. Scmidt, J. Liu and L.W. Molenkamp, Electronics Letters, 36, 983 (2000)
7. Y. Xia and G.M. Whitesides, Annual Review of Material Science. 28, 153 (1998)
8. « Gouttes, Bulles Perles et Ondes », P.G. de Gennes, F. Brochard-Wyart, D. Quéré, Belin Ed., 2002. ISBN 2-7011-3024-7.
9. E. Delamarche, H. Schmid, H.A. Biebuyck, B. Michel, Advanced Materials 9, 741 (1997).
10. F. Carcenac, C. Vieu, A. Lebib, Y. Chen, L. Manin-Ferlazzo and H. Launois, Microelectronic Engineering 53, 163 (2000)
11. F. Carcenac, L. Malaquin and C. Vieu, Microelectronic Engineering 61-62, 657 (2002)
12. M.R.Brown, Science, 263 (1994)
13. P.Auroy, L.Auvray, L.Léger, Macromolecules 24(18),5158 (1991)
14. H.Brunner, T. Vallant, U. Mayer, H. Hoffman, Langmuir 12, 4614 (1996)
15. L.J. Heyderman, H.Schift, C. David, B. Ketyerer, M. Auf der Maur, J. Gobrecht, Microelectronic Engineering 57-58, 375 (2001)
16. S. Luo, IEE transactions on components and packaging technologies Vol 24, N° 1, 43 (2001)
17. H.C. Scheer, H. Schulz, Microelectronic Engineering 56, 311 (2001)
18. C. F. Kane and R. R. Krchnavek, IEEE photonics technology letters **7**, 535-537 (1995)
19. G.S. Ferguson, M.K.Chaudhury, H.A. Biebuyck, G.M. Whitesides, Macromolecules, 26, 5870 (1993)
20. K.Fateh-Alavi, M.E. Mùñez, S.Karlsson, U.W. Gedde, Polymer degradation and stability to be published (2002)
21. H.Hillborg, J.F.Ankner, U.W.Gedde, G.D. Schmid, H.K.Yasuda, K.wikström, Polymer 41, 6851 (2000)
22. M. Ouyang, R.J. Muisener, A. Boulares, J.T. Koberstein, Journal of Membrane science 177, 177 (2000)
23. M.Ouyang, C.Yuan, R.J.Muisener, A.Boulares, J.T. Koberstein, Chem. Mater. 12, 1591 (2000)
24. S. Zankovych, T. Hoffmann, J. Seekamp, J.U. Bruch and C.M. Sotomayor-Torres, Nanotechnology 12, 91 (2001)
25. H. Schift, L.J. Heyderman, M. Auf der Maur and J. Gobrecht, Nanotechnology 12, 173 (2001)
26. M. Tormen, T. Borzenko, B. Steffen, G. Schmidt and L.W. Molenkamp, Microelectronic Engineering 61-62, 469 (2002)
27. T. Fujii, Microelectronic Engineering 61-62, 907 (2002)
28. A. Pépin, P. Youinou, V. Studer, A. Lebib and Y. Chen, Microelectronic Engineering 61-62, 927 (2002)
29. V. Studer, A. Pépin, Y. Chen and A. Ajdari, Microelectronic Engineering 61-62, 915 (2002)

9

Molecules for Microcontact Printing

Jurriaan Huskens, Maik Liebau and David N. Reinhoudt
University of Twente, Enschede, The Netherlands

9.1. INTRODUCTION

Microcontact printing (μCP), pioneered by Whitesides and coworkers,[1-4] is a convenient technique for the replication of patterns on surfaces. Figure 1 shows the general principles of μCP. First, an elastomeric stamp is prepared from an already existing patterned surface, called the master, by covering the surface with a liquid prepolymer (usually poly(dimethylsiloxane), PDMS) followed by curing. The hardened elastomeric stamp is peeled off the master and inked with a solution of an ink, i.e. the molecule of choice to be printed. The solvent is evaporated off and the stamp is brought into contact with the substrate thus transferring the ink from the stamp to the surface at the contacted areas only. The ink forms a self-assembled monolayer (SAM) at these areas. This leaves a patterned surface on which the master patterns have been replicated.

Already in the very first papers,[5-8] various molecules with functional endgroups and various substrates have been described proving its vast applicability. In this way, the polarity of parts of the surfaces could be tuned and this was applied in guiding deposition of e.g. polymers and biomolecules.

An important observation was the early finding[5-7] that patterned SAMs are resistant towards some etch solutions thus allowing the selective etching of unpatterned areas. This has developed into a flowering field of microfabrication.

This chapter gives an overview of molecules which have been used in microcontact printing. Part of the literature in this field deals with the quality and resolution attainable by microcontact printing. Another focuses on its use in lithography i.e. as a step preceding etching, and a large portion describes functionalization of surfaces, e.g. for sensor or biomaterials applications. This review will not cover papers that only use standard inks (e.g. an alkanethiol on gold or a trichloroalkylsilane on silicon oxide), e.g. for protecting

Alternative Lithography: Unleashing the Potentials of Nanotechnology
Edited by Sotomayor Torres, Kluwer Academic/Plenum Publishers, 2003

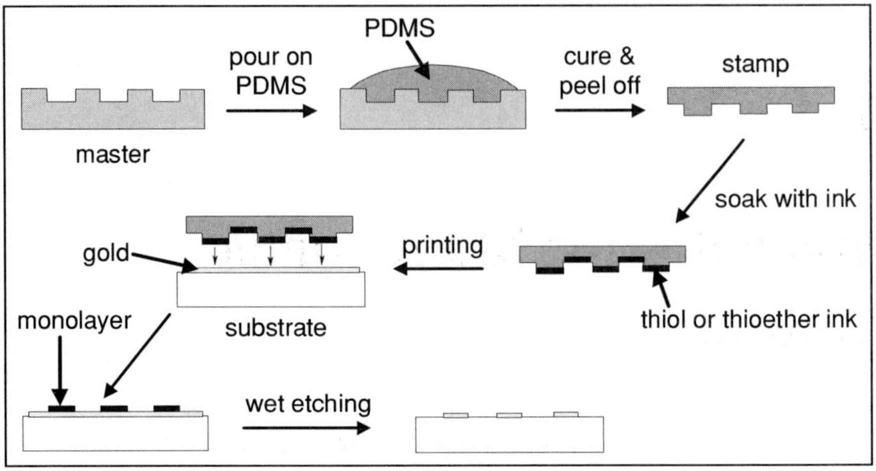

FIGURE 9.1. The microcontact printing process.

part of the surface from deposition of materials applied in a subsequent solution step. Only when they describe advances in resolution, e.g. for attaining submicron etched patterns, or a comparison between SAMs prepared from solution and printed by μCP such papers are included. Other papers included describe the printing with novel, functional molecules, e.g. direct printing of polymers or biomolecules, or printing on non-standard surfaces, e.g. on reactive SAMs.

9.2. QUALITY OF PRINTED SAMS

Self-assembled monolayers[9-12] prepared from solution have been characterized with a multitude of techniques. Spectroscopic techniques give information on the layer thickness (ellipsometry), the orientation and ordering of the molecules in the SAM (IR), or the elemental composition of the SAM (X-ray photoelectron spectroscopy, XPS). Electrochemical techniques provide information on the layer thickness (electrochemical impedance spectroscopy, EIS: capacitance data) and on the accessibility of the underlying metal surface to a redox couple in solution (cyclic voltammetry, CV, or EIS) reflecting both the ordering of the molecules in the SAM and the presence of defects. Scanning probe methods (atomic force microscopy (AFM), scanning tunneling microscopy (STM)) provide local information on outer surface features, e.g. the polarity of functional groups, the presence of defects, and packing of the molecules. The local nature of such probe techniques makes them ideal for studying patterned monolayers prepared by μCP,[13,14] e.g. for imaging patterns or for studying edge resolution.

Process conditions for growing SAMs from solution or for printing SAMs by μCP differ considerably, as printing relies on diffusion of the ink through the polymeric stamp material without wetting of the surface.[15] Therefore, various techniques have been applied to study differences in the quality of solution-grown and printed SAMs because

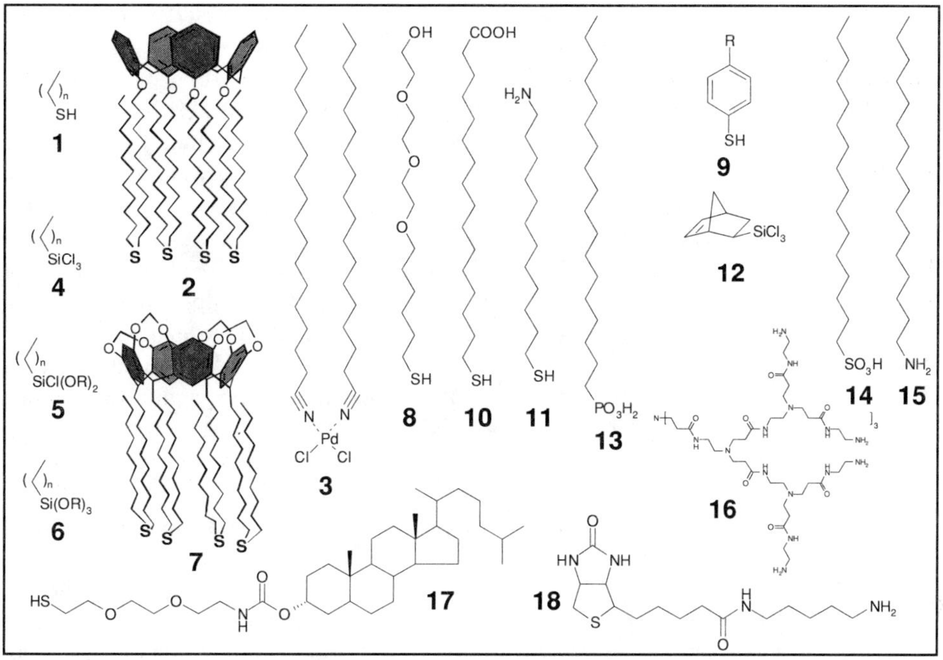

FIGURE 9.2. Molecules used as ink in microcontact printing, discussed in this study.

the quality of printed SAMs is important for further processing, e.g. for subsequent etching steps or solution deposition of other (functional) molecules.

Fischer et al.[16] concluded from a near edge X-ray absorption fine structure spectroscopic (NEXAFS) study that the number of defects and conformations of the alkyl chains of thiols on gold do not differ for printed or solution-grown SAMs. This held for thiols with various chain lengths (Figure 2, **1**) and for ink concentrations of 2 mM. For 0.1 mM ink solutions, significantly less ordered monolayers were found. A friction difference was observed by AFM and attributed to different amounts of domain boundaries. A detailed study (contact angles and STM) by Larsen et al.[17] shows that below 0.1 mM of ink (dodecanethiol; **1**, n = 11) layers become incompletely filled. Above 0.1 mM differently crystallized domains are observed ratio of which changes with concentration. Above 10 mM, printed SAMs are practically indistinguishable from solution-grown SAMs. The concentration dependence of the quality of printed layers was also observed by STM, where the higher concentrations provided monolayers of indistinguishable quality from solution-grown SAMs[18] Printed layers of decanethiol (**1**, n = 9) on gold appeared to be better ordered than those grown from solution as shown by grazing incidence X-ray diffraction and AFM.[19] This was mainly attributed to larger domain sizes for printed SAMs.

Printing with the mostly used hexadecanethiol (**1**, n = 15) on gold (1 mM) appeared to give more gauche defects than grown from solution as shown by IR/Vis sum frequency generation.[20] Contact time between stamp and surface was 5 s and contact angle measurements showed that advancing and receding contact angles became only comparable to solution SAMs for contact times >20 s. Recent results[21] confirm the larger number of defects present in short-time printed SAMs, as concluded from lower electrochemical resistance values and lower contact angles. These studies were done for both decanethiol (**1**, n = 9) and a novel calix[4]arene tetrakis(thioether) ink (**2**).

The sharpness of edges of printed features can be partly attributed to autophobic pinning as shown for printing of alkanethiols on gold.[8] Ink diffusion, however, both through the gas phase and laterally across the surface, poses a problem for edge resolution. Gas phase deposition from a stamp has been observed even at 1 mm distance between stamp and surface for dodecanethiol (**1**, n = 11),[17] and proves the necessity for heavier inks. A partial solution to this problem is found in the application of inking stamps via an "ink pad", a flat piece of PDMS which is inked via normal wet inking.[22] This ensures that the ink is mainly localized at the areas of the stamp that will consecutively be pressed onto the substrate. It also allows monolayer quality to be tuned by changing the concentration of the ink solution applied to the stamp pad. The improved quality of structures printed using this stamp pad procedure was also shown by CFM.[23]

This idea can even be reversed, that is, a patterned stamp can be used to ink a flat stamp which is consecutively pressed onto the substrate for pattern transfer. This has been shown[24] for eicosanethiol (**1**, n = 19) on gold but suffered from too little adhesion and thus high diffusion of the thiol along the stamp surface with concomitant loss of resolution. It was shown to be beneficial, however, for more strongly stamp-adhering inks such as a hydrophobic palladium complex (**3**; used as a catalyst for electroless copper deposition; see below) and heavy inks such as proteins (see below).

Monolayers of silanes (**4-6**) on silicon oxide are less well controlled than thiols on gold because the process involves covalent bond formation on the surface without error correction. The process appears to be crucially dependent on the presence of a small amount of water at the surface, and the kinetics of adsorption depends strongly on the type of silane used, i.e. whether a trichlorosilane (**4**), a monochlorosilane (**5**), or a trialkoxysilane (**6**) is used. Also, the adsorption leads not only to a covalent bond between molecule and surface, but also between different molecules.

A detailed study on the quality of printed silane monolayers has shown excellent surface coverages with a contact time of 30 s using 10 mM octadecyltrichlorosilane (**4**, n = 17) as the ink applied to the stamp.[25] The alkyl chains were predominantly all-trans as concluded from reflection-absorption IR spectroscopy (RAIRS) data. Monolayers prepared from solution (1 mM) appeared to reach high surface coverages only after 15-30 min. Too high ink concentrations led to higher-than-monolayer coverages. As reactive spreading was found to limit the pattern resolution, it was concluded that higher molecular weight inks could contribute to improved pattern resolution. Longer chain silanes (docosyltrichorosilane; **4**, n = 21) give an improvement in printing resolution over the commonly applied octadecyltrichlorosilane owing to reduced diffusion.[26] Printing quality is shown to be dependent on temperature (slightly elevated preferred), on contact time (>3 min necessary, no over-inking observed), and humidity (higher humidity beneficial). An annealing step (1 h, 75 °C) appears to improve the quality of subsequently etched patterns.

Stamp pads have been shown to be useful for application in the printing of silanes on silicon oxide as well, as the control over the ink concentration allowed printing of submicron lines.[27] Edges of the lines were thus of equal height as the main part of the printed features, whereas direct inking of the patterned stamp led to edges of several nm high, probably of polymerized silanes due to over-inking.

9.3. MICROCONTACT PRINTING OF ETCH RESISTS

Microcontact printing of thiols on gold (or silver) provides SAMs that are resistant to wet chemical etching of the underlying metal surface as was described in the very first papers on μCP.[5-7] Etch results depend critically on the order of the SAM, and thus on the chain length of the thiol, on the composition of the etching solution, and on the etching time. Etching results in patterned metal layers, and thus the master pattern is replicated via a patterned SAM and etching into the metal pattern. This metal pattern can subsequently be used as an etch mask in multilevel processing, e.g. for patterning silicon.[6] The technique is so readily applied that even reliable micrometer printing and etching on large, plastic, gold-coated surfaces has been performed with high accuracy.[28,29]

Etching of gold areas protected by hexadecanethiol (**1**, n = 15) SAMs appear to etch 10^6 times slower than unprotected sites,[18] and normally only a few pits are visible of about 30 nm originating from defects in the original monolayer. By STM is was possible to desorb a line of just a few molecules wide of a SAM, self-assembled from solution on a 10 nm thick gold substrate, which resulted in 10-20 nm wide etched lines.

The length of the alkyl chain of the thiol ink appeared to be crucial in the resolution of the obtained structures after etching.[15] The best results so far (90 nm gold dots) were obtained using eicosanethiol (**1**, n = 19) as the ink, and it was concluded that the resolution is limited by gas phase and lateral diffusion of the ink and the necessity of keeping the stamp in place until a well ordered SAM was formed. This finding has led to the development of new inks with multiple thioether moieties, such as calix[4]arene (**2**) and cavitand tetrakis(thioether)s (**7**).[21] These inks have been shown to yield monolayer qualities comparable to thiol SAMs as well as good resolution of the obtained micrometer-sized gold structures after etching (Figure 3), while their molecular weight was about 5 times that of eicosanethiol.

Alternatively, diffusion processes have been used in an advantageous way by performing μCP of hexadecanethiol (**1**, n = 15) on gold under water followed by etching.[30] The water eliminates gas phase diffusion from the non-contacting stamp areas to the surface and reduces the speed of reactive spreading of the ink along the surface from the edges of the contacting stamp areas so that it can be controlled. In this way, 100 nm wide etched lines have been prepared from 50 nm thick gold by contacting with the stamp for 4 min.

Atom transfer radical polymerization has been used to amplify the etch resistance of SAMs of an initiator disulfide adsorbate.[31] This technique led to polymer layers of up to 60 nm thick which resisted wet chemical etching better than the starting disulfide. However, a comparison with the etch resistance of normal thiols was not provided.

Etching of thiol-protected silver patterns was much more rapid than equally prepared patterned SAMs on gold when hexadecanethiol was used as the ink and a combination of ferro/ferricyanide and thiosulfate as the etchant.[32] Also, edge resolution and number of

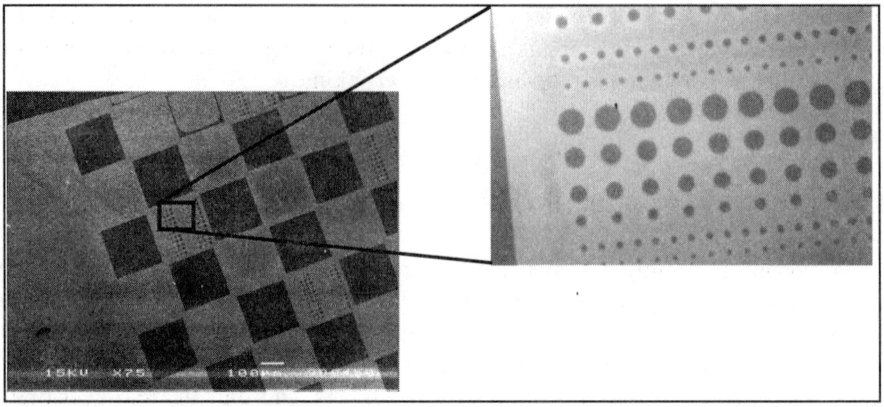

FIGURE 9.3. Etched structure in gold after microcontact printing of cavitand ink **7** and subsequent wet etching [21]. The scale bar on the left indicates 100μm.

defects in the resulting structures were better than for gold. Electroless silver can be used for printing and etching as well, allowing a more general and versatile microfabrication process.[33] The grain structure of electroless silver limits resolution because of reactive spreading of the thiol ink.

Printing of trichlorosilanes on silicon oxide also leads to monolayers that are stable to etching. Printing of octadecyltrichlorosilane (**4**, n = 17) on the native oxide of a silicon wafer led to SAMs stable to etching with HF only upon prolonged contact times between stamp and surface.[34] Etching with HF removed the oxide and etching with KOH allowed patterning of the underlying silicon. Short printing times led to non-resisting SAMs but the change in polarity was enough to cause a difference in deposition rates of etch-resistant polymers,[35] thus still allowing the patterning of the silicon oxide with HF and the underlying silicon with KOH. Etching was also studied for the oxides of crystalline and amorphous silicon.[36]

Multilayer processes have been developed to refine further the etching of silicon. Gold (preceded by titanium or nickel for adhesion) has been evaporated onto silicon wafers, patterned by μCP with hexadecanethiol, and etched.[37] The resulting gold patterns resist etching of the native silicon oxide by HF and the silicon by KOH. The gold patterns are removed after etching in silicon by aqua regia. Alternatively, similarly prepared gold patterns have been used as resists in reactive ion etching into silicon using CF_4/O_2.[38] Feature sizes of 300 nm and depths of 1 μm have thus been achieved. Gold patterns can also be created on a PMMA layer spun on top of a silicon wafer, and the gold patterns resist reactive ion etching into PMMA.[39] This allows standard lift-off technology, and thus deposition of other materials (see Figure 4) on the PMMA patterns and patterning of already patterned silicon substrates.

Gold patterns prepared by printing of octadecylthiol (**1**, n = 17) and etching have also been used to prepare patterned SAMs of octadecyltrichlorosilane on silicon oxide at the areas not covered by gold.[40] It was claimed that patterns obtained in this way were better than those prepared by direct printing of silanes on silicon oxide and that process conditions can be tuned more easily.

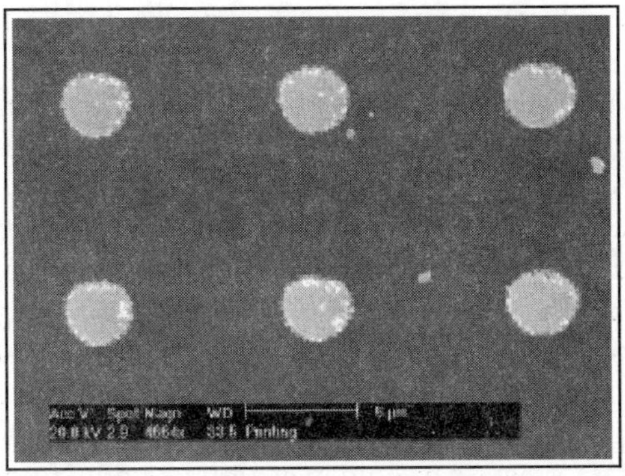

FIGURE 9.4. Cr/Au dots obtained using a multilevel processing technique involving a microcontact printing step [39]. The scale bar denotes 5 μm.

9.4. PRINTING FUNCTIONAL ADSORBATES

From the very beginning, μCP has been used to create patterns of two different adsorbates, one by printing and the other by filling the unprinted regions by self-assembly from solution.[6] For example, it allows different wetting behavior,[41] specific adsorption, or patterned crystallization. The patterns are easily visualized by friction force AFM. Differences between bromide and hydroxyl functionalities have even been demonstrated by AFM height images.[42]

Printing with a tris(ethyleneoxy) thiol adsorbate (**8**) has been achieved to obtain patterned surfaces of a crown ether derivative.[43] Imaging was achieved by friction force AFM and sodium binding in the crown ether was shown to result in a repulsive interaction.

As an alternative to α,ω-functionalized alkanethiols, *p*-functionalized thiophenols (**9**, R: carboxy, hydroxy, nitro, methyl, amine) have been used to prepare patterned surfaces on evaporated silver.[44–46] Micrometer patterns were prepared and used to prove the applicability of surface-enhanced Raman scattering imaging as a pattern imaging technique with a 2 μm resolution limit.

Patterned self-assembled monolayers prepared by μCP have been used to get preferential adsorption of polymers and polymer multilayers. Hammond et al.[47] have printed a carboxylic acid-terminated thiol (**10**) and filled the gaps with an oligoether thiol. Polymer multilayers were obtained by alternate deposition of a polymeric polycation (polydiallyldimethylammonium chloride) and a polymeric polyanion (sulfonylated polystyrene). The film thickness increased linearly with the number of deposited bilayers.

A printed monolayer of 12-aminododecanethiol (**11**) resisted the anodic oxidative deposition of polyaniline which took place only at the uncovered areas.[48] In contrast, an alkylthiol did not resist the deposition, and could therefore be used to limit the surface diffusion of the aminoalkylthiol by filling up the uncovered areas.

Whitesides et al.[49] printed an ω-carboxylic functionalized thiol (**10**) on gold and filled up the gaps by washing with a trimethyl(thioalkyl)ammonium derivative to obtained patterns of opposite charge. Micron-sized gold disks terminated with phosphonate groups were shown to deposit selectively but randomly onto the positively charged quaternary ammonium regions.

The same procedure was used to prepare patterns of carboxylic acid groups and alkyl chains, but here[50] the carboxylic acid groups were converted to anhydrides to allow the covalent attachment of poly(ethylene imine) (PEI). This procedure was extended by attaching an anhydride polymer, again followed by PEI to obtain multilayers.

In another study,[51] the same carboxylic acid thiol **10** was employed to attract iron ions which resisted polymer deposition by CVD. Preferential polymer adsorption onto carboxylate-terminated SAMs has also been used to get isolated crossing wires in a two-layer structure.[52] In contrast, carboxylate-terminated SAMs resisted the adsorption of gold colloids which preferred adsorption to amine-terminated intermediate areas, and which even allowed the formation of gold nanoparticle multilayers.[53]

Whitesides has described the printing of palladium nanoparticles onto a silicon oxide/silicon surface silane-functionalized with functional groups for better adhesion of the nanoparticles and the metal layer deposited thereafter.[54] These palladium colloids served as catalysts in the electroless deposition of copper which resulted in 0.5-1 μm thick metal features. When glass was used as the substrate, free standing metal features could be obtained by dissolution of the glass by HF. The feature sizes of these metal structures could be reduced to about 1 μm by printing onto a polymeric structure that was shrunk before metallization.[55] Optical fibers were treated in a similar way.[56] By printing with a palladium complex (**3**), 170 nm wide lines have been obtained.[57]

An iron salt ($Fe(NO_3)_3.9H_2O$ in ethanol) has been printed onto SiO_2/Si wafers for catalyzing the formation of multiwalled carbon nanotubes in a sequential decomposition step of acetylene.[58] The stamp was hydrophilized in an oxygen plasma before applying the ink solution. Aging of an ethanolic iron salt-containing solution gave a gel-like catalyst which showed a better performance.[59]

Braun et al. printed an amino-terminated thiol (**11**) onto gold which served as an initiator for the formation of a polypeptide.[60] The polymerization led to a layer thickness of about 30 nm.

Another study[61] employed the attachment of the first norbornene monomer to a silicon oxide surface, followed by the attachment of the ruthenium catalyst capable of ring opening metathesis polymerization of norbornene monomers from solution leading to 50 nm thick polymer layers. Direct μCP of the norbornene silane precursor (**12**) was unsuccessful due to slow reactivity. Therefore OTS (**4**, $n = 17$) was printed followed by solution reaction of the norbornene precursor.

9.5. PRINTING ON OTHER SUBSTRATES

Microcontact printing is mostly applied on gold (or the closely related silver) and on silicon oxide (be it the native oxide of silicon, quartz, or glass). Self-assembled monolayers have been prepared, however, on a variety of other substrates, including copper, aluminum, and other oxidic materials, such as superconductors. This section gives an overview of microcontact printing onto such substrates as well as on polymeric substrates and reactive SAMs. Biologically oriented studies are described in the next section.

Although monolayers of thiols on copper behave differently from those on silver or gold, printing of hexadecanethiol (**1**, n = 15) on copper appeared to work as well as on silver and gold.[62] The weaker resistance of copper towards etching was used to prepare mixed patterns of copper and silver features.

Octadecyltrichlorosilane (**4**, n = 17), widely used for printing on silicon oxide, has also been applied for pattern formation on Al_2O_3/Al, TiO_2/TiN, and indium tin oxide (ITO).[63] The patterns were shown to resist chemical vapor deposition of copper. In a similar procedure, deposition of oxidic materials was guided by the stamped patterns.[64] Printing on ITO was studied in more detail by Marks et al.[65] and appeared to be less trivial as initial attempts at room temperature gave irreproducible results. Only after sonication of ITO in acetone and ethanol, and heating to 80-100 °C during printing provided a reliable process.

The formation of SAMs of phosphonic acids on metal oxides has been applied in the μCP of octadecylphosphonic acid (**13**) on Al_2O_3/Al.[66] The ink (applied as an isopropanol solution) appears to be adhering to the outside of the stamp. Shorter alkyl chains offered insufficient protection to etching with aqueous phosphoric, acetic, and nitric acids. Aluminium patterns obtained after etching were conducting and isolated from neighboring aluminum features. This procedure was applied in a multilevel process for obtaining rectifier circuits.[67]

As polymer substrates are interesting for potentially cheap electronic applications, various attempts of μCP on polymeric substrates have been reported. Jeon et al. reported[63] the printing of OTS on plasma treated polyimide, but little detail was provided.

Photoresist patterns on glass or polyimide have been obtained by stamping hexadecanesulfonic acid (**14**) onto the polysiloxane resist to obtain silicon oxide patterns resisting lift-off.[68] In this way, submicron patterns have been obtained and pattern transfer into polyimide was achieved.

Ghosh and Crooks oxidized polyethylene (PE) with aqueous CrO_3/H_2SO_4 to obtain a carboxylic acid-functionalized PE surface which was subsequently converted to the anhydride using ethyl chloroformate.[69] Printing with hexadecaneamine (**15**) resulted in passivated areas, whereas the other areas were used for attachment of hyperbranched polymers.

Hydrogel stamps were applied in the transfer of a poly(amido amine) dendrimer (generation 2; **16**) onto a poly(vinylbenzyl chloride) surface to obtain physisorbed patterns.[70] Fluorescent dyes were covalently attached to the dendrimers in a subsequent step in order to visualize the patterns using fluorescence microscopy.

Conceptually a very versatile approach is the stamping of inks onto a reactive SAM (see Figure 5). Whitesides et al. prepared a monolayer of a carboxylic acid-terminated thiol on gold which was converted to the anhydride by using trifluoroacetic anhydride.[71] The reactive anhydride groups were reacted by μCP with hexadecaneamine (**15**), resulting in amide patterns (Figure 5). The remaining anhydride groups were reacted with an amine from solution. In a similar fashion, poly(ethylene imine) was covalently attached by printing onto such an anhydride surface,[72] as well as poly(t-butyl acrylate).[73]

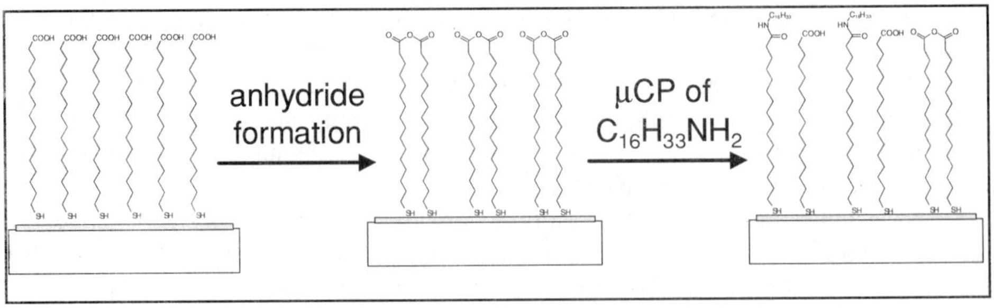

FIGURE 9.5. Microcontact printing on a reactive SAM[71]

9.6. MICROCONTACT PRINTING IN BIO-APPLICATIONS

The ease of use of μCP has triggered its application in a wide range of fields, also including biomaterials. A few key examples will be shown regarding the guided deposition of biomolecules, the direct printing of proteins, and the guided deposition of cells.

Evans et al. have shown the printing of a thiol-functionalized cholesterol derivative (**17**).[74,75] It allowed the deposition and anchoring of a lipid bilayer structure prepared from vesicles. After incorporation of ion channels, they showed that the membranes were still functional.

A mixed SAM of a carboxylic acid and a polyethylene glycol was converted into the corresponding active ester by coupling with pentafluorophenol.[76] Stamping with an amine-terminated biotin unit (**18**) resulted in adhesion of streptavidin and, subsequently, immunoglobulins.

Spencer et al.[77] studied the printing of an N-hydroxysuccinimide ester ω-functionalized disulfide. It allows the covalent attachment of peptides via reaction with their free amino groups.

Biebuyck and Bernard et al.[78,79] have printed proteins (IgGs) onto glass using μCP. In contrast to conventional thiol inks, the proteins are adsorbed at the outer surface of the stamp where they form a monolayer that resists washing. Nevertheless they are transferred rapidly leading to patterns down to 600 nm. Because of the virtually zero diffusion of these heavy proteins, resolution is probably not limited to this value. Individual proteins have been observed in the patterns by AFM and biological activity was ascertained. Proteins appear to adhere so strongly that their interactions with antibody-functionalized AFM tips can be studied by AFM force measurements.[80]

Whereas the examples mentioned above rely on electrostatic interactions between the protein and the substrate, Patel et al.[81] have studied the printing of avidin onto a polymer substrate derivatized with biotin units leading to strong specific interactions. The free sites of avidin appeared still to be accessible for other biotinylated species.

A similar study showed the surface activation of poly(ethylene terephthalate) in order to allow functionalization with biotin and the subsequent adhesion of streptavidin.[82,83] Plasma-oxidized PDMS stamps were used for transfer of the biotin reagent.

Batt et al. studied the printing of antibodies directly onto the native oxide of silicon using an oxygen plasma treated stamp.[84] The patterns were optically visualized after attachment of E. coli cells to the antibodies. Printing of thiol-functionalized antibodies on gold led to better defined patterns than without functionalization and E. coli adhesion was further studied by optical diffraction as well.[85]

Microcontact printing of hexaethylene glycol thiol was followed by adhesion of oligopeptides at the non-patterned areas to obtain areas resisting and promoting cell adhesion, respectively.[86] Various cell types adsorbed specifically onto variously shaped peptide patterns.

Polylysine has been printed onto glass employing the electrostatic interaction between the proteins and the substrate.[87] The stamps were hydrophilized in an oxygen plasma before printing. Neuron cells were attached to such polylysine areas patterned onto microelectrode arrays.[88]

Wheeler et al.[89] have obtained activated ester-functionalized glass substrates onto which proteins (polylysine, BSA, laminin) were printed using μCP. The unreacted areas were passivated using an amino-functionalized poly(ethylene glycol) which resists protein adsorption. Neuron cells were grown onto these patterns and their protein specificity was studied.

In a similar study, such neurons were shown to be active for more than a week and lines of a few micron promoted single cell spreading.[90] Polylysine-conjugated laminin appeared to be a very efficient promotor for neuron adhesion.[91]

The same surface derivatization as described above[89] was applied by Offenhäuser et al.[92] and was followed by μCP of the synthetic peptide PA22-2, promoting neuron cell growth. Their alignment and activity was investigated in detail. A similar procedure was used to obtain neurons immobilized on microelectronic devices.[93]

The cell growth-inhibiting diblock copolymer poly(ethylene glycol)-poly(lactic acid) was printed onto the cell growth-promoting poly(lactic-co-glycolic acid) in order to obtain specific adsorption of retinal pigment epithelial cells.[94] The reverse was also performed and cell morphology appeared to depend on the printed pattern.

Kam and Boxer printed fibronectin onto a glass support in 5 μm lines and the empty areas were covered by a lipid bilayer using vesicles.[95] The fibronectin patterns promoted the adhesion of endothelial cells which appeared not in contact with underlying bilayer areas as was shown by the mobility of bilayer-incorporated albumin.

9.7. CONCLUSIONS

The quality of thiol or thioether SAMs printed on gold depends on the ink chain length, ink concentration, stamp-substrate contact time, and temperature. SAM quality can be comparable to SAMs prepared from solution. Good order and low numbers of defects are crucial for good pattern transfer into metals using etching. Similar considerations hold for silane printing onto silicon oxide, but this process is more empirical.

Various molecules besides thiols and silanes have been used as inks, including thioethers, phosphonates, thiols with various functionalities, proteins, and polymers. For small molecules, diffusion is the main factor limiting resolution, whereas for large molecules such as polymers and proteins, stamp properties (e.g. elastic deformation) are more important.

ACKNOWLEDGEMENTS

Support from the European Commission for the SPINUP project (ESPRIT MEL-ARI and IST FET) is gratefully acknowledged.

REFERENCES

1. Xia, Y.; Whitesides, G. M. *Angew. Chem. Int. Ed.* **1998**, *37*, 550-75.
2. Michel, B.; Bernard, A.; Bietsch, A.; Delamarche, E.; Geissler, M.; Juncker, D.; Kind, H.; Renault, J.-P.; Rothuizen, H.; Schmid, H.; Schmidt-Winkel, P.; Stutz, R.; Wolf, H. *IBM J. Res. Dev.* **2001**, in press.
3. Xia, Y.; Zhao, X.-M.; Whitesides, G. M. *Microelectr. Eng.* **1996**, *32*, 255-68.
4. Kane, R. S.; Takayama, S.; Ostuni, E.; Ingber, D. E.; Whitesides, G. M. *Biomater.* **1999**, *20*, 2363-76.
5. Kumar, A.; Whitesides, G. M. *Appl. Phys. Lett.* **1993**, *63*, 2002-4.
6. Kumar, A.; Biebuyck, H. A.; Whitesides, G. M. *Langmuir* **1994**, *10*, 1498-511.
7. Wilbur, J. L.; Kumar, A.; Kim, E.; Whitesides, G. M. *Adv. Mater.* **1994**, *6*, 600-4.
8. Biebuyck, H. A.; Whitesides, G. M. *Langmuir* **1994**, *10*, 4581-7.
9. Bain, C. D.; Whitesides, G. M. *Angew. Chem. Int. Ed. Engl.* **1989**, *28*, 506-12.
10. Ulman, A. *Introduction to Thin Organic Films*, Academic Press, Boston, **1991**.
11. Dubois, L. H.; Nuzzo, R. G. *Annu. Rev. Phys. Chem.* **1992**, *43*, 437-63.
12. Schreiber, F. *Progr. Surf. Sci.* **2000**, *65*, 151-256.
13. Wilbur, J. L.; Biebuyck, H. A.; MacDonald, J. C.; Whitesides, G. M. *Langmuir* **1995**, *11*, 825-31.
14. Bar, G.; Rubin, S.; Parikh, A. N.; Swanson, B. I.; Zawodszinski Jr., T. A.; Whangbo, M.-H. *Langmuir* **1997**, *13*, 373-7.
15. Delamarche, E.; Schmid, H.; Bietsch, A.; Larsen, N. B.; Rothuizen, H.; Michel, B.; Biebuyck, H. *J. Phys. Chem. B* **1998**, *102*, 3324-34.
16. Fischer, D.; Marti, A.; Hähner, G. *J. Vac. Sci. Technol. A* **1997**, *15*, 2173-80.
17. Larsen, N. B.; Biebuyck, H.; Delamarche, E.; Michel, B. *J. Am. Chem. Soc.* **1997**, *119*, 3017-26.
18. Biebuyck, H. A.; Larsen, N. B.; Delamarche, E.; Michel, B. *IBM J. Res. Dev.* **1997**, *41*, 159-70.
19. Eberhardt, A. S.; Nyquist, R. M.; Parikh, A. N.; Zawodzinski, T.; Swanson, B. I. *Langmuir* **1999**, *15*, 1595-8.
20. Böhm, I.; Lampert, A.; Buck, M.; Eisert, F.; Grunze, M. *Appl. Surf. Sci.* **1999**, *141*, 237-43.
21. Liebau, M.; Huskens, J.; Reinhoudt, D. N. *Adv. Funct. Mater.* **2001**, *11*, 147-50.
22. Libioulle, L.; Bietsch, A.; Schmid, H.; Michel, B.; Delamarche, E. *Langmuir* **1999**, *15*, 300-4.
23. Fujihira, M.; Furugori, M.; Akiba, U.; Tani, Y. *Ultramicrosc.* **2001**, *86*, 75-83.
24. Geissler, M.; Bernard, A.; Bietsch, A.; Schmid, H.; Michel, B.; Delamarche, E. *J. Am. Chem. Soc.* **2000**, *122*, 6303-4.
25. Jeon, N. L.; Finnie, K.; Branshaw, K.; Nuzzo, R. G. *Langmuir* **1997**, *13*, 3382-91.
26. Finnie, K. R.; Haasch, R.; Nuzzo, R. G. *Langmuir* **2000**, *16*, 6968-76.
27. Pompe, T.; Fery, A.; Herminghaus, S.; Kriele, A.; Lorenz, H.; Kotthaus, J. P. *Langmuir* **1999**, *15*, 2398-401.
28. Rogers, J. A.; Bao, Z.; Dodabalapur, A.; Makhija, A. *IEEE Electron Dev. Lett.* **2000**, *21*, 100-3.
29. Rogers, J. A.; Bao, Z.; Baldwin, K.; Dodabalapur, A.; Crone, B.; Raju, V. R.; Kuck, V.; Katz, H.; Amundson, K.; Ewing, J.; Drzaic, P. *Proc. Natl. Acad. Sci.* **2001**, *98*, 4835-40.
30. Xia, Y.; Whitesides, G. M. *Langmuir* **1997**, *13*, 2059-67.
31. Shah, R. R.; Merreceyes, D.; Husemann, M.; Rees, I.; Abbott, N. L.; Hawker, C. J.; Hedrick, J. L. *Macromol.* **2000**, *33*, 597-605.
32. Xia, Y.; Kim, E.; Whitesides, G. M. *J. Electrochem. Soc.* **1996**, *143*, 1070-9.
33. Xia, Y.; Venkateswaran, N.; Qin, D.; Tien, J.; Whitesides, G. M. *Langmuir* **1998**, *14*, 363-71.

34. St. John, P. M.; Craighead, H. G. *Appl. Phys. Lett.* **1996**, *68*, 1022-4.
35. Xia, Y.; Mrksich, M.; Kim, E.; Whitesides, G. M. *J. Am. Chem. Soc.* **1995**, *117*, 9576-7.
36. Wang, D.; Thomas, S. G.; Wang, K. L.; Xia, Y.; Whitesides, G. M. *Appl. Phys. Lett.* **1997**, *70*, 1593-5.
37. Kim, E.; Kumar, A.; Whitesides, G. M. *J. Electrochem. Soc.* **1995**, *142*, 628-33.
38. Whidden, T. K.; Ferry, D. K.; Kozicki, M. N.; Kim, E.; Kumar, A.; Wilbur, J.; Whitesides, G. M. *Nanotechnol.* **1996**, *7*, 447-51.
39. Chen, Y.; Lebib, A.; Carcenac, F.; Launois, H.; Schmidt, G.; Tormen, M.; Muller, G.; Molenkamp, L. W.; Liebau, M.; Huskens, J.; Reinhoudt, D. N. *Microelectr. Eng.* **2000**, *53*, 253-6.
40. Walheim, S.; Müller, R.; Sprenger, M.; Loser, E.; Mlynek, J.; Steiner, U. *Adv. Mater.* **1999**, *11*, 1431-3.
41. Kumar, A.; Whitesides, G. M. *Science* **1994**, *263*, 60-2.
42. Kim, Y.; Kim, K.-S.; Park, M.; Jeong, J. *Thin Solid Films* **1999**, *341*, 91-3.
43. Flink, S.; Schönherr, H.; Vancso, G. J.; Geurts, F. A. J.; Van Leerdam, K. G. C.; Van Veggel, F. C. J. M.; Reinhoudt, D. N. *J. Chem. Soc., Perkin Trans. 2* **2000**, 2141-6.
44. Yang, X. M.; Tryk, D. A.; Hashimoto, K.; Fujishima, A. *Appl. Phys. Lett.* **1996**, *69*, 4020-2.
45. Tryk, D. A.; Yang, X.; Hashimoto, K.; Fujishima, A. *Bull. Chem. Soc. Jpn.* **1998**, *71*, 31-9.
46. Yang, X. M.; Tryk, D. A.; Hashimoto, K.; Fujishima, A. *J. Raman Spectrosc.* **1998**, *29*, 725-32.
47. Clark, S. L.; Montague, M.; Hammond, P. T. *Supramol. Sci.* **1997**, *4*, 141-6.
48. Sayre, C. N.; Collard, D. M. *J. Mater. Chem.* **1997**, *7*, 909-12.
49. Tien, J.; Terfort, A.; Whitesides, G. M. *Langmuir* **1997**, *13*, 5349-55.
50. Huck, W. T. S.; Yan, L.; Stroock, A.; Haag, R.; Whitesides, G. M. *Langmuir* **1999**, *15*, 6862-7.
51. Vaeth, K. M.; Jackman, R. J.; Black, A. J.; Whitesides, G. M.; Jensen, K. F. *Langmuir* **2000**, *16*, 8495-500.
52. Black, A. J.; Nealey, P. F.; Thywissen, J. H.; Deshpande, M.; El-Zein, N.; Maracas, G. N.; Prentiss, M.; Whitesides, G. M. *Sens. Actuat. A* **2000**, *86*, 96-102.
53. Musick, M. D.; Keating, C. D.; Lyon, L. A.; Botsko, S. L.; Peña, D. J.; Holliway, W. D.; McEvoy, T. M.; Richardson, J. N.; Natan, M. J. *Chem. Mater.* **2000**, *12*, 2869-81.
54. Hidber, P. C.; Helbig, W.; Kim, E.; Whitesides, G. M. *Langmuir* **1996**, *12*, 1375-80.
55. Hidber, P. C.; Nealey, P. F.; Helbig, W.; Whitesides, G. M. *Langmuir* **1996**, *12*, 5209-15.
56. Rogers, J. A.; Jackman, R. J.; Whitesides, G. M.; Wagener, J. L.; Vengsarkar, A. M. *Appl. Phys. Lett.* **1997**, *70*, 7-9.
57. Kind, H.; Geissler, M.; Schmid, H.; Michel, B.; Kern, K.; Delamarche, E. *Langmuir* **2000**, *16*, 6367-73.
58. Kind, H.; Bonard, J.-M.; Emmenegger, C.; Nilsson, L.-O.; Hernadi, K.; Maillard-Schaller, E.; Schlapbach, L.; Forró, L.; Kern, K. *Adv. Mater.* **1999**, *11*, 1285-9.
59. Kind, H.; Bonard, J.-M.; Forró, L.; Kern, K.; Hernadi, K.; Nilsson, L.-O.; Schlapbach, L. *Langmuir* **2000**, *16*, 6877-83.
60. Kratzmüller, T.; Appelhans, D.; Braun, H.-G. *Adv. Mater.* **1999**, *11*, 555-8.
61. Kim, N. Y.; Jeon, N. L.; Choi, I. S.; Takami, S.; Harada, Y.; Finnie, K. R.; Girolami, G. S.; Nuzzo, R. G.; Whitesides, G. M.; Laibinis, P. E. *Macromol.* **2000**, *33*, 2793-5.
62. Xia, Y.; Kim, E.; Mrksich, M.; Whitesides, G. M. *Chem. Mater.* **1996**, *8*, 601-3.
63. Jeon, N. L.; Nuzzo, R. G.; Xia, Y.; Mrksich, M.; Whitesides, G. M. *Langmuir* **1995**, *11*, 3024-6.
64. Jeon, N. L.; Clem, P. G.; Nuzzo, R. G.; Payne, D. A. *J. Mater. Res.* **1995**, *10*, 2996-9.
65. Koide, J.; Wang, Q.; Cui, J.; Benson, D. D.; Marks, T. J. *J. Am. Chem. Soc.* **2000**, *122*, 11266-7.
66. Goetting, L. B.; Deng, T.; Whitesides, G. M. *Langmuir* **1999**, *15*, 1182-91.
67. Deng, T.; Goetting, L. B.; Hu, J.; Whitesides, G. M. *Sens. Actuat. A* **1999**, *75*, 60-4.
68. Marzolin, C.; Terfort, A.; Tien, J.; Whitesides, G. M. *Thin Solid Films* **1998**, *315*, 9-12.
69. Ghosh, P.; Crooks, R. M. *J. Am. Chem. Soc.* **1999**, *121*, 8395-6.
70. Martin, B. D.; Brandow, S. L.; Dressick, W. J.; Schull, T. L. *Langmuir* **2000**, *16*, 9944-6.
71. Yan, L.; Zhao, X.-M.; Whitesides, G. M. *J. Am. Chem. Soc.* **1998**, *120*, 6179-80.
72. Yan, L.; Huck, W. T. S.; Zhao, X.-M.; Whitesides, G. M. *Langmuir* **1999**, *15*, 1208-14.
73. Ghosh, P.; Lackowski, W. M.; Crooks, R. M. *Macromol.* **2001**, *34*, 1230-6.
74. Jenkins, A. T. A.; Bushby, R. J.; Boden, N.; Evans, S. D.; Knowles, P. F.; Liu, Q.; Miles, R. E.; Ogier, S. D. *Langmuir* **1998**, *14*, 4675-8.
75. Jenkins, A. T. A.; Boden, N.; Bushby, R. J.; Evans, S. D.; Knowles, P. F.; Miles, R. E.; Ogier, S. D.; Schönherr, H.; Vancso, G. J. *J. Am. Chem. Soc.* **1999**, *121*, 5274-80.
76. Lahiri, J.; Ostuni, E.; Whitesides, G. M. *Langmuir* **1999**, *15*, 2055-60.

77. Zaugg, F. G.; Spencer, N. D.; Wagner, P.; Kernen, P.; Vinckier, A.; Groscurth, P.; Semenza, G. *J. Mater. Sci., Mater. Med.* **1999**, *10*, 255-63.
78. Bernard, A.; Delamarche, E.; Schmid, H.; Michel, B.; Bosshard, H. R.; Biebuyck, H. *Langmuir* **1998**, *14*, 2225-9.
79. Bernard, A.; Renault, J. P.; Michel, B.; Bosshard, H. R.; Delamarche, E. *Adv. Mater.* **2000**, *12*, 1067-70.
80. Garrison, M. D.; McDevitt, T. C.; Luginbühl, R.; Giachelli, C. M.; Stayton, P.; Ratner, B. D. *Ultramicrosc.* **2000**, *82*, 193-202.
81. Patel, N.; Bhandari, R.; Shakesheff, K. M.; Cannizzaro, S. M.; Davies, M. C.; Langer, R.; Roberts, C. J.; Tendler, S. J. B.; Williams, P. M. *J. Biomater. Sci. Polym. Ed.* **2000**, *11*, 319-31.
82. Yang, Z.; Chilkoti, A. *Adv. Mater.* **2000**, *12*, 413-7.
83. Yang, Z.; Belu, A.; Liebmann-Vinson, A.; Sugg, H.; Chilkoti, A. *Langmuir* **2000**, *16*, 7482-92.
84. St. John, P. M.; Davis, R.; Cady, N.; Czajka, J.; Batt, C. A.; Craighead, H. G. *Anal. Chem.* **1998**, *70*, 1108-11.
85. Morhard, F.; Pipper, J.; Dahint, R.; Grunze, M. *Sens. Actuat. B* **2000**, *70*, 232-42.
86. Zhang, S.; Yan, L.; Altman, M.; Lässle, M.; Nugent, H.; Frankel, F.; Lauffenburger, D. A.; Whitesides, G. M.; Rich, A. *Biomater.* **1999**, *20*, 1213-20.
87. James, C. D.; Davis, R. C.; Kam, L.; Craighead, H. G.; Isaacson, M.; Turner, J. N.; Shain, W. *Langmuir* **1998**, *14*, 741-4.
88. James, C. D.; Davis, R.; Meyer, M.; Turner, A.; Turner, S.; Withers, G.; Kam, L.; Banker, G.; Craighead, H. G.; Isaacson, M.; Turner, J.; Shain, W. *IEEE Trans. Biomed. Eng.* **2000**, *47*, 17-21.
89. Wheeler, B. C.; Corey, J. M.; Brewer, G. J.; Branch, D. W. *J. Biomech. Eng.* **1999**, *121*, 73-8.
90. Klein, C. L.; Scholl, M.; Maelicke, A. *J. Mater. Sci., Mater. Med.* **1999**, *10*, 721-7.
91. Kam, L.; Shain, W.; Turner, J. N.; Bizios, R. *Biomater.* **2001**, *22*, 1049-54.
92. Scholl, M.; Sprössler, C.; Denyer, M. C.; Krause, M.; Nakajima, K.; Maelicke, A.; Knoll, W.; Offenhäusser, A. *J. Neurosci. Meth.* **2000**, *104*, 65-75.
93. Sprössler, C.; Scholl, M.; Denyer, M. C.; Krause, M.; Nakajima, K.; Maelicke, A.; Knoll, W.; Offenhäusser, A. *Synth. Met.* **2001**, *117*, 281-3.
94. Lu, L.; Nyalakonda, K.; Kam, L.; Bizios, R.; Göpferich, A.; Mikos, A. G. *Biomater.* **2001**, *22*, 291-7.
95. Kam, L.; Boxer, S. G. *J. Biomed. Mater. Res.* **2001**, *55*, 487-95.

10

Microcontact Printing Techniques

Massimo Tormen

National Nanotechnology Laboratory (NNL), TASC-INFM at Elettra Synchrotron, Basovizza, Trieste, Italy

10.1. INTRODUCTION

Three elements come together to constitute the microcontact printing technique. The first is the capability of a class of molecules to self-assemble, i.e. to arrange spontaneously, as a dense, ordered and stable monolayer on suitable surfaces. Generally, these molecules contain a chemically reactive head that anchors onto the surface, and a linear flexible chain of methyl groups which orients itself approximately with the normal to the surface as the most convenient packing configuration in the monolayer. The second element is that such a monolayer, about 2-3 nm thick, represents a barrier to the permeation of several chemical species from the environment (both in air and in aqueous solution), providing an effective protection against oxidation and corrosion of the underlying surface. The third element is due to the intuition of Xia and Whitesides who invented the μCP technique.[1] Besides known methods of inducing molecules to self-assemble onto a surface from the liquid or vapor phase, they discovered that self-assembly can also occur when a polymeric membrane impregnated with the same kind of molecules is brought into contact with a convenient substrate. The membrane, acting as a reservoir of molecules, allows their localized delivery to the contacted areas of the substrate.

In synthesis, the principle of μCP consists of forming self-assembled monolayers in regions of a substrate that come into contact with the raised features of a polymeric stamp and to utilize the patterned monolayer as a resist mask to selectively remove material from non-protected areas, due to the strong contrast in the etching rate between protected and unprotected areas.

Microcontact printing is a very efficient, fast, and cost effective technique for patterning Au,[1,2] Ag,[2] Cu,[3] Pd,[4] semiconductors as Si,[5] GaAs,[6] ZnSe,[7] oxides as SiO_2, TiO_2, Al_2O_3[8,9] (either as bulk material or as native oxides of Si, Ti, and Al), transparent electrically conductive materials as indium tin oxide (ITO), and indium zinc oxide

Alternative Lithography: Unleashing the Potentials of Nanotechnology
Edited by Sotomayor Torres, Kluwer Academic/Plenum Publishers, 2003

(IZO).[10] Moreover, using a pattern etched into a thin film of any of the above materials as a mask for dry or wet chemical etching, it is possible to transfer the pattern to other kinds of substrates, thus extending the field of potential applications of μCP.

Most of the demonstrations of microcontact printing are based on the alkanethiols as a molecular "ink" due to the favourable characteristics of their self-assembly on noble metals (Au, Ag, Cu, Pt, Pd). In fact, simple contact of a few seconds with an inked stamp is sufficient to leave on those surfaces a monolayer-fingerprint which is capable of lasting for the time necessary to etch unprotected areas to a depth of 10-100 nm. However, for printing on oxides alkanethiols, are useless (they do not react chemically with these surfaces), and other kinds of molecules are needed, alkyl-trichlorosilanes[8] or alkanephosphonic acids.[9,10] Unfortunately, the assembly process of these molecules occurs on a longer time-scale (minutes to hours) than alkanethiols on Au (Ag, Cu). It also requires more strict control of the environmental conditions (temperature and moisture), and gives rise to a less ordered and protective monolayer. All of these factors make the SAM/oxide systems more interesting for chemical patterning (i.e. μCP without the etching step) than for lithographic applications. Chapter 9 deals specifically with the properties molecules need to base for microcontact printing.

A general classification divides the different micro- and nano-fabrication approaches into top-down and bottom-up techniques. Conceptually, microcontact printing cannot be considered as either, representing instead a synthesis of both. In fact, both aspects contribute concurrently to form a patterned nm-scale resist, top-down, delimiting areas at the scale of the microstructures present in the stamp; bottom-up, introducing a two-dimensional crystalline order in the monolayer that extends over mesoscopic areas. This naturally divides the presentation of μCP in two parts devoted to these different views of the process. Since chapter 9 was devoted to the chemistry of self-assembly, this chapter will focus on the top-down elements of the technique: the fabrication and use of stamps, the preparation of substrates, the conditions for printing and the etching process.

At the end of this chapter, I will present experimental results about a kind of stamp that looks promising for the future of the technique and a demonstration of multilevel processes involving accurate alignment of the μCP step.

Necessarily, the chapter will not include new important developments of μCP in fields other than lithography. In fact, the flexibility of the patterning principle on which μCP is based has allowed an extension of the initial concept and of the field of applicability of the technique. Abandoning the exclusive use of self-assembling molecules as ink, μCP can be seen simply as a method to transfer and arrange molecules in a pattern on a substrate. For instance, μCP has allowed the direct patterning of biomolecules (e.g. proteins,[11-13]DNA,[14] cells).[15] Also μCP has been used to form template of catalysts for the growth of nanotubes,[16] the electroless metal plating,[17] and the local initiation of thin polymer films cross-linking.[18-19] Further examples are the use of microcontact printing for locally defining both the chemical reactivity (also by multiple functionalisation)[20]and the surface energy with application in wetting-dewetting processes.[19]

Among the prerogatives of μCP we can include its high throughput, the saving of chemicals and materials, compared to the highly wasteful techniques based on the patterning of polymeric resists, and the low level of investments required. In some cases the advantage over other lithographies goes beyond the economic factors, introducing technical possibilities that are not available to other methods. This is true for the patterning of biomolecules that denature in harsh processes involving solvents, pressure and heating, and for which the direct patterning by microcontact printing represents an ideal solution.

Before embarking on our technical discussion on the microcontact printing techniques we must recognize that the topic is very broad and only a part can be covered by this contribution. Therefore, we will often refer the reader to existing publications, starting from the review articles that already exist on this and related techniques.[22-26]

10.2. THE SELF-ASSEMBLY OF ALKANETHIOLS

Self-Assembled Monolayers (SAMs) of n-alkanethiols are fairly well characterised systems. Their structure has been the object of a multitude of studies by techniques such as X-ray diffraction,[27] Infrared Spectroscopy,[28] NEXAFS,[29] ellipsometry,[30] XPS[31] LEED,[32] HREELS[33-34] He diffraction,[35] AFM,[36] STM,[37-39] and chemical[40] and lateral force micoscopy.[41-42] These and other techniques have been fundamental for studying the role of relevant parameters in the formation of SAM's. Among the most important parameters are the length of the alkyl chain (number of methyl units),[38] the nature and morphology of the substrate,[43] the degree of surface coverage,[44] the assembling technique (evaporation or sublimation in UHV chamber, deposition from solution, μCP),[43] and the presence of functional groups at the end or inside the alkyl chain.[45-46]

For practical purposes, however, many details relative to the complex structure of SAMs appear to play a minor role in μCP, provided that the printed monolayers are dense, ordered, and have a sufficiently low density of defects, to effectively contrast the permeation of ions from the etching solution. This fact represents an element of robustness of the μCP technique that allows it to perform well on many different substrates, with different inks, and with rather large process latitude with respect to other process parameters (e.g. concentration of ink in the stamp, temperature of the substrate, duration of contact printing). Nevertheless, in order to make use of the technique at its present state, it is essential to gain better insight into the behavior and the properties of SAMs. We summarize here some of the most relevant points.

A great contribution to the understanding of the mechanism of alkanethiol self-assembly comes from an in-situ STM study.[47] In that investigation, by imaging the surface at consecutive stages of coverage, Poirier and Pylant showed that self-assembled monolayers on Au(111) are formed during a two-step process. The first step is the formation of crystalline islands in which the molecules lie with their axes parallel to the surface. At saturation coverage of this phase, the monolayer undergoes a phase transition to a more dense phase which involves the realignment of the molecular axes with the surface normal. The realignment is caused by the molecule-molecule interaction. This lateral stress that drives the alkyl chains to realign and to increase the packing density is confirmed also by measurements made by Berger et al., using a method they developed to monitor the build-up of lateral stress in the monolayer.[48] Several studies point out the existence of domains with different lattice structures on the monolayer. The presence of different phases may be observed in some cases as a change of the substrate protection provided by the chemisorbed layer. Often in our investigations of μCP we observed in the same printed sample several distinct levels of substrate etching, with abrupt boundaries between these areas. However, the distinction between these intermediate levels is lost for long etching time.

At full surface coverage, the molecular packing in the monolayer is dominated by the spacing between the sulphur atoms on the substrate.[49] Alkanethiol monolayers on Au(111), for instance, exhibit long-range crystalline order commensurate to the substrate with predominant $(\sqrt{3} \times \sqrt{3})R30^\circ$ lattice structure. In the presence of different chemical functionalities at the end or inside the alkyl chain, the packing and orientation of the molecules can be strongly influenced by the interactions between the functional groups (steric effect),[46,51] and the resulting monolayer may no longer be impermeable to ions from the etching solution. The presence of a chemical functionality can lead to even more dramatic effects on the structure of the monolayer. This is the case of α-ω-alkane**di**thiols, where the presence of an SH functionality at both ends of the alkyl chain leads to monolayers with molecules lying on the surface, which makes this kind of molecule not suitable for microcontact printing.[52]

We mentioned above that molecules in the SAMs on noble metals tend to stand on the substrate. However, in general their orientation is not perpendicular to the surface, but they form a tilt angle, and organize into domains in which the molecules crystallize with the same orientation of the alkyl chain. The alkyl chain of the SAMs on gold, for example, form a tilt angle with the surface normal of $\sim 30^\circ$, whereas the orientation of the chains in monolayers formed on silver is much closer to the surface normal (tilt angle $\sim 12^\circ$).[53-54] The origin of this fact is related to the structure of the substrate. In particular, the different spacing of sulphur atoms on the Ag and Au surface is compensated by a different film thickness (and thus by a different tilt angle) to give bulk densities that are close to each other in the films and nearly that of bulk n-alkanes.[55] Therefore, SAMs on Ag appear thicker and more compact than the corresponding monolayers on Au.

An important aspect for the technological exploitation of the properties of SAM's is their stability under various environmental conditions. Monolayers of alkanethiols on coinage metals are stable in air at room temperature for weeks. At elevated temperatures the monolayers also appear substantially stable (up to 100 °C) on the time-scale of a day. However, at this temperature the molecules start to desorb (with a low evaporation rate) and diffuse. Annealed monolayers show less surface coverage, but nonetheless a high degree of order and perfection characterized by larger domains than non-annealed SAMs.[56-57] In general, the stability of the monolayer tends to increase with the number of methyl units in the alkyl chain (as shown by voltammetric studies reporting a shift of the peak corresponding to the reductive desorption potential on the order of -15 or -20 mV for every methyl unit added to the chain).[58-60]

One of the most important aspects in relation with the μCP technique is the diffusivity of the molecules both in air and on the surface where they assemble. In fact, this aspect determines the resolution achievable by printing. The best results in terms of resolution have been reported for long length chain thiols (eicosanethiols) due to their lower diffusivity with respect to short chain alkanethiols. Unfortunately, from the literature no information is available for even longer thiols (>20 methyl groups), and thus it is not clear whether the trend towards higher resolution continues with heavier alkanethiols.

10.3. THE STAMP

Whitesides and collaborators have pioneered new concepts in micro- and nano-fabrication by using elastomeric stamps as a key element for patterning. These approaches, collectively known as "Soft Lithography",[61] exploit different physical and chemical principles combined with the use of a stamp to create some kind of chemical or topographic contrast on a substrate. Besides, microcontact printing, elastomeric stamps have been used in several ways. These include patterning curable polymers by molding, defining channels and capillaries at the interface stamp-substrate in microfluidic applications,[62-65] using the stamps as amplitude, phase-shift, and reflective masks in photolithography,[66-70] or as removable and multi-use resist masks for dry etching, electroplating and lift-off processes.[71] Moreover, elastomeric stamps have allowed the patterning of charges onto dielectric polymers such as PMMA,[72] the definition of 2D structures on non-planar substrates[73] the development of strategies for fabricating 3D metallic microstructures,[74,75] in process exploiting wetting-dewetting effects,[76] or as micromechanical systems.[77,78] It is beyond the scope of the present chapter to review the wide and rapidly growing field of soft lithography. Nonetheless, it is worthy to point out that mastering the technology of elastomeric stamps gives access not only to the technique of microcontact printing, but also to a variety of other patterning methods and to the exploration of new principles in micro- and nano-fabrication. Almost all of the remarkable developments in soft lithography have been achieved using stamps made of siloxane elastomers. Stamps are usually fabricated by molding. A fluid elastomer precursor is poured onto a master (mold), and cured thermally or by exposure to UV light. The resulting membrane is then carefully peeled from the mold, relying on small reversible deformations of the material. The usefulness of polydimethylsiloxanes (PDMS) as a material for stamps has been shown since the earliest demonstrations of microcontact printing.[1] A list of the properties of PDMS will contribute to clarify the reasons of its utility.

10.4. PROPERTIES OF POLY(DIMETHYL SILOXANE) ELASTOMERS

10.4.1. Chemical formulation

Poly(dimethyl siloxane) is a linear polymer consisting of a backbone chain of Si-O units and two methyl side groups connected to each Si atom. PDMS is very inert, so reactive functionalities must substitute for the methyl groups at the ends or along the chain in order to form curable polymer blends.

Most of the research on microcontact printing has been carried out using a commercial two-component siloxane polymer, called Sylgard 184 (Dow Corning). This has the advantage of easy preparation and satisfactory elasticity and durability. However, materials with improved mechanical stability (i.e. with a higher Young's modulus) are required for high-resolution (< 500 nm) printing.[79] Blends of functional siloxanes with vinyl and hydride groups cross-link via the platinum catalyzed hydrosilylation reaction (Figure 1). The molecular weight of the two polymer components, the ratio between

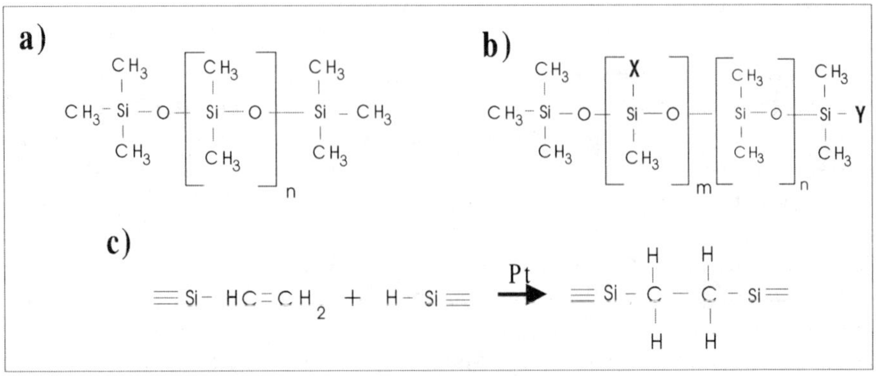

FIGURE 10.1. **a)** Molecular structure of Poly(dimethyl siloxane). **b)** Functional siloxanes can be obtained replacing methyl groups by reactive groups either in side (X) or terminal (Y) positions. **c)** Hydrosilylation reaction of siloxanes with vinyl and hydrosilane functionalities.

inert and reactive groups, the amount of catalyst, and the temperature of curation, can be chosen to tailor the properties of siloxane elastomers according to what is needed.[79] Alternative cross-link methods are possible, such as exposure to UV or to electron beam.[80]

10.4.2. Elastic properties

Elastic properties of PDMS networks are of fundamental importance for the fabrication of high quality μCP stamps. In general, two conflicting requirements have to be fulfilled, namely a high mechanical stability of the microstructures on the stamp and their capability to conform to the morphology of the substrates down to the molecular scale during the contact printing. However, as we will demonstrate in the experimental part at the end of the chapter, the importance of this last aspect seems to be overestimated (in other words, even materials with lower compliance could be used, with advantages in terms of mechanical stability of the stamp).

Young's modulus PDMS elastomers ranges between ~0.1 and ~10 MPa and Poisson's ratio is very close to 0.5,[84] which tells us that PDMS elastomers are almost incompressible. However, only the most stiff elastomers, i.e. those with Young's modulus of ~10 Mpa, provide sufficient mechanical stability for the high-resolution, high aspect-ratio (1-2) structures as required for printing features below ~500 nm. Useful criteria for producing elastomers with a compression modulus up to ~10 MPa (cf. the commercial Sylgard 184 with E =~3 MPa) are reported by Schmid and Michel.[79] Their study report also how the compression modulus, the elongation at break and the surface hardness correlate with the chemical formulation of the PDMS elastomers and the curing conditions. In particular, materials of high compression modulus and great surface hardness are very brittle,[79] leading to stamps that are rapidly spoiled under repeated stamping conditions, particularly in non clean conditions, by cracking due to concentrated stress in in the points of contact with dust particles. Concerning this point, Schmid et al.[79] report that bimodal networks containing a distribution of short and long chains between cross-links lead to materials with a higher fracture energy. This fact

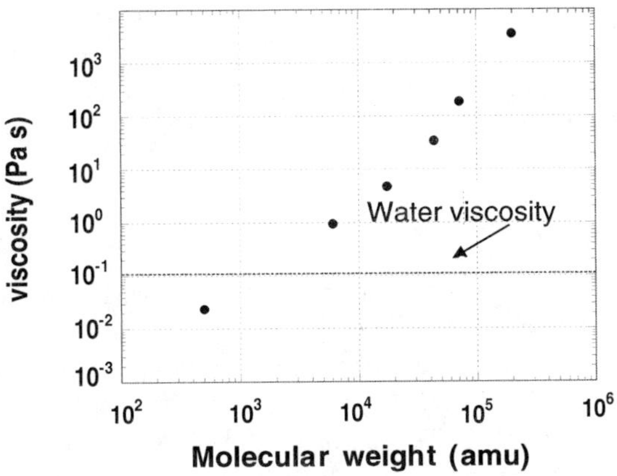

FIGURE 10.2. Viscosity of PDMS polymers as a function of the molecular mass. DMS oligomers may exhibit viscosity even below that of water.

can be exploited in order to obtain less brittle materials without reducing its Young's modulus.

Young's modulus and Poisson's ratio of PDMS elastomers have been studied analyzing the stress-strain relationship in contact with a surface using various measurements schemes.[82,83] Phenomenological models describing the strain energy density function[84] can be very useful for predicting the behavior of stamps under various load conditions.

10.4.3. Viscosity of PDMS precursors

The fluid dynamic properties of the polymer precursors are relevant for the fabrication of μCP stamps by molding. PDMS precursors are always in the melted state at room conditions (the glass transition temperature T_g is ~-120°C), and their viscosity in the range of the molecular weight commonly employed for μCP stamps ranges between 10^{-3} and 10^3 Pa*s (Figure 2).[85] Low molecular weight siloxanes are specially suited for the fabrication of thin elastomeric films (on the order of hundreds of nanometers) due to their low viscosity. In the case of mixtures of siloxane polymers, as in the case of two component curable systems, the viscosity is given to a good approximation by

$$\textbf{\textit{Log}}\ \eta(x) = x\textbf{\textit{Log}}\ \eta_1 + (1\text{-}x)\textbf{\textit{Log}}\ \eta_2, \qquad (1)$$

where η_1 and η_2 are the viscosities of the two components of the blend, and x and $(1\text{-}x)$ the relative mass concentrations.

10.4.4. Thermal properties

Poly(dimethylsiloxane) exhibits a rather strong volume (linear) thermal expansion, on the order of 10^{-3} K^{-1}(~ 3 10^{-4} K^{-1}) in the range of temperatures between -55 and 150 °C. Therefore, stamps cured at temperatures well above room temperature are subject to non-negligeable shrinkage after demolding and cooling down, which should be pre-compensated for in the pattern design. However, even with a corrected pattern design, PDMS membranes are too sensitive to thermal fluctuations to ensure good dimensional stability. This makes printing with free standing membranes unsuitable for applications that need high position accuracy.

10.4.5. Surface properties

In microcontact printing, the surface properties of PDMS play an important role for the process of wetting an elastomeric stamp with an ink solution and in the spreading of the front of conformal contact at the stamp/substrate interface during the contact printing.

Generally, non-modified PDMS elastomers exhibit hydrophobic surfaces (due to the absence of polar groups in the polymer) with a correspondingly low surface energy of ~22 mN/m.[85] As a consequence, low polarity solvents more easily wet PDMS stamps. For example, ethanol is a rather polar solvent, but less than water. Therefore, it more easily wets the PDMS surface and penetrates inside the polymer transporting the molecular ink in the polymer bulk. However, inks based on water do not wet the hydrophobic surface of PDMS stamps and do not permeate its bulk. This fact should make problematic the application of microcontact printing to a broad class of molecules such as inorganic complexes (colloids),[86] macromolecules[87] and especially biomolecules[88] like proteins and cells, that cannot be introduced in solvents other than water without being denatured. Since, such molecules do not normally diffuse into the bulk of PDMS, some degree of affinity to the surface of the stamp is required for their immobilization during the inking process. At the same time, however, the interaction with the stamp should not overcome the attractive force to the surface on which these molecules have to be printed, otherwise the transfer during the contact printing would be prevented. All these facts call for methods to tailor the surface energy of the elastomers used for the stamps.

An easy method for making hydrophobic PDMS surfaces hydrophobic is exposing them for a short time to an oxygen plasma (a few seconds at low power in a RIE machine is sufficient) or to an ultraviolet/ozone (UVO) cleaning process. These treatments lead to surfaces that are OH-terminated. Unfortunately, the surface modification is not permanent, and in a normal environment the surface becomes hydrophobic again within ~1 week. To deal with this problem, strategies for permanent surface modification of PDMS have been developed using self-assembled monolayers of functionalized alkanetrichlorosilane molecules grafted onto OH-terminated PDMS surfaces. Depending on the end functional group present the on the SAM, the surface can acquire a hydrophilic[89] or an even more hydrophobic character.[90] Therefore, the mechanism of self-assembly represents a possibility for "fine tuning" of the surface properties of the stamps (even by utilizing mixed SAM's of different functionalities).

Another process that can be applied to silanol-terminated PDMS surfaces is that of "irreversible sealing". PMDS surfaces presenting silanol groups (SiOH) at the surface can be bonded to each other or to other OH-terminated surfaces simply by bringing them into contact, due to the following reaction

$$\equiv \mathrm{SI} - \mathrm{OH} + \mathrm{HO} - \mathrm{SI} \equiv \; \longrightarrow \; \equiv \mathrm{SI} - \mathrm{O} - \mathrm{SI} \equiv + \mathrm{H_2O} \qquad (2)$$

One application of this mechanism is in microfluidic systems for irreversibly sealing the microfluidic circuitry by binding a patterned membrane to a substrate.[62,91]

On the other hand, some circumstances require an easy release of the PDMS from a surface, such as the release of a stamp from a mould. Materials that are commonly used to fabricate molds (such as novolak resists and PMMA, which can be patterned by photolithography and electron-beam, respectively) naturally exhibit low adhesion forces to PDMS. Others, like silicon, need a surface treatment to improve the demolding process (an example is given in the experimental part of this chapter). Calculations of the conditions for the formation and the spreading of the (conformal) contact as a function of the work of adhesion of the PDMS to different substrate are available in the literature.[92,93] Work of adhesion describes also the separation of the stamp at the end of the printing process.

10.4.6. Permeability to chemicals

In microcontact printing the stamp acts as a reservoir of molecules. This behavior is ensured by the high solubility and permeability of poly(dimethyl siloxane) to gases and especially organic molecules. These characteristics make PDMS one of the most attractive materials both for the separation of volatile organic compounds from air in anti-pollution and environmental applications,[94,95] and for the controlled release of chemicals and drug delivery.[96] PDMS is also permeable to most organic solvents, with various degrees of swelling induced in the polymer network.[97]

10.4.7. Chemical stability

PDMS is stable in most chemical environments except for the most aggressive ones (HF, H_2SO_4 and KOH containing solutions). In oxygen plasma PDMS shows a very low etching rate. This is due to the formation of a thin layer of SiO_x on the polymer surface, which is very slowly etched only by the physical component of the process (i.e. milling). This strong resistance to plasmas and in particular to oxygen plasma has been exploited in some applications such as dry etching mask.[98,99] In order to enhance the etching rate, combinations of oxygen with fluorocarbons (CHF_3) can be used. Finally, PDMS is stable in biological environments and biocompatible. In fact, PDMS is used also for contact lenses and prostheses.[100]

10.4.8. Optical and dielectric properties

PDMS is a transparent material within the spectrum of visible to ultraviolet light (800-300 nm). Its refraction index is ~1.4. This has made possible the development of an optics of the elastomers,[101] based on devices fabricated by moulding.

Moreover, the transparency of the material allows the realisation of transparent μCP stamps on glass or quartz. This makes possible the use of the stamp in mask aligners for the development of multilevel lithography schemes.

PDMS is a good dielectric with dielectric constant and dielectric strength of ~2.5 and ~15 MV/m, respectively.

10.4.9. Elastomer actuation

Mechanisms for elastomer actuation offer interesting technological possibilities in micromechanical systems. Nothing excludes that in future the contact of the stamp with the substrate could be obtained by actuation. Essentially, the possibilities for elastomer actuation are electric, thermal and magnetic. Electric actuation with siloxane elastomers offer a much larger actuation effect than piezoelectric ceramics (in some cases this can go up to 100% strain) at the price of slower response (~1 ms) and lower generated stresses (~MPa).[102] We have already mentioned the large thermal expansion coefficient of PDMS, which is the root of dimensional stability problems of μCP stamps. In other cases, this effect has been used as a resource of the material for the fabrication of thermally actuated devices, in particular in combination with its optical properties (e.g. in thermally actuated interferometric sensors).[103-104] Finally, magnetic actuation becomes possible by filling the PDMS prepolymers with magnetic nanoparticles or introducing extended elements of ferromagnetic materials.[105]

10.5. STAMPS FOR MICROCONTACT PRINTING

10.5.1. Membrane-stamps

The earliest stamps employed in μCP were flexible membranes of millimeter-scale thickness with protruding structures of micrometer-scale height.[1] Such "membrane-stamps", rather easy to fabricate and to use, can be usefully applied to single-level patterning on planar as well as non-planar substrates. However, they suffer from a low dimensional stability that makes impractical their use in the fabrication of complex multi-level devices that require high placement accuracy. Indeed, strain caused by handling (Figure 3), adhesion forces towards the substrate, thermal expansion, and swelling produced by the solvent contained in the inking solution,[106] cause serious problems in the registration of subsequent steps at an accuracy level comparable to the intrinsic resolution limit of μCP. Moreover, defects come out when printed patterns contain large empty areas, due to the sagging of unsupported regions of the membrane (Figure 3).

One technique for the assessment of pattern distortions in microcontact printing based on the detection of the Moiré patterns between a stamped and a reference pattern. This technique has shown that strong distortions on the order of hundreds of μm/cm can be reduced to below 1 μm using stamps fabricated as thin film (~0.1 mm) on a rigid substrate.[107] In order to eliminate or at least reduce the above effects, alternative kinds of stamp have been developed, which make the process of printing more reliable and accurate.

10.5.2. Stamps on substrates

The basic idea for preventing long-range pattern distortions is that of bonding the elastomeric stamp to a stiff backplane[79,107] in order to introduce a lateral constraint against in-plane strain. Using a stamp of a 90 mm diameter and 1 mm thickness fabricated directly by molding the commercial elastomer Sylgard 184 onto a rigid quartz plate, Schmid et al.[79] reported a maximal pattern distortion between two prints of

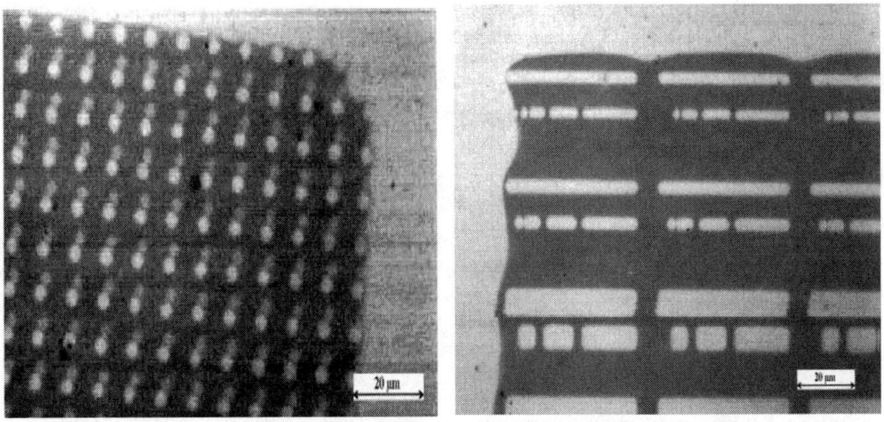

FIGURE 10.3. **Left:** A membrane-stamp of 20x20 mm^2 was used to print. After the initial contact between the stamp and the membrane, an edge of the stamp one millimetre away from the area shown was first lifted up and then allowed to recontact the substrate. The pattern produced in the second contact shows a shift with respect the first one. **Right:** Sagging effect. The membrane relaxes outside the printed field printing on undesired areas.

1.5 nm. The residual distortion was ascribed to the tendency of such a 1 mm thick elastomeric layer to be non-uniformly distorted. Analogously, Burgin et al.[108] reported a runout of 0.3 nm/cm and an improvement in the linewidth control and pattern uniformity using an elastomeric stamp of 100 microns supported by a glass plate. Multilayered stamps[79] combining elastomers with different Young's moduli also have been proposed to solve divergent mechanical requirements of the stamps. In fact, both mechanical stability of high resolution features and compliance for compensating the long range warp of the substrates are necessary for high quality printing. Multilayered elastomeric materials enlarge the space of parameters for matching the required stamp properties. This possibility is obtained at the price of a more complex fabrication process. Thin stamps on rigid substrates allow solving several problems of μCP, but poses also new difficulties.

The impossibility of a sequential contact of the substrate due to the rigidity of the stamp causes air trapping at the stamp-substrate interface. In the case of sequential contact the air is pushed continuously forward beyond the contact front. This results in defects and large non-printed areas on the substrate. A second problem, which arises during the fabrication of the stamp, is that rigid backplanes prevent the sequential release of the stamp from the mold, creating a high risk of damaging one or both of them during the separation.

An intermediate approach considered by Schmid et al.[79] is able to solve both of the mentioned problems. This solution consists of using a flexible glass plate as a substrate for the elastomeric stamp. The flexible glass backplane allows the sequential release of the stamp from the mold preventing the damage of the stamp. Moreover, the sequential propagation of the contact front during μCP avoids the formation of air pockets, and the long-range warp of the substrate to be printed is easily compensated by the compliance of the glass plate.

Stamps fabricated on metal foils offer a technologically interesting possibility. These stamps can be wrapped around a cylinder and used for continuous printing, analogously to the technology for printing newspapers. Metal foils avoid the introduction of large uncontrolled strain in the stamp pattern, though it seems very difficult to control the process and the mechanical system for highly accurate alignment in printing with rolling stamps. However, this approach would allow the patterning of large areas as required for some specific applications, e.g. photovoltaic cells, and thus it is currently intensively pursued by some large-scale electronic industries. It is worth mentioning that interesting possibilities for patterning on special topographies have been also explored. Whitesides et al. showed for example strategies for patterning metals on surfaces with high curvature, with a topology different from that of the plane, or in the cavities of hollow objects.[73]

Geissler et al.[109] have investigated the possibility of printing flat stamps (i.e. unstructured membranes) for microcontact printing. In this case the pattern is defined by the procedure of contact inking, using a patterned inking pad of PDMS. However, this approach seems somewhat unconvenient for several reasons. The contact between a flat stamp and a substrate suffers from the problem of air trapping at the membrane/substrate interface, as is always the case when large contact areas have to be created. Furthermore, problems of membrane sagging are shifted from the printing step to inking, but not solved. Finally, any re-inking would require a re-alignment between the stamp and inking pad, or a thorough cleaning of the flat membrane from ink residues due to previous inking.

Finally, we mention an approach that has been pursued at Wuerzburg University with the aim of drastically reducing all sources of pattern distortion by fabricating ultrathin (down to sub-micron thickness) stamps on rigid substrates. This technique will be described in the section "microcontact printing with ultrathin stamps".

10.6. STAMP FABRICATION

Stamps for microcontact printing are commonly fabricated by molding. This is a highly accurate process that can reliably invert a 3D profile of a master into a thermosetting polymer with resolution well below 100 nm. In fact, the resolution limit appears to be set by the intrinsic properties of the material (the combination of elasticity and surface tension may cause the rounding of sharp corners) more than by the molding process itself. Moreover, molding offers the possibility of replicating a stamp several times from a single master.

Therefore, the resort to primary lithography tools such as electron and ion beam[110] lithography, or even high resolution laser printing (for a process of rapid prototyping of devices requiring a low pattern resolution ($\sim$100 μm),[111] is limited to the fabrication of the mold.

Different materials can be lithographically structured and used as moulds. These materials include films of photoresist,[112] PMMA,[113] metals, SiO_2 and Si_3N_4on Si, glass or metal substrates. In most cases the structures are engraved directly into the substrates (Si (001) is the common choice). In practice, there is quite a large freedom in choosing the material for the mould. Any material that can be patterned down to the desired scale could be utilized, provided that both the structures of the mould are stable during the stamp fabrication, and that the adhesion forces between the elastomer and the mould be sufficiently low to allow for an easy release of the stamp from the mold. Structures in polymeric molds, for instance, might get spoiled by the thermal process of PDMS cure or by intermixing with the elastomer precursor. Often, in order to favor the release of a stamp, the surface of the mold needs to be treated in advance to lower its surface energy. In most cases, such surface treatments are obtained by self-assembly of n-alkyltrichlorosilane or fluorinated alkyltrichlorosilane monolayers on hydroxy-terminated surfaces. Alternative anti-adhesive coatings, such as Teflon sputter deposited on the masters, have also been used to favor the release of stamps.[114]

A relevant parameter during the contact printing is the surface roughness of the stamp on the top of the raised regions (those that contact the printing substrate). Since these raised parts correspond to grooves into the master, it is important to minimize the roughness at the bottom of recessed areas in the master. Subtractive processes, such as plasma etching, tend to create some roughening of the surface at the bottom of the etched grooves. This effect can be avoided by etching the structures of the mould into the first layer of a bilayer (e.g. SiO_2/Si), where the second material acts as an etching stop. Alternatively, additive processes, may be used. In this case the roughness of the recessed areas of the stamp corresponds to that initially present on the substrate on the substrate, which can be more easily controlled.

10.7. THE MICROCONTACT PRINTING PROCESS

10.7.1. Inking of the stamp

Since aspects related to the choice of the molecular ink for microcontact printing have been the object of a detailed discussion in chapter 9, I do not come back to that point. I simply remind the reader that, at present, a consensus exists that the best molecular inks to create patterned monolayer resists by μCP are long chain alkanethiols (18-20 methyl groups). In contrast, short chain alkanethiols tend to be less protective and more diffusive, with a consequent reduction of the process resolution.[115]

The common method of inking a μCP stamp consists of impregnating it by dip into an ethanolic solution of the desired molecules and, after waiting a few minutes, drying it under a nitrogen stream.

Although ethanol is not the only solvent permeating PDMS elastomers,[116] it represents the most convenient choice due to the low degree of swelling induced into the polymer network, which is much lower than e.g. toluene, acetone, tetrahydrofuran, heptane.[97] On the contrary, aqueous solutions are not convenient for the inking process because of the low solubility of alkanethiols in water and the low permeability coefficient of water in PDMS However, this problem can be overcome by a surface treatment of the PDMS, as described in the section "Surface properties" on the PDMS material part.

A "contact inking" method, as opposed to the previous "wet inking" method, has been used to avoid any wetting of the stamp with solvents and the consequent swelling effects.[117] Contact inking makes use of an unstructured PDMS membrane impregnated with ink as an inking pad for the stamp. The amount of ink transferred to the stamp during the inking process can be controlled by changing the concentration of the thiol solution used for impregnating the pad. A further advantage of contact inking is that it localizes the ink delivery to the surface of the elevated features of the stamp, reducing the amount of ink entering in vapour phase or diffusing from the walls of the features during the contact printing. The concentration of thiols in a stamp, although not known in absolute and expressed in terms of the concentration in the ink solution, strongly affects the quality of the printed pattern, the maximum achievable resolution and level of surface protection (pin-hole density). The optimum concentration of thiols in ethanol ranges roughly between 0.1 and 1 mM, depending on the resolution of the structures on the stamp (low concentration is preferable for printing small features).

After inking, a stamp can be repeatedly used for printing without re-inking. This is due to the fact that the ink impregnates the bulk of the stamp, from which slowly diffuses out. This is clearly not true for inks that localize at the stamp surface, that are almost completely transferred at every print.[118] However, contact an measurements show that the quality of the printed monolayers progressively decreases as a function of the number of prints performed after the inking process,[119] and after ~10 prints the stamp has to be re-inked.

10.8. SUBSTRATES FOR MICROCONTACT PRINTING

Although in principle microcontact printing might be applied to any given substrate/SAM system, in practice, its interest as a patterning technique the research has been focused on few systems only from the wide literature on self-assembling. The reason is that for a lithographic use of μCP, the SAMs need to provide an effective barrier against wet chemical etching. This means that only a very low defect density in the monolayer can be tolerated. Moreover, for applications, it is desirable that the process of self-assembling be fast, and take place at room temperature without need for strict environmental control. Because of the above, microcontact printing is mostly performed with alkanethiols on Au, Ag, Cu, and more recently Pd,[120] which exhibit these favourable characteristics. I also will focus the discussion on these systems. Nevertheless, it is worth to mention that alkylsilanes that bind covalently to hydroxylated surfaces (such as SiO_2, glass, TiO_2, Al_3O_2), and alkanephosphonic acids ($CH_3(CH_2)_nPO(OH)_2$) that graft on ZrO_2, TiO_2, Al_3O_2, [121,7] are gaining interest. This is due to the technological importance of the class of oxides, even though the process appears more difficult to handle, leads to patterns defined at lower resolution, and requires contact time of several minutes up to ~1 hour more than the corresponding process with alkanethiols.

10.8.1. Preparation of the substrates for μCP with alkanethiols

The present section deals with methods of preparing the substrates for printing monolayers of alkanethiols, also addressing the question of how their properties (in particular the morphology) affect the performances of the microcontact printing step and that of the subsequent wet chemical etching.

Thin films of Au (or Ag, Cu, Pd) used for μCP can be deposited onto a substrate by different methods (e-beam or thermal evaporation, sputtering, electroplating, electroless plating). In most cases, however, the above metals do not adhere sufficiently to the substrate and tend to flake off during some of the process steps (release of the stamp after contact printing, wet etching). Therefore, a thin layer (or sub-monolayer), typically 1-10 nm of metals such as Cr, Ti, Ni is introduced between the substrate and the metallic film supporting the thiol self-assembly in order to improve the adhesion.

After transferring a printed pattern into Au (Ag, Cu, Pd) by chemical etching, most applications also require the etching of the adhesion layer left on the bottom, e.g., in electronic device fabrication in order to cut the electric connectivity. However, this might be difficult to obtain by a second step of wet chemical etching, because of electrochemical effects, such as redox reactions occurring between the top layer and the adhesion layer, which essentially forms a galvanic cell. In fact, electrochemical effects might lead to etching processes difficult to control. Often this results in the under-etching of the adhesion layer and the lift-off of the top layer. A possible solution is represented by the use of dry etching processes for the removal of the adhesion layer. However, a more elegant method exists for binding Au (Ag, Cu, Pd) films to a substrate without any metallic interlayer, namely, by functionalising the substrate with thiol groups on the free surface. This method has been used in the case of hydroxylated substrate surfaces, by self-assembly of a monolayer of 3-mercaptopropyltrichlorosilane prior to the metal evaporation.[122,43]

We now address the question of how the substrate morphology affects the results of μCP. Losic et al.[43] studied the influence of surface morphology on the integrity of alkanethiol SAMs assembled from solution and by microcontact printing, comparing six gold substrates differing in method of preparation and surface roughness. Electrochemical methods were used to asses the quality of the SAMs and their ability to provide an effective barrier to ion permeation (in particular to $[Fe(CN)_6]^{3-}$, which is a component of the Au etching solution). They found that the density of defects and ion permeation are correlated to the roughness of the substrate. The lowest defect densities were found for very smooth substrates such as gold films on cleaved mica and large atomically flat gold substrates, the latter ones obtained by stripping a gold film from the cleaved mica on which they had been evaporated. The effect of substrate morphology on the level of protection provided by the SAM was found both in monolayers assembled from solution and by microcontact printing, even if printed monolayers always exhibit a lower quality.

The substrate preparation affects not only the self-assembly of the monolayer onto the substrate, but also the wet chemical etching. In particular, in polycrystalline Au the etching rate in the grains and at the grain boundaries appear different. Grains (typical size 15-30 nm for Au) result either totally protected or completely dissolved,[123] thus contributing to the edge roughness of printed features.

Although best results by μCP can be obtained on smooth substrates even rather rough surfaces may be nicely printed and may be useful for applications. For example, microcontact printing has been performed with good results on electroless plated silver substrates.[124-126] Using electroless plating or electroplating as a deposition technique avoids the resort to UHV deposition techniques. The advantage is the possible use of μCP for a variety of cost effective applications and the ability to coat non-planar substrates (e.g. surface of a fiber[127,128] or the cavities of a hollow object),[124] which can be patterned afterwards by μCP with an appropriately designed stamp.

Finally, we report on methods to prepare substrates containing structures resulting from previous lithographic processes. A multilayer resist scheme can be used conveniently in this case, and it can be also used for solving material incompatibilities (e.g. Au on Si in electronic applications) by introducing interlayers of compatible materials, or for enhancing the low aspect ratio of μC printed structures by transfer with anisotropic plasma etching. As an example we mention a process in which substrates coated with a thin film of a thermoplastic (PMMA) as a planarization layer. Subsequently this was covered by Cr (adhesion layer) and Au and printed. The pattern formed in gold was used as dry etching mask for producing high aspect ratio features in the underlying polymer layer by oxygen plasma etching.[129,130] The resulting high aspect ratio polymer pattern may be used in turn to define structures in other materials by metal lift-off or electroplating.

10.9. PRINTING CONDITIONS

The parameters relevant in the microcontact printing process with alkanethiols are, apart from the length of the alkyl chains of the alkanethiols as already mentioned, their concentration in the solvent and the contact time of the stamp with the substrate. The temperature may also affect the process, in particular affecting the diffusive behaviour of the molecules. Liebau et al.[131] report the need of increasing the temperature to 60 °C for printing ordered and etch protective monolayers of more complex and heavier inks (thioether derivatives with multiple anchoring sites). However no systematic investigation about temperature effects on microcontact printing has been reported.

Other factors that in principle could affect the printing process with alkanethiols such as the humidity of the environment and the pressure exerted by the features on the substrate seem to be irrelevant. Typically, the concentrations of long chain alkanethiols in ethanol as inking solution range between 0.2 to 2 mM, and contact times range between 10 sec and 1 min. A longer contact time and higher concentration lead to more protective SAMs with a lower density of pin-holes, at the price of a predictable broadening of the features due to diffusive spreading of the ink.[132] Losic et al.[133] investigated electrochemically the quality of the monolayers obtained by microcontact printing as a function of the thiol concentration in the ink solution, finding that only at concentrations of ~50 mM, printed monolayers appear indistinguishable from those assembled directly from solution.

10.9.1. Conformal contact

Conformal contact of the stamp, i.e., the adaptation to the shape of the substrate, at all scales, from macroscopic down to the nanoscale roughness of the substrate, has been emphasized in several papers as an essential condition for a reliable printing. However, concerning this specific point it is possible to formulate some criticisms. First of all, no counterexample has been provided in the literature showing the failure of the conformal contact. The formation of gaps between stamp and substrate caused by the inability of the stamp to conform to the substrate should have been observed as non-continuos "leopard-skin"-like monolayer on the surface. Only in the case of large voids left between stamp and substrate the failure in the monolayer continuity becomes manifest. In our investigations on μCP we observed such an effect using an extremely thin and rough elastomeric stamp (~300 nm thick elastomeric stamp with a roughness of 25 nm fabricated on a rigid substrate) due to the inability of compensating the substrate roughness. However, such a problem of conformal contact disappeared already with elastomeric stamps of ~1 μm thickness having surface roughness below 10 nm, as I will show in the last section of this chapter. A second argument is the experimental evidence that, due to the ink diffusion, the quality and the protection of the monolayer is even enhanced in tiny regions (few tens of nm) of non-contact around the contact areas. This, we believe, can be explained by the fact that the matrix of the PDMS, not contacting these areas, does not interfere with the self-assembly process. Two papers[114,93] have appeared recently, which analyze the mechanics of the contact between an elastomeric stamp and a substrate, and propose guidelines for matching the mechanical properties of stamps to the requirements of their use. In particular, Bietsch et al. introduced a criterion for the spontaneous deformation of the stamp profile mediated by the adhesion force that brings the elastomeric surface to conform to a rough substrate. According to this criterion conformal contact occurs when the work of adhesion "gained" in the contact is sufficient to compensate the work that has to be provided to elastically deform the stamp. In this case the contact tends to spread all over the surface.

A second important point analyzed in both the above papers is the condition for the vertical collapse of the areas of the stamps, which are not supported by features. This represents a serious problem for printing with membrane-stamps and thin (~100 μm) stamps on a rigid substrate. However, the case corresponding to the approach based on ultrathin stamps (~1 μm) is not considered. At the end of the chapter it will be demonstrated experimentally that ultrathin stamps completely solve the problem of sagging. The theoretical analysis by Hui et al.[93] considers additional effects that cause distortions in the profile of the features of the elastomeric stamp such as surface tension (responsible for the rounding of sharp corners in elastomeric stamps after release it from the mold), distortion of thin features (buckling) due to the weight of the stamp or to an external load, lateral collapse of two neighboring punches adhering to each other by surface forces, residual stresses in the stamp due to thermal and chemical shrinkage, pull-off forces, effects due to surface roughness, etc. According to calculations by Hui et al.[94], the surface tension effect in PDMS sets a lower limit on the radii of curvature of the edges of ~50 nm.

10.10. CHEMICAL ETCHING

Monolayers of alkanethiols possess the ability to inhibit the corrosion and the oxidation of coinage metal[134,135] as well as other metals such as Fe,[136,137] Pt[138] and semiconductors.[139] This was a well known property before being exploited in the μCP technique as a convenient way to perform the pattern transfer based on wet chemical etching. The origin of this property is the non-permeability of the ordered and compact alkanethiol monolayer to gases such as O_2 or to ions, which physically isolates the substrate from the reactants. A typical etching bath for Au, Ag, and Cu used in μCP contains CN^- and O_2 chemical species. Other chemical or electrochemical conditions under which Au dissolves while SAM/Au system appears stable are obtained in solutions containing Cl- , Br-,[140] I- , ionic species. One formulation of the etching bath containing CN^- and O_2 is an aqueous solution of KOH and KCN (1M, 0.1 M respectively). However, in order to avoid the hazardous KCN component, the above solution has been replaced by a different one consisting of $K_3Fe(CN)_6$, $K_4Fe(CN)_6$, $K_2S_2O_3$, and KOH in water. The optimal concentration of the different chemicals (0.001 M, 0.01 M, 0.1 M, 1 M respectively) was chosen in order to produce the lowest density of defects into a gold substrate protected by a SAM.[141,142]

Several studies have reported a lower protection provided by microcontact printed monolayers with respect to those obtained from solution. On one side, high protection is required in order to avoid the intrusion of the ions of the etching bath through the pits in the monolayer. This would need the inking of the stamp with a highly concentrated solution of alkanethiols (>10 mM). On the other, high resolution patterning requires low concentration inks (~0.1 mM) to reduce diffusion effects.

A solution to the conflict between these aspects seems impossible by a simple optimization of the ink. However, Geissler et al. have recently made some progress in the etching process.[143] In order to overcome the problems due to monolayer imperfections they used new defect-tolerant strategies for the etching. The principle is to heal defects in a monolayer with additives present in an etch bath. Additives (octanol in the experiments performed by Geissler et al.) with a high affinity for defects in the SAMs enter the pits and remain there due to the interaction with the neighboring alkyl chains. However, the additives are not able to assemble themselves on the bare substrate due to the absence of thiol groups. The additives prevent the penetration of the etching solution by plugging the holes.

An alternative strategy for a defect-tolerant etching process is based on the use of chemical species of large size in the etching solution. Defects below the dimension of the largest molecules of the etching bath are not affected by the etching process, due to the inability of these molecules to penetrate through the openings in the monolayer. Both approaches have proven highly effective in improving the protection and the edge roughness of the etched pattern into the substrate.

10.11. MICROCONTACT PRINTING USING ULTRATHIN STAMPS

In this part of the chapter, I will describe an approach to microcontact printing with stamps consisting of a thin patterned elastomeric film of sub-μm thickness bonded to a flat rigid substrate. In spite of their extremely reduced thickness, the deformable layers can still compensate for the substrate's roughness. Moreover, such stamps are capable both of printing reliably high-resolution patterns over macroscopic areas, presently demonstrated over ~1 cm^2, and to solve effectively the main drawbacks of the technique represented by distortions at the macroscopic scale, e.g. stretching and thermal expansion, as well as at the microscopic scale e.g., sagging of unsupported areas, buckling and lateral collapse.

Fabrication, use, problems, and relative solutions of ultrathin stamps will be discussed. The advantages that ultrathin stamps offer over analogous stamps with thicker (~100 μm) elastomeric layers and over free-standing patterned membranes will appear

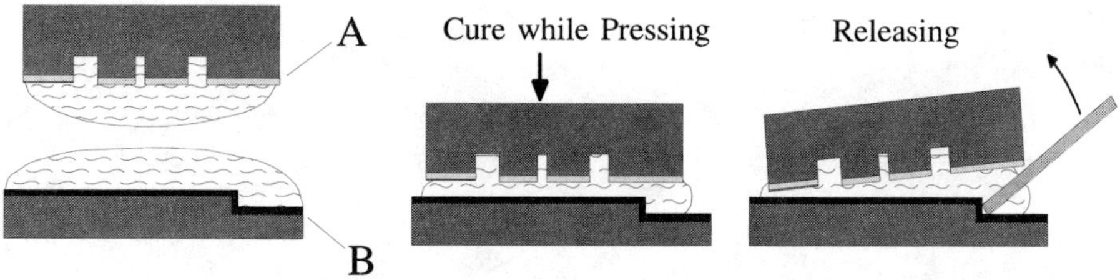

FIGURE 10.4. Process steps for fabrication of ultrathin stamps. A: mold B: stamp substrate.

from the experimental results. The technique that is described here was developed at the University of Wuerzburg in the frame of the European Union Project SPINUP. The author of this chapter was involved in such research.

10.11.1. Fabrication of ultrathin stamps

A method to fabricate thin elastomeric stamps on a substrate is schematically shown in Figure 4.

A liquid elastomer precursor is sandwiched between a mould and a flat substrate. Pressure is applied for thinning the precursor layer allowing the extra material to flow out. Depending on the formulation of the two-component elastomer precursor and the amount of catalyst, the curing can be performed at room temperature or speeded up thermally, typically at ~50-100 °C. During the cross-linking reaction the precursor viscosity increases. The pressure is maintained until the polymer flow stops and the polymerization is completed. Separation of a stamp with a very thin elastomeric film from the mold may be difficult, even for molds provided with an anti-adhesion coating. A practical way to facilitate this process is to pry off the mold with a blade inserted at an edge of the stamp (a step engraved on the substrate may be essential for inserting the blade). Generally after the removal of the mould, a residual layer remains on the stamp substrate. The thickness of this layer depends on the (time-dependent) viscosity of the precursor, the applied pressure, and the diameter of the mould/substrate interface Figure 5.

The dynamics of two plates pressed against each other with a viscous liquid in between, known as Reynolds problem,[144] is described by the equation

$$F = -\frac{3 \cdot \eta(t)}{2} \frac{\pi R^4}{h^3} \frac{dh}{dt}, \tag{3}$$

where $\eta(\mathbf{t})$ and $\mathbf{h(t)}$ are the (time-dependent) viscosity and the layer thickness of the liquid between the plates. **R** is the radius of the plates, and **F** is the applied force. The final thickness is given approximately by the thickness of the film after a "curing time" t_c, assuming that the precursor viscosity remains constant until t_c and becomes infinite afterwards. The continuous and dotted lines represent, respectively, the thinning of a precursor (viscosity $7\ 10^{-2}$ Pa·s) used to fabricate sub-μm thick stamps and that of a standard material (Sylgard184, viscosity 4 Pa·s). The initial thickness of the polymer

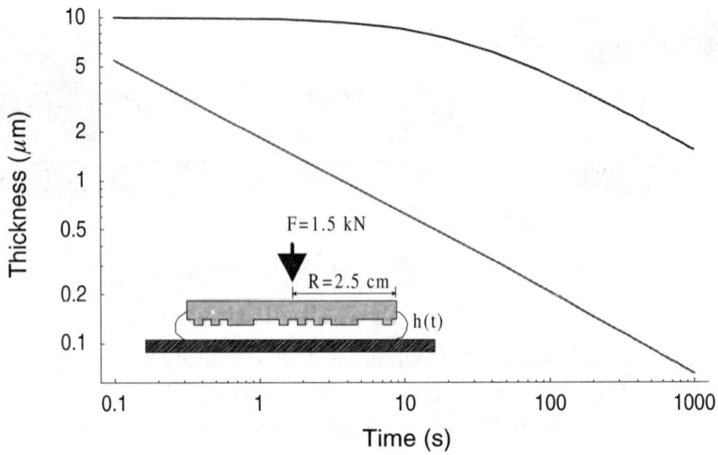

FIGURE 10.5. Calculated thickness of the elastomeric film as a function of the time elapsed from the beginning of pressing, for two precursors of different viscosities.

layer is assumed to be 10 μm. The values of the mold radius R = 2.5 cm and that of the force F = 1.5 kN correspond to those used in the fabrication of stamps described in the text .

To enhance the adhesion of PDMS to the glass and silicon substrates, the substrates were hydroxy-terminated by dipping in H_2O_2: H_2SO_4 (1:3), an extremely aggressive solution to biological tissues. Subsequently, vynil-octodecyltrichlorosilane was assembled onto the surface from the liquid phase. The vynil group of the grafted molecules which takes part in the cross-linking reaction that occurs in the bulk PDMS strongly binds the elastomer to the substrate.

For anti-adhesion treatments of Si or glass molds the same process scheme can be used as for enhancing the adhesion, aside from the chemical nature of the self-assembling molecules, which in this case must provide a strong hydrophobic character to the surface. This can be obtained using molecules such as n-alkyltrichlorosilane or perfluo-n-alkyltrichlorosilane, preferably assembled from gas phase.

10.11.1.1. Formulation of the elastomeric material. As shown in the analysis of the thinning of viscous liquid films (Figure 5), precursors of low viscosity are preferable to those of high viscosity for molding sub-μm thick stamps on substrates. In our experiments we used mixtures of two functional polymers, a vinyl-terminated poly (Dimethylsiloxane) and a MethylHydrosiloxane-Dimethylsiloxane copolymer. In particular, we formulated the elastomeric material by choosing the vinyl component from PDMS with commercial codes DMS-V00, DMS-V05, DMS-V21, with viscosity of 0.7, 5, 100 mPa·s respectively, and the hydrosiloxane component between HMS-301 and HMS-501,with viscosity of 30 and 13 mPa·s, respectively, purchased from ABCR, Germany. The time of cross-linking could be adjusted between a few seconds and several hours by varying the amount of catalyst (the Pt complex SIP6831.0, ABCR) within the range of 100-10000 ppm. The composition was optimized in order to obtain cured elastomers in about 1

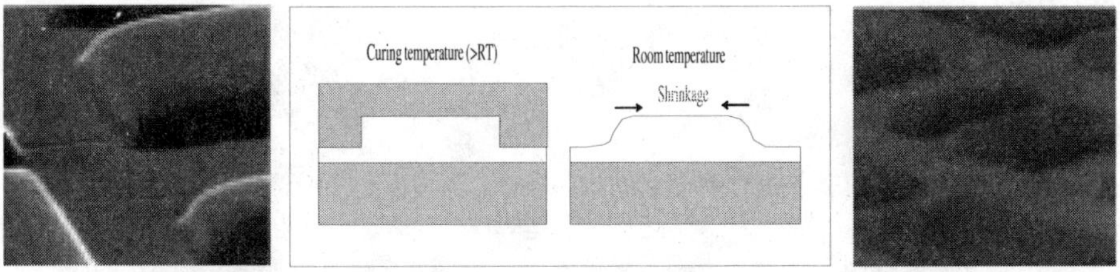

FIGURE 10.6. Left Stamp cured at 70 °C. Thermal shrinkage induces distortions in the shape of the features. Center Effect of thermal shrinkage on the shape of the edges. In membrane-stamps the shrinkage cannot be appreciated on the local scale, being an isotropic effect, while the presence of a constraint on the surface introduces a bending in the walls of the features. Right Stamp cured at room temperature.

hour at room temperature. Room temperature cross-linking prevents thermal shrinkage effects on the shape of the features (Figure 6).

In preparing the elastomer precursor we followed some practical rules useful for material handling. These are:

1. Diluting the Pt catalyst into vinyl-functionalized PDMS, and then mixing the obtained solution with the hydrosiloxane part afterwards. On the contrary, adding the catalyst to the hydrosiloxane component or to the two PDMS components already mixed would induce a fast and localized reaction. Small stable droplets of catalyst form before the catalyst can be dispersed uniformly in the prepolymer.
2. Filtering the prepolymers. In our study we used 200 nm pore size filters.
3. Evacuating the trapped air after mixing the prepolymers. A low vacuum of ~10 mbar is sufficient.

10.11.1.2. Contacting thin elastomeric stamps to a substrate. Three major problems have plagued the formation of defect-free contact between thin and ultrathin stamps with substrates in a series of μCP experiments. These problems include the formation of air pockets at the interface between elastomeric stamp and substrate, the non-uniformity of the elastomer thickness, and the need of compensating long wavelength warp of the substrates to be printed. Experiments with thin (1-100 μm) unstructured elastomeric films on glass substrates clearly have shown that the contact front during μCP does not propagate as a straight line, but bends, folding into close perimeters. The air enclosed by these perimeters at the stamp-substrate interface prevents contact and causes print defects. A solution to this problem was obtained by introducing some restrictive conditions on the design rules of the stamp pattern, such as avoiding large uninterrupted contact regions (approximately 500 × 500 μm^2), and creating paths topologically connected to the periphery of the stamp. Regarding the second problem, i.e., the non uniformity of the elastomer film thickness, one solution is the use of extremely flat and rigid substrates both for the mold and for the stamp substrates. This avoids the bending that usually takes place during the fabrication of the stamp. Finally, the long wavelength warp of the substrates for printing, the planarity of which cannot be ensured with high accuracy, if μCP has to be used in cost effective applications, is eliminated by applying an external pressure during microcontact printing.

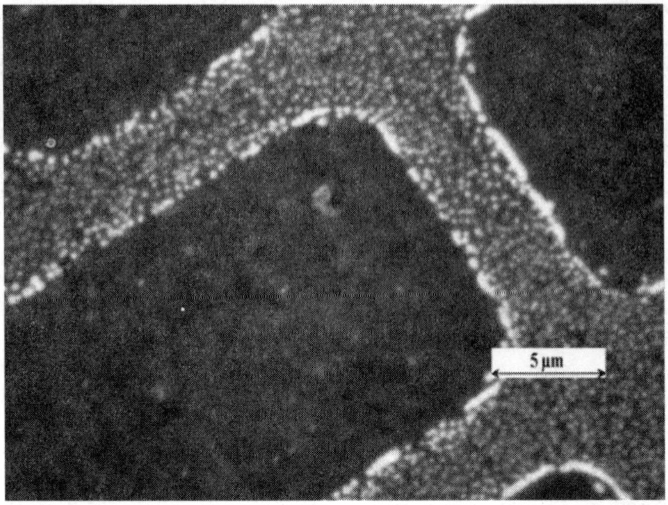

FIGURE 10.7. Failure of conformal contact between an ultrathin stamp (~300 nm) and substrate results in non-continuous protection of the metal film. The pattern was etched in a 20 nm thick Au film.

Generally, the local roughness of the elastomeric stamp and of the substrate for printing, both usually below 10 nm, does not pose problems. In fact, the failure of conformal contact was observed (Figure 7) only in one case using an extremely thin elastomeric stamp (~300 nm) which had a rather large surface roughness (~25 nm). Already with a 600 nm thick elastomeric stamp, with a surface roughness of ~10 nm, a continuous monolayers can be printed.

10.11.1.3. Experimental process. Three stamps with different elastomeric film thickness and different substrates (glass and silicon) have been fabricated using the same mold. The mold consisted of a glass disk of 5 cm diameter, 1 cm thickness, and very flat (the deviation from planarity below 30 nm), supporting a 600 nm thick PMMA 950K film patterned by electron beam lithography at 30 KeV. In order to avoid charging effects 20 nm Au were thermally evaporated on the PMMA layer before electron beam exposure. The Au film was stripped and the PMMA was developed in toluene:iso-propyl alcohol (1:5). Feature sizes between 100 nm and 40 μm have been defined over a 12×12 mm^2 area with an overall filling factor of 50%.

The characteristics of the stamps are summarized in Table I. For all stamps the elastomer precursor was a mixture of two functional polymers, the DMS-V05 and HMS-301,Vinyl-terminated PDMS and MethylHydrosiloxane - Dimethylsiloxane copolymer, with viscosity of 6 and 30 mPa·s, respectively, from ABCR, Germany. The concentration of the catalyst (the Pt complex SIP6831.0, ABCR) in the mixture was chosen to ensure the curing of the elastomer in ~1 h at room temperature. In order to enhance the adhesion of PDMS to the glass and silicon substrates, these were treated in advance by hydroxy-terminating their surfaces and assembling vynil-octodecyltrichlorosilane.

The thinning of the precursor film was accomplished by applying a force of 1.5 kN during the fabrication of the stamps A and B (Table I), resulting in films with a residual

TABLE 10.1. Parameters relative to three stamps (A,B,C) used in the experiments

Stamp	Mold	Pressure / Curing Temperature	Stamp Substrate	PDMS Base Layer Thickness/ Ununiformity	Feature Height/ Roughness	Pressure Applied during μCP
A		P = 2 MPa	Optical flat,			
	Material:		1 cm thick	200 ± 100 nm		4 MPa
	PMMA	Room	glass, 5 cm Ø			
	600nm	Temperature				
B	Patterning:				600 / 10 nm	
	EBL			500 ± 500 nm		2.5 MPa
			Silicon wafer			
	Substrate:					
C	1 cm thick	P = 250 Pa	650 μm thick			
	5 cm Ø glass.	Room		200μm		1 kPa
		Temperature				

thickness of 200 nm and 500 nm, respectively. Stamp C was fabricated without additional weight on the mold, leading to a final PDMS thickness of ~200 μm. The height of the protruding features were ~600 nm for all three stamps, corresponding to the height of the structures in the mould.

10.11.2. Experimental results

The results of printing depend critically on the properties of the stamps. This is well illustrated by the comparison of the different μCP performances of three stamps with different substrates and elastomeric film thickness (Table 1).

10.11.2.1. Stamp A. The printing process was performed by applying a pressure of 4 MPa on the stamp in contact with the printing substrate (650 μm thick Si covered by Cr/Au 5/20 nm). The quality of the pattern etched in the Au layer appears uniform over the entire printed area (~1 cm^2). The pattern is well defined both for low and high resolution features. The smallest printed spacing between lines in the pattern is ~100 nm (Figure 8).

The stamp can withstand the pressure of 4 MPa without showing any loss of resolution in the pattern and without being spoiled. No sagging effect is observed even at a distance of hundreds of μm from the nearest printed field. In this case the applied pressure compensates for the deviations from perfect planarity of the printing substrate by flattening it on the highly controlled planarized stamp surface.

10.11.2.2. Stamp B. The pressure used for printing was 2.5 MPa. As in the previous case (stamp A) the thin elastomeric film of stamp B prevents sagging. However, the uniformity of the PDMS layer is not controlled as it was for stamp A. In the present case the lower rigidity of the silicon substrate used for the stamp causes problems in two steps of the process. The first, during the stamp fabrication when a non-uniform PDMS layer arises as a result of the bending of the substrate. The second is during the

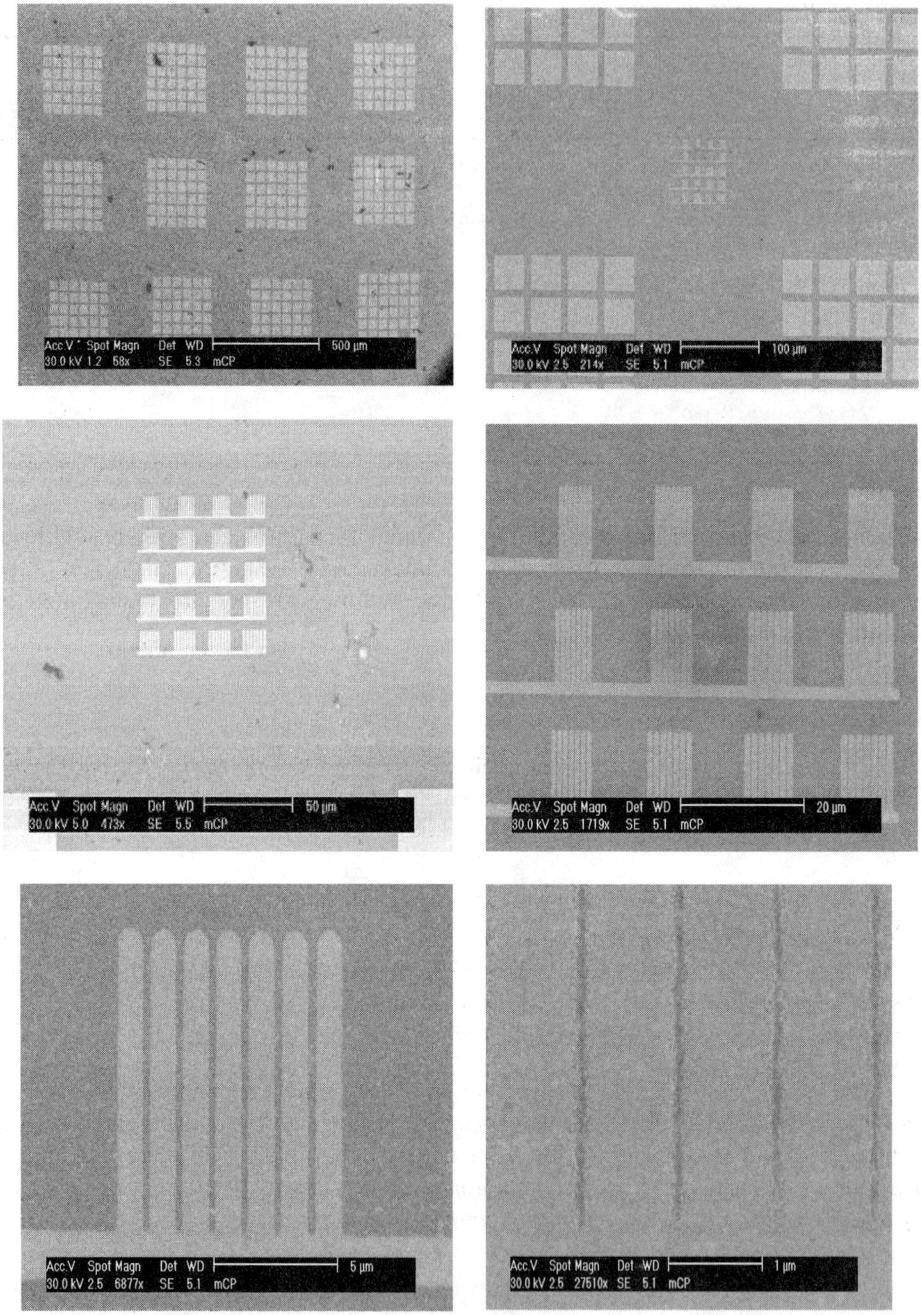

FIGURE 10.8. Scanning electron microscope images of a pattern transfered into a 20 nm thick Au film on Ti/Si substrate by μCP (stamp A) and wet chemical etching. The sequence zooms from mm scale area down to the 100 nm scale. The pressure applied to the stamp during μCP was 4 MPa. The printed pattern is uniform on areas of 10 × 10 mm2. No sagging effect is present.

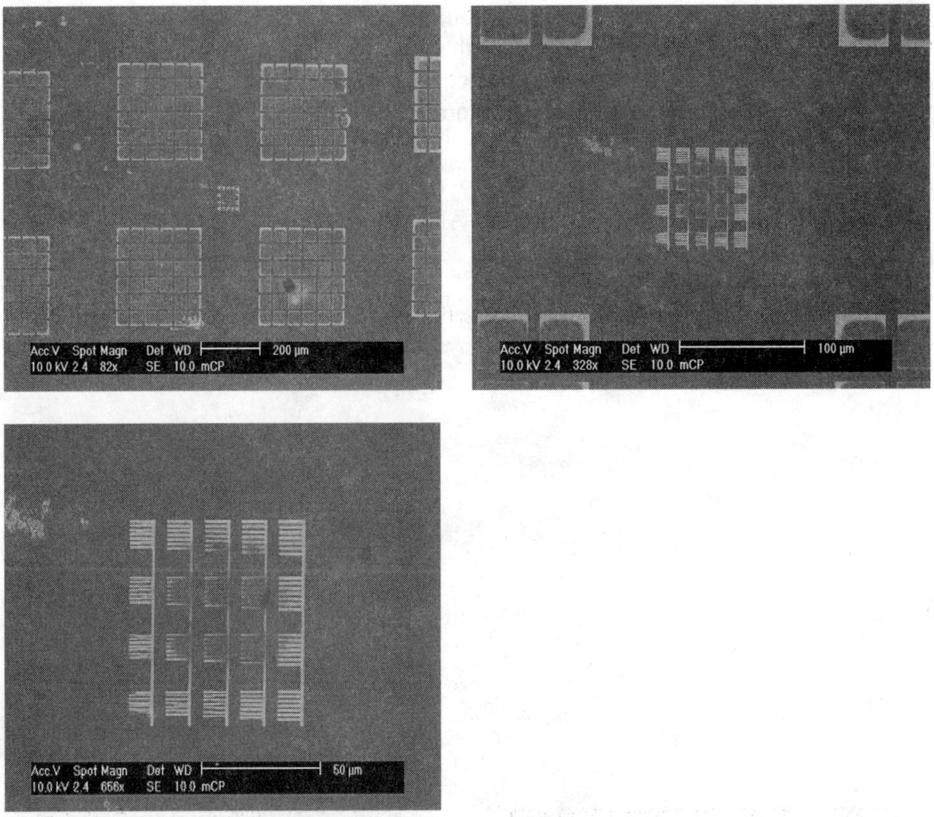

FIGURE 10.9. SEM images of patterns transferred into a 20/5 nm thick Au/Ti film on Si substrate by μCP and wet chemical etching. The printing pressure is 2.5 MPa. The printed pattern is present on areas of 10 × 10 mm2 and do not exhibit any sagging effect. The presence of concentrated forces corresponding to the contact fields gives rise to a bending of the substrate and a pressure modulation across each printed field. This is reflected in a non uniform protection of the Au layer. The resolution is comparable to that obtained with stamp A and C.

printing process where again the substrate can easily bend due to the concentrated forces applied by the structures of the stamp. It is interesting to notice that in this case the long scale bending of the substrate gives rise to a pressure modulation. This is enhanced at the borders of the printed fields; we see this in the printed structures as a contrast in the level of protection of the Au layer (Figure 9). However, the thinness of the base layer prevents any sagging effect even in the presence of a small bending of the substrate, and the resolution is similar to that obtained with stamp A (and C, see below).

10.11.2.3. Stamp C. The pressure used in this case was ~1000 Pa, which ensures the contact of the complete pattern with the substrate. However, the sagging effect in the printed results becomes severe. Larger forces result in a complete protection of the Au substrate. The resolution of the pattern transfer is similar to that obtained with stamps A and B, indicating that the application of pressure up to some MPa does not affect the resolution of the μCP process.

The comparison between the results obtained with stamps A, B, and C shows that the thickness of the stamps affects mainly the long scale uniformity and the reliability (e.g. sagging) of the printing process, without affecting the quality and the resolution of the smallest features, here 600 nm and 100 nm.

10.12. A MULTILEVEL PROCESS: MIX AND MATCH WITH ACCURATE ALIGNMENT OF THE μCP

One of the main reasons for the introduction of thin μCP stamps on rigid substrates is to perform the registration of a series of consecutive steps with accuracy comparable

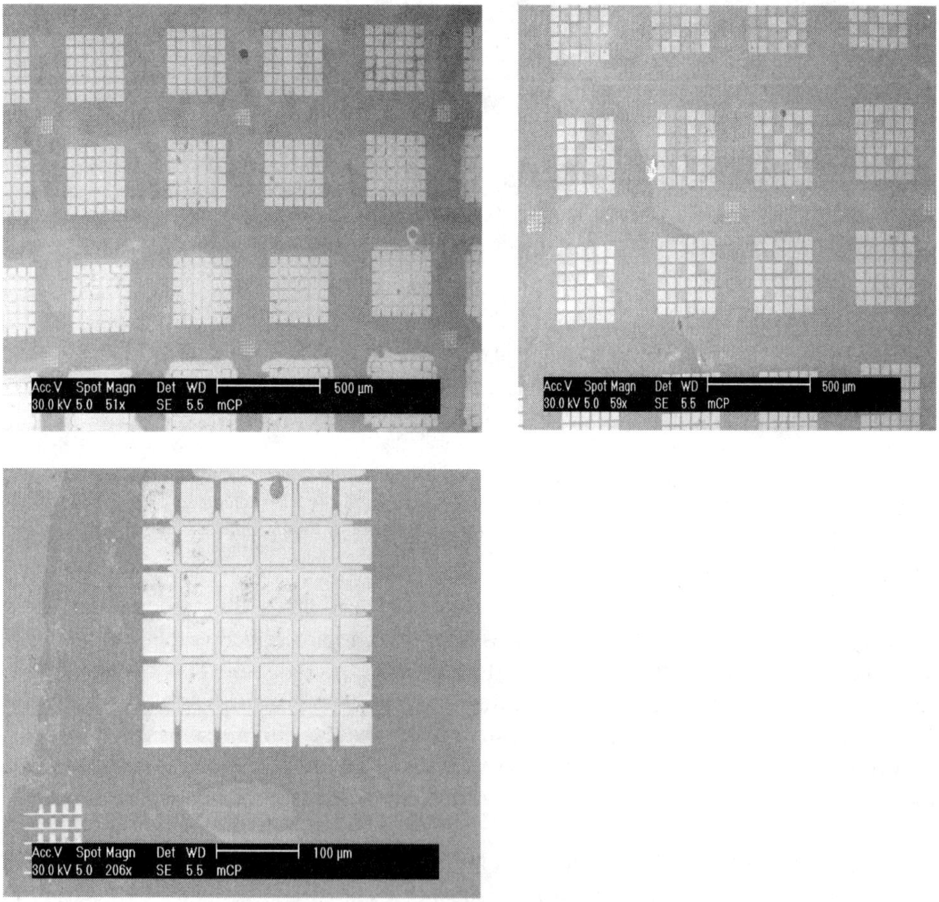

FIGURE 10.10. SEM images of a pattern transfered into a 20/5 nm thick Au/ Ti film on Si by μCP (stamp C) and wet chemical etching. The printing pressure was 1000 Pa (the weight of the stamp on the sample). The printed pattern is present on the entire area (10 × 10 mm2). A strong sagging effect has been found in all the printing results even at no externally applied pressure.

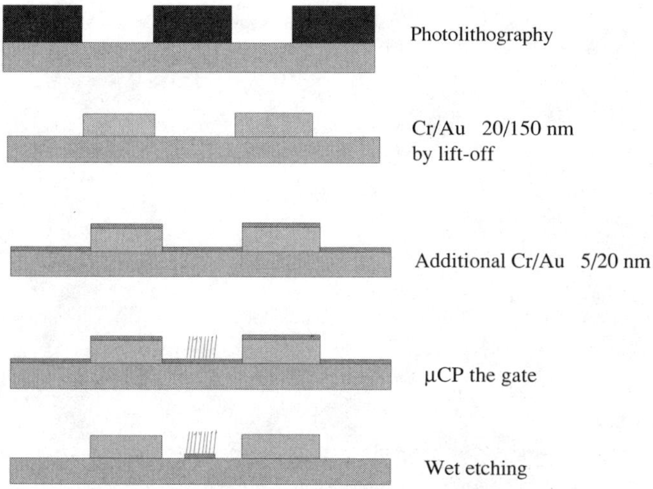

FIGURE 10.11. Scheme of the mix and match process.

to the resolution of the technique.[109] The distortions suffered by the membrane-stamps make them unsuitable for printing features separated by distances of centimeters with sub-μm placement accuracy. The technique of ultrathin stamps on transparent substrates combined with the use of a home made μCP aligner has allowed us to demonstrate that it is possible to obtain μm registration accuracy in mix and match processes in which the μCP is performed as a second lithographic process.

A homemade alignment system was built for the demonstration of a mix and match process. The system had the following characteristics:

1. Stage with 6 degrees of freedom.
2. Vacuum chuck mounted on the piezoelectric actuator to attach the sample.
3. Supporting stage for the quartz plate of the stamp.
4. Microscope with optical resolution better than 1 μm.

A photomask containing two sets of complementary patterns translated a fixed distance (800 μm) with respect to each other was fabricated by electron-beam lithography and used to produce a mold for μCP. A thin stamp (~3 μm) was fabricated on a quartz plate. The scheme used for the mix and match process is depicted in Figure 10. The same pattern was defined onto a Si substrate by photolithography and lift-off of Au/Cr (150/20 nm) using the same photomask that was used for the mold. Subsequently, the substrate of the sample and the metallic pattern lying on it were coated with an additional Au/Cr (20/5 nm) film. For the μCP step, the quartz plate supporting the stamp was mounted onto the aligner, registered to the pre-existing structures on the sample, and brought gently into contact with the sample. Sub-μm accuracy (Figure 11) was obtained in the global alignment of structures present on a 5×5 mm^2 area. Moreover, areas as large as 1×1 mm^2 without features do not exhibit the sagging effect. Figure 12 shows a larger field.

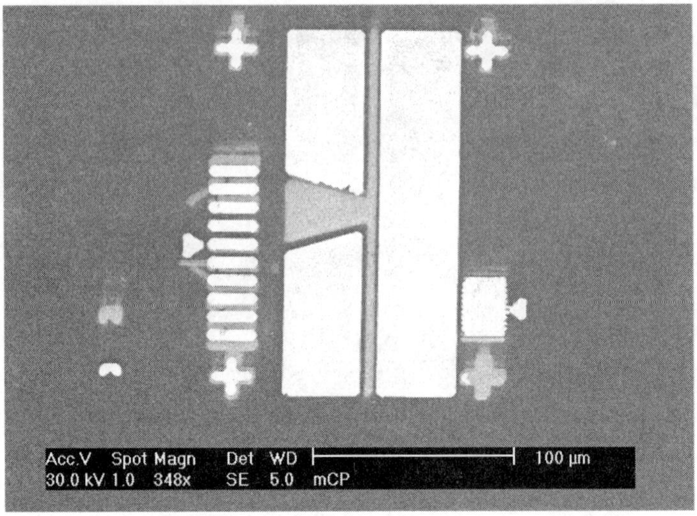

FIGURE 10.12. Aligned structures defined by photolithography (bright) and μCP (intermediate gray).

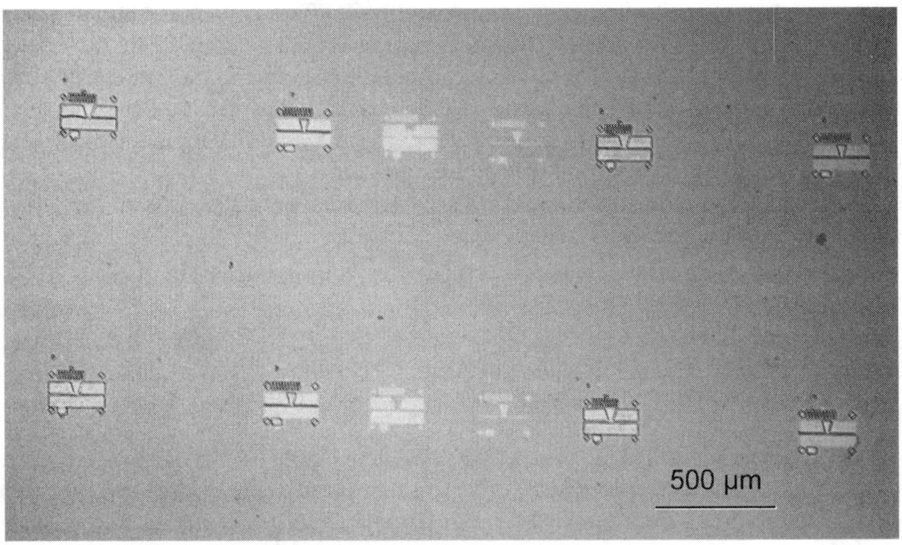

FIGURE 10.13. Field of aligned microstructures.

CONCLUSIONS

The interest exited in the scientific community by μCP is stimulating an intense research activity on the different aspects of this technique. The simplicity of the principle, that makes possible the introduction of μCP in a lab at low investment costs, contrasts with the problems that are usually encountered for an advanced utilization of the technique. This chapter tries to provide the reader with concepts, hints, and references that seem essential to achieve the best performances at the present status of the technique, and to furnish a solid background for facing challenges that its further developments will pose.

ACKNOWLEDGEMENTS

The experimental results reported in the section "microcontact printing with ultrathin stamps" have been obtained at the University of Wuerzburg in the group Experimentelle Physik III, with financial support from European Union ESPRIT basic research project SPINUP. I am grateful to my colleagues T. Borzenko, Georg Schmidt, Bernd Steffen and Laurens Molenkamp.

REFERENCES

1. A. Kumar and G. Whitesides, *Appl. Phys. Lett.* 63 (1993) 2002.
2. Y. N. Xia, X. M. Zhao, G. M. Whitesides 32 (1996) 255
3. Y. N. Xia, E. Kim, M. Mrksich, G. M. Whitesides *Chemistry of materials* 8 (1996) 601.
4. A. Carvalho, M. Geissler, H. Schmid, B. Michel, and E. Delamarche *Langmuir* 18 (2002) 2406.
5. Y. Jun, D. Le, and X.-Y. Zhu *Langmuir* 18 (2002) 3415.
6. E. Kim, G. M. Whitesides, M. B. Freiler, M. Levy, J. L. Lin, R. M. Osgood, Nano technology 7 (1996) 266.
7. A. R. Noble-Luginbuhl and R. G. Nuzzo *Langmuir* 17 (2001) 3937.
8. P. M. St. John and H. G. Craighead *Appl. Phys. Lett,* 68 (1996) 1022.
9. L. B. Goetting, T. Deng, and G. M. Whitesides *Langmuir* 15 (1999) 1182.
10. T. L. Breen, P. M. Fryer, R. W. Nunes, M. E. Rothwell *Langmuir* 18 (2002) 194
11. C. D. James, R. C. Davis, L. Kam, H. G. Craighead, M. Isaacson, J. N. Turner, and W. Shain 741 *Langmuir* 14 (1998) 741.
12. H. B. Lu, J. Homola, C. T. Campbell, G. G. Nenninger, S. S. Yee B. D. Ratner, *Sensors and Actuators B* 74 (2001) 91.
13. E. Ostuni, C. S. Chen, D. E. Ingber, and G. M. Whitesides , *Langmuir* **2001,** 17,2828.
14. E. A. Smith, M. J. Wanat, Y.Cheng, S. V. P. Barreira, A. G. Frutos, and R. M. Corn 2502 *Langmuir* **2001,** 17,2502-2507
15. R. S. Kane, S. Takayama, E. Ostuni, D. E. Ingber, G. M. Whitesides, Biomaterials 20 (1999) 2363
16. H. Kind, J.-M. Bonard, Ch. Emmenegger, L.-O. Nilsson, K. Hernadi, E. Maillard-Schaller, L. Schlapbach, L. Forró, and Klaus Kern *Adv. Mater.* 11 (1999) 1285.
17. Wolfe D. B., J. C. Love, K. E. Paul, M. L. Chabinyc, G. M. Whitesides *Appl. Phys. Lett.* 80 (2002) 2222.
18. N. L. Jeon, I. S. Choi, G. M. Whitesides, N. Y. Kim, P. E. Laibinis Y. Harada, K. R. Finnie, G. S. Girolami, and R. G. Nuzzo 4201 Appl. Phys. Lett., Vol. 75, No. 26, 27 December 1999
19. L. Yan, W. T. S. Huck, X.-M. Zhao, and G. M. Whitesides *Langmuir,* 15, (1999) 1208.
20. H. D. Inerowicz, S. Howell, F. E. Regnier, and R. Reifenberger, *Langmuir* 18 (2002) 5263.
21. Y. Xia, D. Qin, Y. Yin *Current Opinion in Colloid & Interface Science* 6 (2001) 54.
22. J. Tien, Y. Xia, G. M. Whitesides, *G.M. Microcontact Printing of SAMs*, in *Self Assembled monolayers of thiols* Edited by A. Ulman, Academic Press, San Diego, CA, 1998).
23. H. A. Biebuyck, N. B. Larsen, E. Delamarche, and B. Michel IBM J. RES. & DEV. 41 (1997) 159.
24. Y.Xia and G. M. Whitesides, *Angew. Chem. Int. Ed.* 37, (1998), 550.
25. B. Michel, A. Bernard, A. Bietsch, E. Delamarche, M. Geissler, D. Juncker, H. Kind, J.-P. Renault, H. Rothuizen, H. Schmid, P. Schmidt-Winkel, R. Stutz, H. Wolf *IBM J. RES. & DEV.* 45 (2001) 697.
26. G. M. Whitesides, E. Ostuni, S. Takayama, X.Y. Jiang, D. E. Ingber, *Annual Review of Biomedical Engineering* 3 (2001) 335.

27. P. Fenter, A. Eberhardt, P. Eisenberger, *Science* 266 (1994) 1216.
28. P. E. Laibinis, G. M. Whitesides, D. L. Allara, Y.-T. Tao, A. N. Parikh, and Ralph G. Nuzzo *J. Am. Chem. Soc.* 113(1991) 7152.
29. H. Rieley, C. K. Kendall, A. Chan, R. G. Jones, J. Lüdecke, D. P. Woodruff, B. C. C. Cowie *Surface Science* 392 (1997) 143.
30. N. Garg, E. Carrasquillo-Molina, T. R. Lee, *Langmuir* 18 (2002) 2717.
31. S. D. Evans, S. D. Cooper, S. R. Johnson, T. M. Flynn, and A. Ulman *Supramolecular Science* **4** (1997) 247.
32. L. C. Cheng, S. L. Bernasek, A. B. Bocarsly, T. A. Ramanarayanan *Chemistry of Materials* 7 (1995) 1807.
33. Y. Chen, R. E. Palmer, J. P. Wilcoxon *Surface Science* 454-456 (2000) 963.
34. A.-S. Duwez, L. M. Yu, J. Riga, J.-J. Pireaux, J. Delhalle, *Thin Solid Films* 327-329 (1998) 156.
35. N. Camillone, C. E. D. Chidsey, P. Eisenberger, P. Fenter, J. Li, K. S. Liang, G. Y. Liu , G. Scoles *J. Chem. Phys.* 99 (1993) 744.
36. S. Lee, A. Puck, M. Graupe, R. Colorado, Y. S. Shon, T. R. Lee, S. S. Perry *Langmuir* 17 (2001) 7364.
37. E. Delamarche, B. Michel *Thin Solid Films* 273 (1996) 54.
38. J. Voets, J. W. Gerritsen, R. F. P. Grimbergen, H. van Kempen, *Surface Science* 399 (1998) 316.
39. K. Kobayashi, T. Horiucgi, H. Yamada, K. Matsushige *Thin Solid Films* 331 (1998) 210.
40. M. Fujihira, M. Furugori, U. Akiba, Y. Tani, *Ultramicroscopy* 86 (2001) 75.
41. Y. Kim, K.-S. Kim, M. Park, J. J. *Thin Solid Films* 341 (1999) 91.
42. E. Barrena, C. Ocal, M. Salmeron, *Surface Science* 482-485 (2001) 1216.
43. D. Losic, J. G. Shapter, and J. J. Gooding *Langmuir* **2001,** 17,3307.
44. G.E. Poirier , E.D. Pylant *SCIENCE* 272 (1996) 1145.
45. Q. Zhang, H. Huang, H. He, H. Chen, H. Shao, Z. Liu. *Surface Science* 440 (1999) 142.
46. K. Tamada, J. Nagasawa , F. Nakanishi, K. Abe, M. Hara, W. Knoll, T. Ishida, H. Fukushima, S. Miyashita, T. Usui, T. Koini, T. R. Lee *Thin Solid Films* 327-329 (1998) 150.
48. R. Berger, E. Delamarche, H. P. Lang, Ch. Gerber, J. K. Gimzewski, E. Meyer, H.-J. Güntherodt *Science.* 276 (1997) 2021.
49. E. Delamarche, B. Michel *Thin Solid Films* 273 (1996) 54.
51. S. Frey, K. Heister, M. Zharnikov, M. Grunze, K. Tamada, R. Colorado Jr, M. Graupe, O. E. Shmakova, T. R. Lee *Israel Journal of Chemistry* 40, (2000) 81.
52. K. Kobayashi , H. Yamada, T. Horiuchi, K. Matsushige *Applied Surface Science* 144-145 (1999) 435.
53. P. E. Laibinis, G. M. Whitesides, D. L. Allara, Y.-T. Tao, A. N. Parikh, R. G. Nuzzo, *J. Am. Chem. Soc.* 113 (1991) 7152.
54. H. Sellers, A. Ulman, Y. Shnidman, and J. E. Eilerss *J. Am. Chem. Soc.* 115 (1993) 9389.
55. N. Camillone III, C. E. D. Chidsey, P. Eisenberger, P. Fenter, J. Li, K. S. Liang, G.-Y. Liu, and G. Scoles, *J. Chem. Phys.* **99** (1993) 744.
57. E. Delamarche, B. Michel, H. Kang, and Ch. Gerber*Langmuir 10 (1994)* 4103.
58. S. Imabayashi, M. Iida, D. Hobara, Z. Q. Feng, K. Niki, T. Kakiuchi, *J. Electroanal. Chem. 428* (**1997**) 33.
59. T. Kakiuchi, H. Usui, D. Hobara, M. Yamamoto, Langmuir; (Article); 2002; ASAP Article.
60. K. Aoki, T. Kakiuchi *J. Electroanalytical Chemistry* 452 (1998) 187.
61. Y.Xia and G. M. Whitesides, *Angew. Chem. Int. Ed.* 37, (1998), 550.
62. D. C. Duffy, J. Cooper McDonald, O. J. A. Schueller, and G. M. Whitesides *Anal. Chem.* 70 (1998) 4974.
63. G. M. Whitesides and A. D. Stroock *Physics Today*, June 2001 42.
64. E. Delamarche, A.Bernard, H. Schmid, A. Bietsch, B. Michel, and H. Biebuyck, *J. Am. Chem. Soc.* 120 (1998) 500.
65. A. Bernard, B. Michel, and E. Delamarche, *Anal. Chem.* 73 (2001) 8.
66. D. Qin, Y. Xia, G. M. Whitesides, *Adv. Mater.* (1997), 9, 407
67. J. A. Rogers, R. J. Jackman, O. J. A. Schueller, G. M. Whitesides, *Appl. Opt.* 35 (1996) 6641.
68. H. Schmid, H.Biebuyck, B. Michel,O. J. F. Martin *Appl. Phys. Lett.* 72, (1998) 2379.
69. J. A. Rogers, K. E. Paul, R. J. Jackman, and G. M. Whitesides, *Appl.Phys. Lett.* 70 (1997) 2658.
70. D. Qin, Y. Xia, A. J. Black, and G. M. *J. Vac. Sci. Technol. B*, 16 (1998) 98.
71. R. J. Jackman, D. C. Duffy, O. Cherniavskaya, and G. M. Whitesides *Langmuir* 15 (1999) 2973.
72. H. O. Jacobs and G. M. Whitesides *Science* 291 (2001) 1763.

73. Y. Xia, N. Venkateswaran, D. Qin, J. Tien, G.M. Whitesides, *Langmuir* 14, (1998) 363.
74. J. A. Rogers, R. J. Jackman, G.M. Whitesides, D. L. Olson, J. V. Sweedler *Appl. Phys. Lett.* 70 (1997) 2464.
75. R. J. Jackman, S. T. Brittain, A.Adams, H.Wu, M. G. Prentiss, S. Whitesides, G. M. Whitesides, *Langmuir* 15, (1999) 826.
76. R. J. Jackman, D. C. Duffy, E. Ostuni, N. D. Willmore, and G. M. Whitesides *Anal. Chem* 70 (1998) 2280.
77. Elastomeric light valves Qin D, Xia YN, Whitesides GM *Adv. Mater.* 9 (1997) 407.
78. J. C. Lötters, W. Olthuis, P. H. Veltink, Bergveld P *Microsystem Technologies* 3 (1997) 64.
79. H. Schmid and B. Michel *Macromolecules* 33 (2000) 3042.
80. J. M. Desimone, G. A. York, J. E. Mcgrath, A. S. Gozdz, M. J. Bowden *Macomolecules* 24 (1991) 5330.
81. Joost C Lötters, Wouter Olthuis y , Peter H Veltink and Piet Bergveld, *J. Micromech. Microeng.* 6 (1996) 52.
82. F. Robbe-Valloire, M. Barquins, *Intern. J. of Adhesion and Adhesives* 18 (1998) 29.
83. M. K. Chaudhury, G. M. Whitesides *Langmuir* 7 (1991) 1013.
84. T. Kawamura, K. Urayama, and S. Kohjiya *Macromolecules* 34 (2001) 8252.
85. P.Y. Longin, C. Verdier, M. Piau, *J. Non-Newtonian Fluid Mech.,* 76 (1998) 213.
86. P. C. Hidber, W. Helbig, E. Kim, G. M. Whitesides, *Langmuir* 12 (1996) 1375.
87. W. T. S. Huck, L. Yan, A. Stroock, R. Haag, G. M. Whitesides, *Langmuir* 15 (1999) 6862.
88. C. D. James, R. C. Davis, L. Kam, H. G. Craighead, M. Isaacson, J. N. Turner, and W. Shain *Langmuir* 14 (1998) 741.
89. Ch. Donzel, M. Geissler, A. Bernard, H. Wolf, B. Michel, J. Hilborn, and E. Delamarche, *Adv. Mater.* 13 (2001) 1164.
90. J. Genzer and K. Efimenko *Science* 290 (2000) 2130.
91. S. Perutz, E. J. Kramer, J. Baney, C.-Y. Hui *Macromolecules* 30 (1997) 7964.
92. A.Bietsch, Bruno Michel *J. Appl. Phys.* 88 (2000) 4310.
93. C. Y. Hui, A. Jagota, Y. Y. Lin, E. J. Kramer *Langmuir* 18 (2002), 1394.
94. W.-I. Sohn, D.-H. Ryu, S.-J. Oh, J.-K. Koo *J. Membrane Science* 175 (2000) 163.
95. T. Aoki *Prog. Polym. Sci.* 24 (1999) 951.
96. L. E. Matheson, Y. Chen, *International Journal of Pharmaceutics* 125 (1995) 297.
97. E. Favre *Eur. Polym. J.* 32 (1996) 1183.
98. R. J. Jackman, D. C. Duffy, O. Cherniavskaya, and G. M. Whitesides *Langmuir* 15 (1999) 2973.
99. M. Tormen, T. Borzenko, G. Schmidt, J. Liu and L. W. Molenkamp Electronics Letters 36 (2000) 983.
100. B. Arkles *Chemtech* 13 (1983) 542.
101. J. L. Wilbur, R. J. Jackman, and G. M. Whitesides, E. L. Cheung, L. K. Lee, and M. G. Prentiss *Chem. Mater. 8* (1996) 1380.
102. R. Pelrine, R. Kornbluh, Q. Pei, J. Joseph, *Science* 287 (2000) 836.
103. B. A. Grzybowski, S. T. Brittain, and G. M. Whitesides, *Rev. Sci. Instrum.* 70 (1999) 2031.
104. J. A. Rogers, O. J. A. Schueller and G. M. Whitesides *Appl. Phys. Lett.* 72 (1998) 1951.
105. M. Khoo, C. Liu *Sensors and Actuators A* 89 (2001) 259.
106. E. Delamarche, H.A. Biebuyck, H. Schmid, B. Michel, *Adv Mater.* 9 (1997) 741
107. J. A. Rogers, K. E. Paul, and G. M. Whitesides . *J. Vac. Sci. Technol. B*, 16 (1998) 88.
108. T. Burgin, V.-E. Choong, and G. Maracas *Langmuir* 16 (2000) 5371.
109. M. Geissler, A. Bernard, A. Bietsch, H. Schmid, B. Michel, E. Delamarche *J. Am. Chem. Soc.* 122 (2000) 6303.
110. R. Hull, T. Chraska, Y. Liu, D. Longo *Materials Science and Engineering C* 19 (2002) 383.
111. Y. Xia and G. M. Whitesides, *Annu. Rev. Mater. Sci.* 28, (1998) 153.
112. H. Kind, M. Geissler, H. Schmid, B. Michel, K. Kern, and E.Delamarche *Langmuir* 16, (2000) 6367.
113. H. A. Biebuyck, N. B. Larsen, E. Delamarche, and B. Michel IBM J. RES. & DEV. 41 (1997) 159.
114. A. Bietsch and B. Michel *J. Appl. Phys.* 88 (2000) 4310.
115. E. Delamarche, H. Schmid, A. Bietsch, N. B. Larsen, H. Rothuizen, B. Michel, and H. Biebuyck, *J. Phys. Chem. B*, 102 (1998) 3324.
116. D. Bhanushali, S. Kloos, C. Kurth, D. Bhattacharyya *J. Membrane Science* 189 (2001) 1.
117. A. Bietsch, H. Schmid, B. Michel, E.Delamarche*Langmuir.* 15 (1999) 300.

118. A. Bernard, J. P. Renault, B. Michel, H. R. Bosshard, and E. Delamarche *Adv. Mater.* 12 (2000) 1067.
119. S. D. Evans, S. D. Cooper, Johnson SR, Flynn TM, Ulman A *Supramolecular Science* 4 (1997) 247.
120. D. B. Wolfe, J. Ch. Love, K. E. Paul, M. L. Chabinyc, and G. M. Whitesides *Appl. Phys. Lett.*. 80, (2002) 2222.
121. W. Gao, L. Dickinson, Ch. Grozinger, F. G. Morin, and L. Reven *Langmuir* 12 (1996) 6429.
122. J. J. Goodding, V. Praig, E. A. H. Hall, *Anal. Chem.* 70 (1998) 2396.
123. B. Michel, A. Bernard, A. Bietsch, E. Delamarche, M. Geissler, D. Juncker, H. Kind, J.-P. Renault, H. Rothuizen, H. Schmid, P. Schmidt-Winkel, R. Stutz, H. Wolf *IBM J. RES. & DEV.* 45 (2001) 697.
124. Y. A. Xia, N. Venkateswaran, D. Qin, J. Tien, and G. M. Whitesides *Langmuir* 14 (1998) 363.
125. J. Tate, J. A. Rogers, C. D. W. Jones, B. Vyas, D. W. Murphy, W. J. Li, Z. A. Bao, R. E. Slusher, A. Dodabalapur, H. E. Katz *Langmuir* 16 (2000) 6054.
126. S. T. Brittain, O. J. A. Schueller, H. K. Wu, S. Whitesides, G. M. Whitesides *J. Phys. Chem. B* 105 (2001) 347. 127. R. J. Jackman, S. T. Brittain, A. Adams, H. Wu, M.G. Prentiss, S. Whitesides, and G. M. Whitesides *Langmuir*, 15 (1999) 826.
128. R. J. Jackman, J. L. Wilbur, G. M. Whitesides *Science* 269 (1995) 664.
129. G. Schmidt, M. Tormen, G. Muller, L. W. Molenkamp, Y. Chen, A. Lebib, H. Launois *Electr. Lett.* 35 (1999) 1731.
130. Y. Chen, A. Lebib, F. Carcenac, H. Launois, G. Schmidt, M. Tormen, G. Muller, L. W. Molenkamp, M. Liebau, J. Huskens, S. N. Reinhoudl *Microelectr. Eng.* 53 (2000) 253.
131. M. Liebau, J. Huskens, and D. N. Reinhoudt *Adv. Funct. Mater.* 11 (2001) 147.
132. L. Libioulle, A.Bietsch, H.Schmid, B. Michel, and E. Delamarche, *Langmuir* 15 (1999), 300.
133. D. Losic, J. G. Shapter, and J. J. Gooding *Electrochemistry Communications* 3 (2001) 722.
134. R. G. Nuzzo, D. L. Allara, *J. Am. Chem. Soc.*, 105 (1983) 4481.
135. M. Ishibashi, M. Itoh, H. Nishihara, and K. Aramaki, *Electrochimica Acta*, 41, (1996) 241.
136. T. Kawai, H. Nishihara, K. Aramaki, *J. Electrochem. Soc.* **1996**, 143, 3866.
137. K Nozawa, H. Nishihara, and K. Aramaki, Corrosion Science 39 (1997) 1625.
138. M. A. Hines, J. A. Todd, P. Guyot-Sionnest, *Langmuir* 11 (1995) 493.
139. C. W. Sheen, J.-X. Shi, J. Martensson, A. N. Parikh, D. L. Allara, *J. Am. Chem. Soc. 11*4 (1992) 1514.
140. F. P. Zamborini and R. M. Crooks *Langmuir* 14, (1998) 3279.
141. X. M. Zhao, J. L. Wilbur, and G. M. Whitesides *Langmuir* 12 (1996) 3257.
142. X. M. Zhao, J. L. Wilbur, and G. M. Whitesides *Langmuir* 12 (1996), 5504.
143. M. Geissler, H. Schmid, A. Bietsch, B. Michel, and E. Delamarche *Langmuir* 18 (2002) 2374.
144. D.J. Achenson. *Elementary Fluid Dynamics.* Oxford University Press, Oxford 1990.

11

Local Oxidation Nanolithography

Mechanism, Feature Size and Applications

Ricardo Garcia
Instituto de Microelectrónica de Madrid, CSIC, Madrid, Spain

11.1. INTRODUCTION

The potential of scanning probe microscopes for the atomic and nanometer scale modification and manipulation of materials was recognized right after the invention of the scanning tunneling microscope STM.[1,2] Atomic scale desorption and fragmentation was demonstrated on Si(111) 7×7 surfaces.[3] A paradigmatic example of single atom manipulation was provided by moving individual atoms of xenon on a nickel surface.[4] Since then, a number of methods have been proposed and applied to manipulate surfaces with atomic and nanometer-scale features.[5-9] However, just a few of the proposed methods are suitable for the development of a scanning probe-based lithography. Ultra high vacuum requirements, low temperatures or unreliable pattern definition restricts considerably the number of methods suitable for lithographic purposes. In this context lithography means a method for large scale patterning of surfaces and device fabrication.

Local oxidation or nano-oxidation lithography is emerging as a powerful and versatile approach to fabricate nanometer-scale devices with an atomic force microscope (AFM). The term local comes from the smallness of the sample region that interacts with the end of a sharp tip (the AFM tip). In a typical local oxidation experiment an AFM tip is brought into close proximity of the sample. In ambient conditions the application of a voltage between tip and sample drives the oxidation of the surface. The present knowledge allows establishing some similarities between local oxidation and conventional anodic oxidation. The AFM tip is used as a cathode and the water meniscus formed between tip and surface provides the electrolyte. The strong localization of the electrical field lines near the tip apex in the presence of oxygen species gives rise to a nanometer-size oxide dot.

Alternative Lithography: Unleashing the Potentials of Nanotechnology
Edited by Sotomayor Torres, Kluwer Academic/Plenum Publishers, 2003

A feature that distinguishes local oxidation nanolithography from other scanning probe-based modification methods is that modification and imaging processes operate independently. Long range van der Waals forces and/or short range repulsive forces are used to image the surface while an external voltage is applied to induce the local oxidation of the surface. Another relevant feature is that it allows *in situ* control of the device electrical and topographic characteristics during its fabrication.

In 1990 Dagata and co-workes modified hydrogen-terminated silicon surfaces with an STM operated in ambient air.[10] Simultaneously, Thundat and co-workers used an STM to smooth tantalum surfaces.[11] Secondary ion mass spectroscopy in the case of silicon and x-ray photoelectron spectroscopy in tantalum showed that the modifications were due to the oxidation of the respective surfaces. The local oxidation method has been improved and optimized through the years. The STM was replaced by the AFM as the instrument of choice for performing the oxidation.[12] Equally important has been the reversal of the tip polarity, from a positive bias in some the earlier experiments[10] to a negative bias in subsequent experiments. Actually, it is a negative biased tip what defines an anodic oxidation process. Dynamic AFM modes have also been applied to perform local oxidation experiments,[13-15] in particular noncontact AFM operation has greatly improved the reproducibility and definition of the motives.[14-15] Remarkably local oxidation experiments have also been performed by using a scanning electron microscope.[16] Selective oxidation of single H-Si bonds has also been demonstrated in ultra high vacuum.[17] Those experiments illustrate the ultimate limits of local oxidation, however, the oxidation mechanism at ambient pressure and UHV are different. Here, we are only concerned with experiments performed at ambient pressure.

Both STM and AFM have been used to locally oxidized surfaces. However, the AFM is more suitable for performing local oxidation oxidation experiments because the feedback process that controls imaging in force microscopy can be applied independently of the exposure mechanism (the electrical field) in local oxidation. As a consequence, high resolution images of the sample can be obtained and used for precise alignment or assessment of the lithography without introducing further modifications.

Local oxidation experiments were first performed on silicon (111) and poly-crystaline tantalum faces by[10] and[11] respectively. Since then a number of materials have been locally oxidized such as compound III-V semiconductors,[18-19] titanium,[20] molybdenum[21] and silicon nitride films[22] as well as and self assembled monolayers[23] and carbonaceous films.[24] The intense activity has revealed some of the relevant factors affecting the oxidation process such as the voltage and pulse duration,[25-27] doping of the substrate,[28-30] the formation of a liquid meniscus[15,31] or the chemical composition of the atmosphere.[32] The fabrication of single electron and metal-oxide transistors,[33-34] data storage memory arrays[15] and machining of silicon structures[35] have illustrated some of its device applications.

11.2. LOCAL OXIDATION MODES

The AFM used for local oxidation has all the elements of a conventional force microscope, cantilever-tip ensemble, laser beam, segmented photodiode, piezo-scanner and software to control the process (Figure 1). In addition the cantilever-tip ensemble

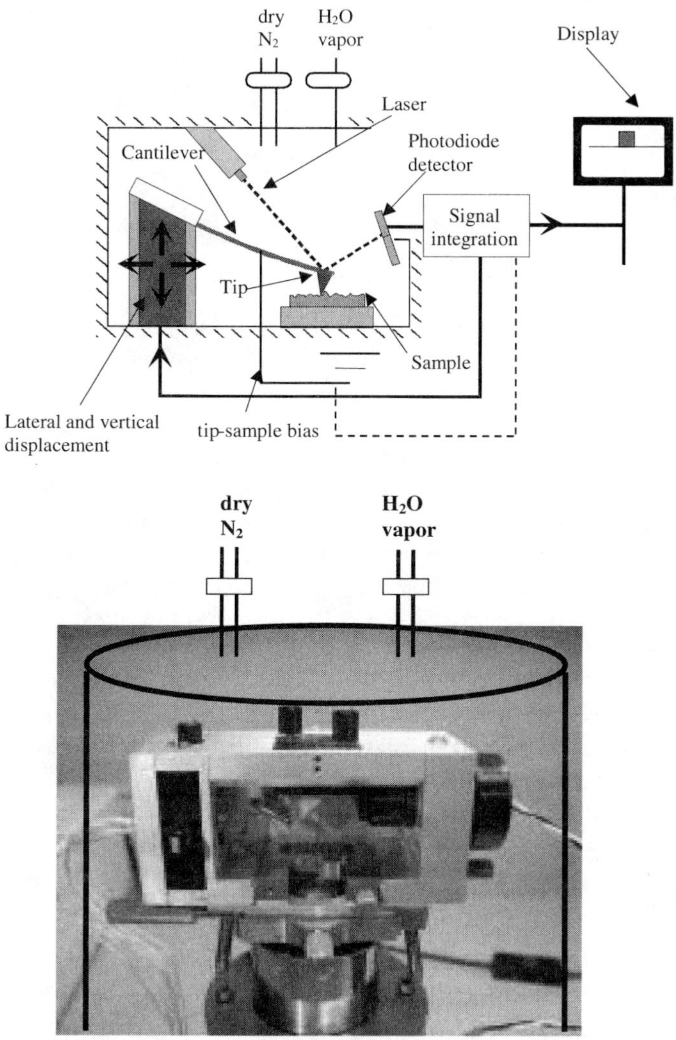

FIGURE 11.1. (a) Scheme of an atomic force microscope suitable for local oxidation experiments. (b) Close up of an AFM head. The humidity control chamber is depicted by continuous lines.

should be conductive and a power supply should be intercalated between tip and sample. The whole system should also be enclosed in a chamber to allow the control of the relative humidity. Finally, to perform local oxidation in a routine way, a special program should be used to control the tip motion. The AFM can be operated in contact mode and also several dynamic modes such as amplitude modulation AFM which allows the operation without tip-sample mechanical contact.[36-37] Both contact and amplitude modulation AFM modes are suitable to perform local oxidation experiments.

Figure 2 shows a scheme of an AFM interface during a local oxidation experiment. In contact AFM (c-AFM) oxidation the tip is in mechanical contact with the sample.

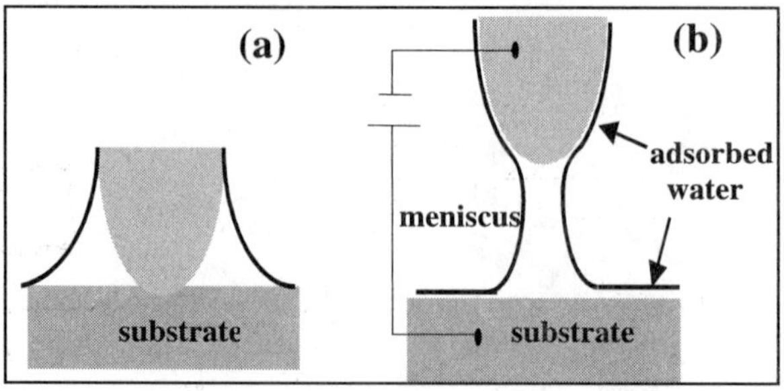

FIGURE 11.2. Schematic of an AFM interface in a humid environment. (a) In c-AFM the mechanical contact between tip and sample in humid air drives the spontaneous formation of a water meniscus. (b) In nv-AFM the formation of a water meniscus is induced by the application of an external electrical field.

In humid air, the mechanical contact of surfaces with submicrometer features drives the spontaneous formation of a water meniscus between tip and sample.[38] The meniscus is the source of the oxyanions needed for the oxidation of the sample surface.

Figure 2(b) shows a scheme of the AFM interface in a non contact AFM (nc-AFM) oxidation experiment. In nc-AFM the formation of the liquid bridge is induced by the application of an electrical field.[15] The protocol to perform local oxidation experiments in non contact AFM involves several steps. First, a suitable cantilever-tip oscillates ~ 5 nm above a region of the sample surface to be modified. The surface of the sample should be covered by a few monolayer thick film of water. Then a voltage pulse between tip and sample is applied. If the voltage is above a certain threshold value a liquid bridge is formed. Then another voltage pulse is applied to induce the oxidation.

To drive the anodic oxidation the sample is positively biased with respect to the tip. The voltage has a double function (i) it produces the hydrolysis of the water molecules within the liquid meniscus and (ii) drives the oxyanions to the silicon-silicon oxide interface (Figure 3).

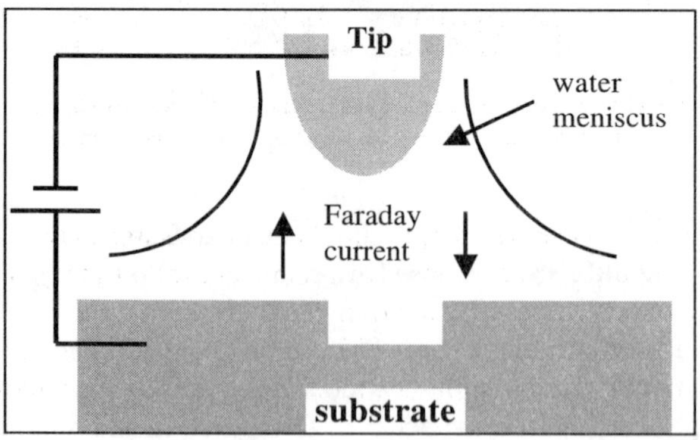

FIGURE 11.3. Scheme of an AFM interface during an anodic oxidation experiment. The tip is negatively biased with respect to the sample. The electrical field produces oxyanions in the liquid meniscus, drives them to metal (semiconductor)/oxide interface and enhances the oxidation growth rate. The oxide grows more rapidly underneath the tip.

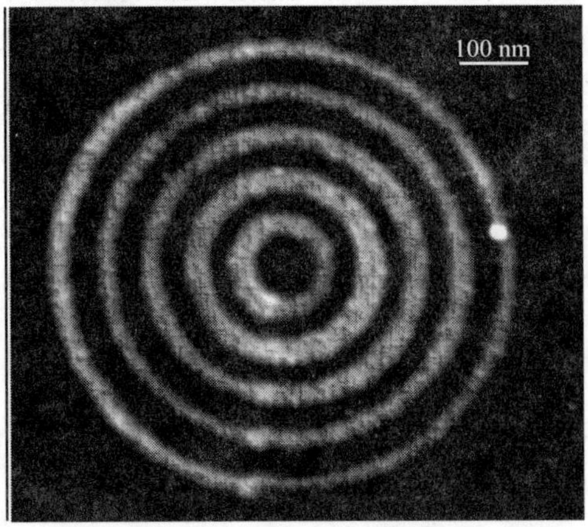

FIGURE 11.4. Concentric rings fabricated by nc-AFM with a sequence of voltage pulses of 24 V for 10 ms. The outer ring has required the application of about 60 pulses.

The electrical field is introduced by applying either a continuous voltage or a pulse. In the first case the tip is displaced along the sample while the voltage is switched on. In the second case, the tip rests still on a fixed position during the oxidation process, then the voltage is switched off and the tip is displaced to another position. In both cases is possible the fabrication of continuous structures. Figure 4 shows a series of concentric rings fabricated by nc-AFM oxidation with the application of voltage pulses of 10 ms and 24 V.

For identical electrical conditions non-contact AFM oxidation produces oxides of higher aspect ratio than contact AFM. This result is a consequence of two effects. First, non contact AFM allows controlling the lateral size of the liquid meniscus. This in turn controls the lateral size of the oxide dot. Second, the vertical growth rate is smaller in contact AFM oxidation. This is a consequence of the mechanical energy needed to deflect the cantilever during the growth of the oxide in c-AFM.[39]

The specialized literature offers a variety of instruments and cantilevers to perform local oxidation nanolithography. In general local oxidation requires some minor modification of the AFM like the incorporation of additional circuits to apply voltage pulses. The microscope is placed into a closed box with inlets for dry and H_2O saturated nitrogen. The relative humidity is kept around 30%-50%. Usually noncontact AFM oxidations are performed with doped n^+-type silicon cantilevers with force constant (k) and resonance frequency (f_0) of about 30 N/m and 314 kHz, respectively. On the other hand, contact AFM experiments are usually performed with Si_3N_4 cantilevers with k of about 1 N/m. Contact Si_3N_4 cantilevers are metallized with a layer 5 nm of Cr and 10 nm of Au. The oxidation is performed by applying a constant force in nN range. The feedback is switched off during the local oxidation process in both contact and nc-AFM operations. Typical applied voltages range between 5 to 24 V.

11.3. LIQUID MENISCUS

The liquid bridge or water meniscus has a two-fold role in AFM oxidation. The bridge provides the oxygen species (mostly OH^-) needed to oxidize the surface. Second,

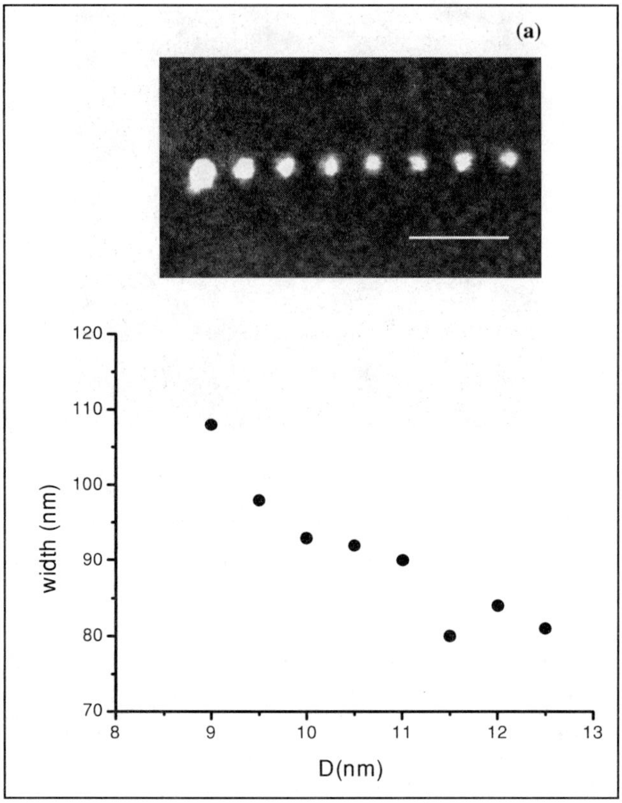

FIGURE 11.5. Dependence of the oxide width on the meniscus length. (a) AFM image of a sequence of oxide dots. The dots have been generated by nc-AFM oxidation by applying voltage pulses of 0.5 ms and 18 V. In each case a liquid meniscus was formed under identical conditions (tip-surafce separation and applied voltage) then the liquid bridge was stretched by increasing the tip surface separation. Scale bar denotes 250 nm. (b) Dot width dependence on the tip-surface average distance. Thinning the liquid bridge reduces the dot width.

it also provides the electrical field pathway to induce the oxidation. More importantly, the difference between the electrical permittivity inside and outside the liquid bridge, 81 and 1 respectively, acts as an effective lens, focusing the fieldlines within the neck limits. As a result the ions involved in the oxidation are confined within the meniscus.[15,40]

Mechanical contact drives the spontaneous formation of the water meniscus in c-AFM oxidation. The lateral size of the meniscus has a relevant role because it influences the dot size (see below). In contact AFM the size of the liquid bridge is determined by the tip size, the hydrophilic character of the surfaces and the water vapor pressure which leaves the observer with few options to reduce the meniscus size without affecting the overall oxidation process. However, in non contact AFM once the liquid bridge is formed its dimensions may be modified by changing the tip-surface separation. Figure 5 illustrates the close relationship between the liquid bridge and local oxide size. The experiment consists in the formation of a series of identical liquid bridges. Each bridge is stretched independently to have a series of bridges of different lengths. Then, a voltage pulse is applied to generate an oxide. The voltage pulse is identical for all bridges. Finally, the width of the oxide is plotted versus the length of the meniscus.[15]

The above experiment has required the application of two different voltage pulses, one to form the liquid bridge and the other to grow the oxide. First, a 0.5 ms pulse of 18 V is applied to build the bridge. Then, tip and sample are separated, i.e., the bridge is stretched. Finally, to grow a dot, another 0.3 s pulse of 8 V is applied. The experimental data (circles) show that the width of the oxide dot decreases by increasing the tip-sample separation (Figure 5(b)). The width is reduced from 108 nm to 81 nm by increasing the separation from 8.5 to 12.5 nm.

11.4. KINETICS

The study of the kinetics and mechanism of local oxidation is of fundamental importance to fulfill its potential for lithographic applications. This has been the subject of a number of contributions in the past few years.[25-29,39,40-45] The effort has revealed the dominant factors affecting the oxidation rate growth, such as the voltage strength, pulse duration, doping of the substrate, space charge build up, relative humidity and cantilever force constant. Several models have also been proposed to explain the oxidation mechanism, although none of the models seem to explain the wide range of experimental data. Because silicon surfaces have been extensively locally oxidized, the rest of the section is devoted its kinetics.

Secondary ion mass, x-ray, photoemission and Auger spectroscopies have confirmed the growth of the oxide under the tip in local oxidation experiments.[10-11,46,47] Local silicon oxides show an amorphous structure according to cross-sectional transmission electron microscopy analysis.[30] Depending on the doping type, local silicon oxides may show density and *stoichiometric* differences with respect to thermal oxides.[30,47]

Faraday currents have also been measured during oxidation of a silicon surface.[25] The net current through the interface was very close to the one expected for the anodic oxidation if calculated from the measured volume growth according to the electrochemical reaction described below

$$\mathrm{Si} + 2\mathrm{h}^+ + 2(\mathrm{OH}^-) \rightarrow \mathrm{SiO_2} + \mathrm{H_2} \tag{1}$$

The above reaction implies the decomposition of water molecules according to

$$2\mathrm{H_2O} \rightarrow 2\mathrm{H}^+ + 2(\mathrm{OH}^-) \tag{2}$$

while the oxidation takes place in the anode according to

$$\mathrm{Si} + 2\mathrm{h}^+ + 2(\mathrm{OH}^-) \rightarrow \mathrm{Si(OH)_2} \rightarrow \mathrm{SiO_2} + 2\mathrm{H}^+ \tag{3}$$

Hydrogen generation occurs at cathode to complete the electrochemical reaction

$$2\mathrm{H}^+ + 2\mathrm{e}^- \rightarrow \mathrm{H_2} \tag{4}$$

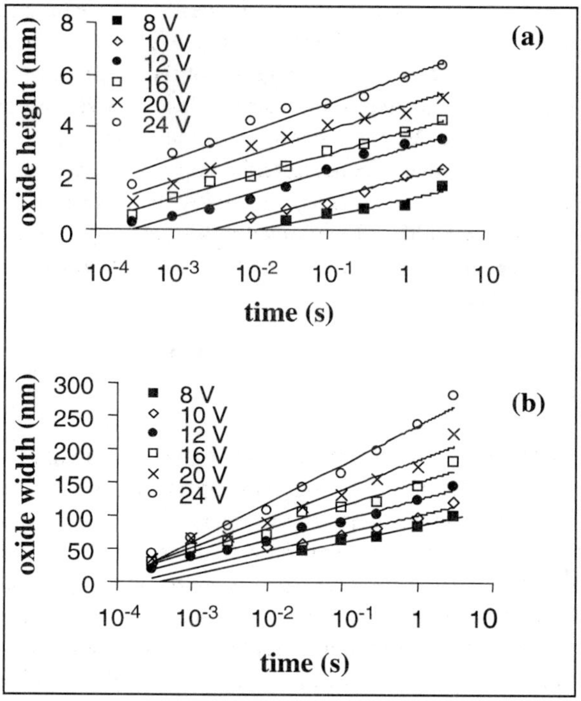

FIGURE 11.6. Silicon oxide dependence on applied voltage and pulse duration in nc-AFM. Height (a) and width (b) dependence on pulse duration for different voltages.

The kinetics of the oxidation can be empirically determined by measuring the oxide size as a function of the voltage and time. Figure 6 shows the dependence of the lateral and vertical dimensions of oxide dots as a function of voltage and pulse duration.[27]

The experiment was performed on a p-type Si(100) surface covered with a native oxide. Remarkably, local oxidation happens in presence of thin native oxides. The data shows a logarithmic dependence of both height and width on time. For a given pulse duration the width and height show an almost linear increase with the applied voltage. The above results imply an oxidation process that is characterized by a high initial growth rate, that decays rapidly and saturates.

Fitting height and width experimental dependencies on time and voltage allows to deduce the following relationships,

$$h = h_0(V) + h_1(V) \ln t/t_0 \quad (5)$$

$$w = w_0(V) + w_1(V) \ln t/t_0 \quad (6)$$

where h_0, h_1, w_0 and w_1 show a dominant linear dependence on V. In the above equations size, voltage and time are expressed in nm, volts and seconds respectively. Similar logarithmic dependencies have been obtained by Avouris et al.[25]and Stievenard et al.[43] However, empirical power of time laws of the type $h \sim \alpha_0 V(t/t_0)^{\gamma}$ have also been reported by Teuschler et al. on Si(111) faces[29] and Dubois and Bubendorff on titanium surfaces.[26] It must be noted that in some cases the data can be approximately fitted by both logarithmic and power law dependencies.

11.5. MECHANISM

Low temperature oxidation of solids is a diffusion process where the key factor is to determine whether the oxidation rate is controlled by processes occurring at the metal-oxide interface or by reactions at the oxide-oxygen interface or by diffusion through the oxide. Once the process has been established, the activation energy can be determined. To explain the direct log form observed experimentally in many local oxidation experiments, Dagata et al.[41] have used Uhlig's model.[48] Uhlig obtained the direct-log form, $h(t) = k''(V)\ log(k_U t+1)$ by assuming that the rate-controlling step involves interaction of oxyanions with either holes or electrons, and defects at a metal (semiconductor)-oxide interface. To deduce the above equation, Uhlig also considered that the oxide tends to acquire excess negative charge whenever the oxide has a greater electron affinity or higher work function than the metal (semiconductor here). The production of charged defects leads to a buildup of space charge within the oxide which inhibits further growth over long pulse times. Dagata noticed that substituting $t^{0.4}$ for t in the Uhlig log form provides a good quantitative fit to several local oxidation experiments. A further analysis of local oxidation data led to the conclusion that a single overall rate does not provide sufficient insight into how oxyanions, trapping sites, and electronic species interact. Dagata et al. proposed a model that includes a temporal crossover of the oxide growth rate from transient to steady-state growth and a spatial crossover from predominantly vertical to coupled lateral growth. Unfortunately, the model involves the solution of a system of coupled equations that can only be solved numerically. Furthermore, the confirmation of this model requires the use of rather long oxidation times (several seconds) which is hard to accomplish experimentally without introducing considerable experimental errors due, among other things, to the mechanical interaction of the microcantilever with the growing oxide.

Dubois and Bubenddorff [26] have proposed a different model to obtain a power law. They assumed an ionic diffusion process described by Cabrera and Mott,[49]

$$\frac{dh}{dt} = u_0 \exp\left(\frac{W}{kT}\right) \sinh\left(\frac{qa\varepsilon}{kT}\right) \tag{7}$$

where W is the energy barrier and q, $2a$ and ε are the charge, the separation between interstitial sites and the total electric field within the oxide respectively. Then they considered that nonstoichiometric states (space charge) close to the bulk/oxide interface strongly limit the oxidation rate. In addition, it was also assumed that the flux of oxyanions through the oxide layer is totally controlled by the region where the electric field is decreased by the presence of space charge. This and other considerations allow them to determine ε and finally the arrive at,

$$h = h_b(\frac{t}{t_0})^{1/(\delta+1)} \quad (8)$$

where δ is a dimensionless parameter characteristic of the material.

11.6. SUBSTRATES

Silicon and tantalum were the first materials to be oxidized by scanning probe microscopy. Initially the oxidation of silicon was demonstrated on a hydrogen terminated Si(111),[10] however, the method was readily extended to other H-passivated silicon surfaces such as Si(100) and Si(110). It was also found that the presence of thin oxide film on the surface did not prevented further local oxidation. This observation and the technological relevance of Si(100) surfaces the material of choice to perform local oxidation experiments. However, local oxidation lithography is not restricted to silicon surfaces. Compound semiconductor surfaces such as GaAs have also been patterned by STM[18] and AFM.[19,50]

Metallic surfaces that are anodically oxidized are also candidates to perform local oxidation nanolithography. Tantalum,[11] titanium,[20,51-52] niobium,[21] aluminum[53] and chromium[54] surfaces have been locally oxidized. Figure 7 shows a series of NbO_x dots. The dot size depends on the oxidation time, however, for identical electrical conditions a rather uniform size is obtained which illustrates the reproducibility of local oxidation to modify Nb surfaces.

Titanium nitride films were converted into titanium oxinitride layers by S. Gwo et al.[55] The kinetics of local oxidation of silicon nitride films was also studied by the

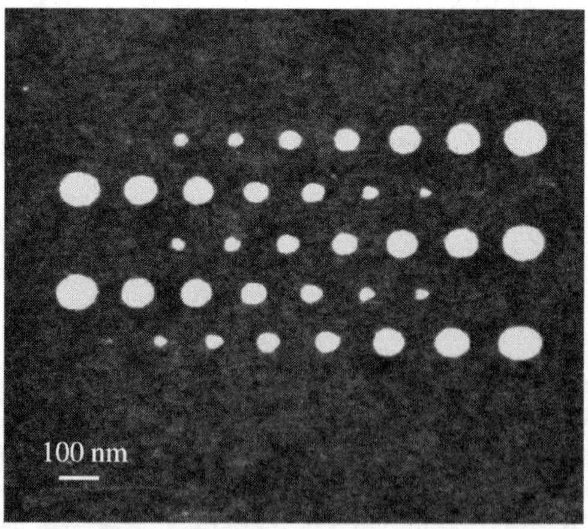

FIGURE 11.7. AFM image of a series of NbOx dots. The size depends on the oxidation time. Rows (top to bottom) 1,3 and 5 and left to right have been generated by applying pulses of 10-3, 3×10-3, 10-2, 0.1, 0.3, 1 and 3 s respectively. Rows 2 and 4 from 3 to 10-3 s.

same group.[56] Recently, Maoz et al. have provided an example of the versatility of local oxidation lithography to modify self assembled monolayers by demonstrating the oxidation of monolayers of methyl terminated silane on silicon surfaces.[23] Other materials such as $YBa_2Cu_3O_{7-y}$ have also been locally oxidized.[57] The variety of materials suitable for local oxidation experiments underlines the general character of this method.

11.7. RESOLUTION

Minimum feature size is a pivotal figure of merit to classify lithographic techniques. Several sub-10 nm motives have already been claimed in local oxidation experiments. García et al. have determined an average linewidth of 10 nm for an array of 4864 dots with a 40 nm spacing on a silicon surface.[15] Gotoh have used single-walled carbon nanotube tips to fabricate TiO_x lines.[58] AFM cross-sectional images of parallel lines give a full width at half maximum (FWHM) of 7 nm. Carbon nanotubes have also been used by Cooper et al. to produce arrays of dots on a titanium surface.[59] Arrays of 25 dots with a 20 nm periodicity were generated. The above results illustrate some sub-10 nm potential, however, in many cases the measurements have been obtained with little consideration to the influence of the tip on the measurements. The nonlinear character of the imaging process in AFM, i.e. different regions of the tip interacting simultaneously with the surface, makes hard to obtain faithful linewidth measurements. In general, an AFM image is a combination of tip and object shapes. That explains the extended use of FWHM measurements as a means to give an estimation of the linewidth. The question is to what extend tip-object convolution effects influence width measurements in local oxidation.

To answer this question we have fabricated silicon oxide structures in the proximity of an object of well defined geometry sexithiophene monolayer islands (T6) to calibrate both tip geometry and object width.[60] Two main conclusions were deduced from this work. (i) The side slope of silicon oxide structures is significantly smaller that the slope of the T6 island, 0.09 vs. 0.23. (ii) Side slopes measured over a four year period show an average value of about 0.1 with a standard deviation of 0.05. Those values seem to be independent of the tip shape and imply shallow angles with the horizontal. As a consequence, *unreconstructed* AFM images give an accurate picture of silicon oxides. Furthermore, it was deduced a minimum feature size of 14 nm at the base for a silicon oxide protruding 1 nm from the surface. However, the above limit only applies for p-type silicon surfaces due to the dependence of the growth rate on the material.

Figure 8 shows an AFM image of 18 interdigitated SiO_x lines. The spatial spectrum of the cross-section shows a dominant peak at 19.5 nm, which demonstrates the reliability of local oxidation to pattern sub 20 nm motives. Higher density packings have also been demonstrated by fabricating patterns with a separation of 13 nm. The shallow angles of the oxide makes possible to pack structures with a pitch smaller than the baseline width of a single structure.

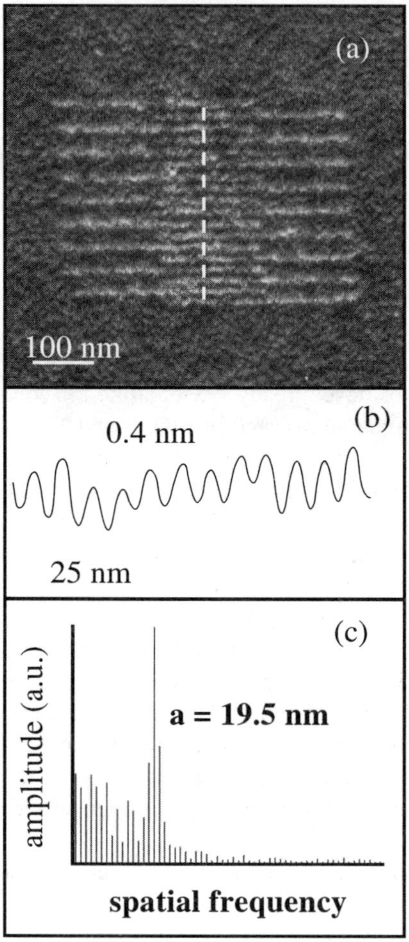

FIGURE 11.8. (a) AFM image of a set of 18 interdigitated lines obtained by nc-AFM oxidation. (b) Cross-section along the dashed line in (a). (c) The Fourier transform of (b) reveals a periodicity of 19.5 nm

11.8. APPLICATIONS

Local oxidation nanolithography has been applied to fabricate several electronic and mechanical devices with nanometer-scale features such as data storage memories, single electron transistors, superconducting interference devices or conducting wires. In many cases local oxidation nanolithography is used in combination of other methods such as photolithography or chemical wet etching. Usually, the critical or most relevant features of the device are fabricated by local oxidation. The masks and devices described in this section should not be considered representative of the many devices fabricated by local oxidation, however, they illustrate the flexibility and versatility of local oxidation for nanometer-scale science studies.

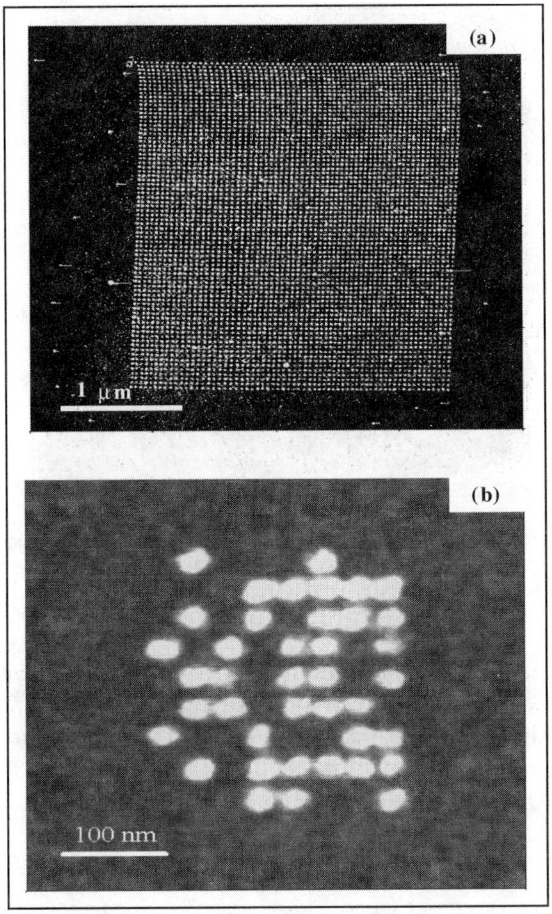

FIGURE 11.9. (a) AFM image of an array of 4864 silicon oxide dots. The dots are 40 nm apart and with an average width of about 20 nm. The same tip has been used to grow the dots and to image them afterwards. (b) AFM image of □ (3.14159265358979323846) written by local oxidation. Each oxide dot is a '1' in binary code. Dot size is 3 and 30 nm in height and width, respectively.

11.8.1. Memories

It is relatively straightforward to pattern arrays of dots by local oxidation. Figure 9(a) shows an array of 4864 SiO_x dots fabricated by nc-AFM.[15] The dots are 40 nm apart and 25 nm in width. The image illustrates the reproducibility and reliability achieved by local oxidation to pattern thousands of structures in a single experiment. The straightforward process to write an oxide dot (bit) by local oxidation has suggested its application for data storage.[15,59] Figure 9(b) shows an AFM image of π (3.14159265358979323846). The number is written in binary code written on a silicon surface. Each dot (SiO_x) represents a '1' while the absence of a dot is identified as '0'. The dots are 3 nm and 30 nm in height and width respectively. The lateral spacing is 35 nm, which means an areal density of 0.1 Tbit/cm^2 (0.7 Tbit/in^2). Using carbon nanotubes as AFM tips, Cooper at al[59] have

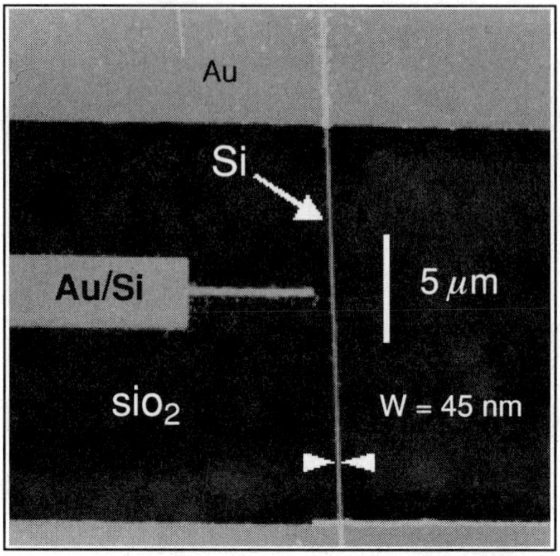

FIGURE 11.10. AFM image of side-gated field effect transistor. The silicon wire bridging the gold electrodes has been fabricated by local oxidation and selective chemical etching. The gate has been fabricated by electron beam lithography.

demonstrated an areal density of 1.6 Tbits/in^2 in the patterning of dots. Although areal densities of magnetic hard disks are one to two orders of magnitude smaller, the written rate ($\sim$5 Kbit/s) and the permanent character of oxide dots (bits) limit the applications for data storage.

11.8.2. Pattern transfer

The capability of a pattern to serve either as a mask or a template is a key feature for a successful lithography. Several examples illustrate the versatility of local oxidation to fabricate a variety of ultra thin oxide masks and templates. Campbell et al. pioneered an approach to fabricate silicon wires by local oxidation AFM that combined masking and chemical etching.[61] An example of this method is illustrated in the fabrication of a side-gate field effect transistor on a silicon on insulator substrate (Figure10).[62] The gate and electrodes have been fabricated by electron beam and optical lithographies respectively while the silicon wire is the result of a two step process. First a silicon oxide line is written between the electrodes, then by using KOH the unexposed silicon is etched away leaving only a silicon wire underneath the SiO_x line.

Nano-machining provides another example of pattern transfer. Machining is understood as a process for generating three-dimensional structures. By combining local oxidation AFM and anisotropic chemical etching Chien et al.[63] have fabricated a variety of silicon nanostructures such as gratings with high aspect ratios (400nm/55nm) (Figure 11) and two-dimensional array of hexagonal pits on a (110)-oriented Si surface (Figure 12).

Selective-area epitaxial growth of Si has been achieved by using an ultrathin mask fabricated by direct writing of nanoscale patterns on a Si_3N_4 layer. Windows for selective epitaxial growth are defined by wet etching of the locally oxidized regions.[64] Similarly oxide masks have been used to pattern platinum silicide PtSi regions.[65]

FIGURE 11.11. SEM image of a Si grating fabricated by local oxidation and wet etching in a KOH solution The vertical wall height is about 300 nm. Image courtesy of S.Gwo.

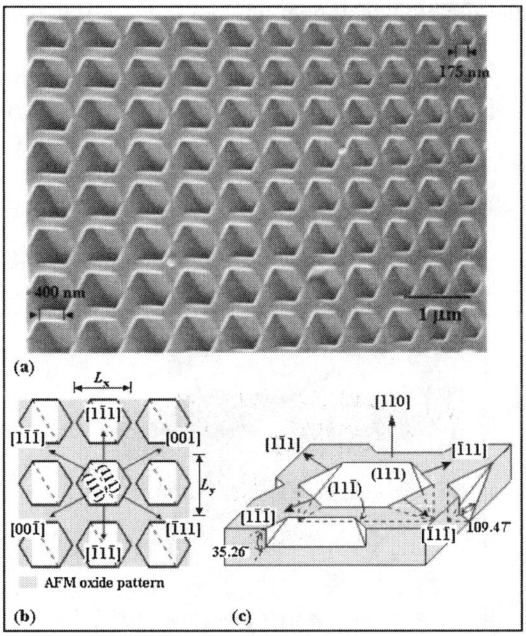

FIGURE 11.12. SEM image of a two-dimensional array of hexagonal pits with varied sizes on the (110)-oriented Si surface. (b) Illustration to indicate crystallographic directions of edges and bottom faces of hexagonal pits. The shadowed area is the oxide grid patterned by AFM nano-oxidation before pattern transfer. (c) Stereoscopic view of a hexagonal pit. Image courtesy of S.Gwo.

11.8.3. Transistors

Single electron transistors are tree terminal devices that are sensitive to the movement of a single electron.[66] The electrical characteristics are dominated by the Coulomb blockade effect. Coulomb blockade is observed in metallic clusters when the charging energy of the island is larger than $k_B T$. To observe Coulomb blockade effects at temperatures close to room temperature requires metallic islands the characteristics dimensions below 10 nm. Matsumoto and co-workers[67] have developed a method to fabricate Nb/Nb oxide-based single electron transistor with a side gate structure. The process includes optical and reactive ion etching process to pattern a 3 nm thick sputtered Nb film on a SiO_2/Si substrate. Local oxidation defines the tunnel barriers and the insulating film that separates the island from the gate electrode. Single electron charging effects were observed at 100 K. By using Ti instead of Nb as the conducting film Coulomb blockade characteristics were observed at room temperature.[33,68]

Snow et al.[34] have explored the potential of a metal-oxide-metal tunnel junction as a key component in the fabrication of transistors. They argue that metal-oxide junctions do not require epitaxial growth processes which in turns would ease the overall fabrication process. A schematic view of the device is show in Figure 12. Local oxidation AFM was applied to define the NbO_x tunnel barrier. Figure 12b shows a 0.2 μm × 0.2 μm AFM image of the tunnel junction region. The transistor operates by using a gate field to modulate the transmission of electrons through a lateral/metal oxide barrier. The transistor has a 30 nm-wide lateral Nb/NbO_x tunnel junction on top of a planar Al_2O_3/Al buried gate.

Figure 13 shows the I-V characteristics of the device at 300 K. An effective modulation of the source-drain current is clearly produced by the gate bias with negligible gate leakage current.

Other type of transistors such as side-gated FET[61-62] and metal-oxide-semiconductor field effect transistors have also been fabricated.[69]

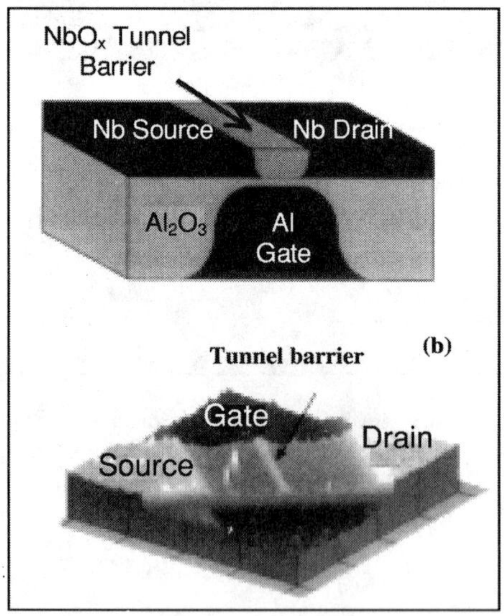

FIGURE 11.13. (a) Schematic of a Nb/NbOx tunnel junction fabricated on top a planar Al/Al2O3 gate. (b) A 2 μm× 2 μm AFM image of the tunnel junction region. The Al gate lies under the depressed area of the image. Image courtesy of E. Snow.

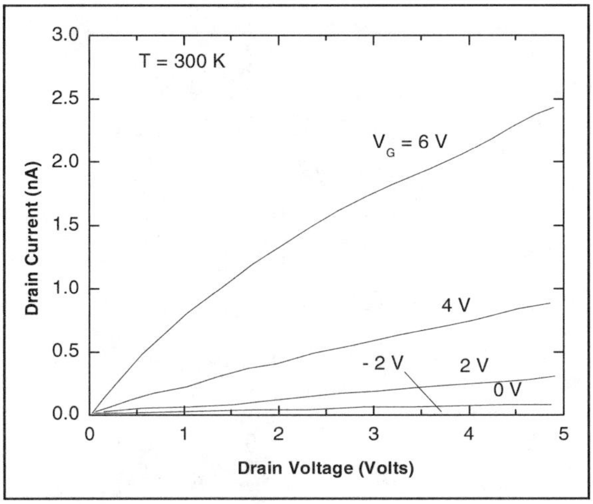

FIGURE 11.14. I-V characteristics of the transistor shown in Fig. 13 for various gate voltages. The gate leakage current was less than 1 pA.[34]

11.8.4. Mesoscopic devices

Bouchiat and co-workers have applied local oxidation to fabricate Josephson junctions and superconducting quantum interference devices SQUID.[70] Their process involves the deposition of a strip of a thin film of Nb on an insulating substrate (αAl_2O_3). Then the substrate is selectively oxidized by AFM to generate tunneling bridges and rings as the ones plotted in Figure 15.

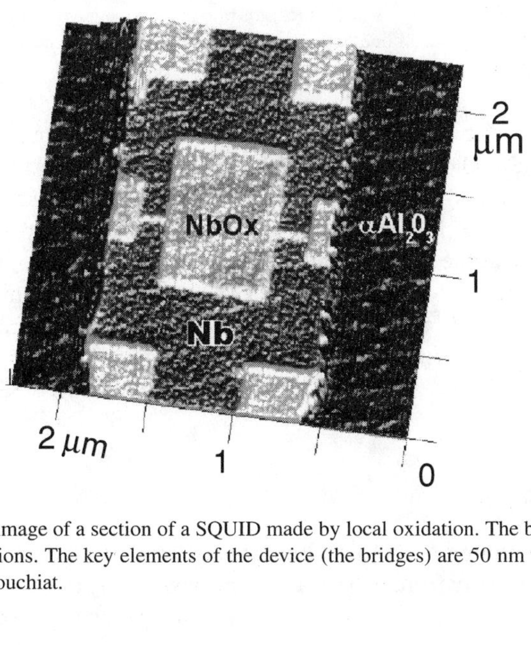

FIGURE 11.15. AFM image of a section of a SQUID made by local oxidation. The brightest regions are the patterned oxide protrusions. The key elements of the device (the bridges) are 50 nm wide and 400 nm long. Image courtesy of V. Bouchiat.

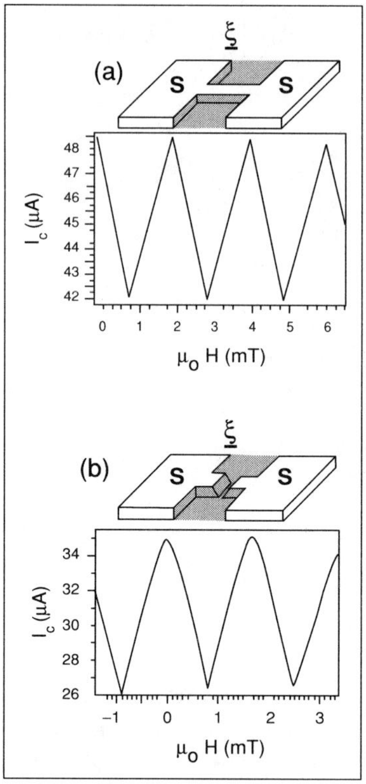

FIGURE 11.16. Dependence of the switching current (maximum value of the Josephson current) on the magnetic field for two kinds of microbridges. (a) For SQUIDs with 0.3 □m long Dayem-like bridges a perfect sawtooth modulation are obtained since the bridges have widths and lengths larger than the superconducting coherence length. (b) For SQUIDs with variable thickness bridges, the magnetic field dependence of Ic shows deviation from linear behaviour (dotted line) which is a signature of Josephson junction of dimension shorter than the superconducting coherence length.[69]

The bridges consist of superconducting lateral constrictions of between 30 and 100 nm in width and 200-1000 nm in length. The modulation of the switching current with the magnetic field (Figure 16(a)) exhibits a sharp symmetric sawtooth shape with a modulation depth of about 15% $(\Delta I_s/I_s)$. The periodicity is around 2 mT, a value that is in good agreement with the superconducting flux quantum of $h/2e$. On the other hand, for SQUIDs with bridges with variable thickness a deviation from linearity is found (Figure 16(b) which reveals a junction of dimensions shorter than the coherence length.

Other low dimensional quantum structures such as quantum dots,[71] antidot lattices[72] and quantum point contacts[73] in GaAs/AlGaAs haterostructures have also been fabricated by local oxidation.

11.8.5. Parallel writing

Presently all scanning probe-base lithographies are based on a sequential process, the tip is displaced laterally from one region to the next during the writing. Consequently, they are intrinsically slow for mass production of mechanical or electronic devices. To overcome this limitation several groups have proposed the parallel operation of arrays of cantilevers for imaging and surface modification[74,75] where each tip of the array operates independently to read or write individual pixels. The optimization of those proposals would have profound implications in the technological applications of local oxidation lithography because the local oxidation of surfaces is compatible with a parallel operation. Minne and co-workers have demonstrated that local oxidation of 1 cm^2 area by operating independently an array of a 50 cantilevers.[74]

SUMMARY

Over the past few years local oxidation nanolithography has emerged as a very flexible and versatile method to fabricate nanometer-scale devices. There a tip of an atomic force microscope is used as counter-electrode to anodically oxide a surface. In most experimental situations the method requires the formation of a liquid meniscus bridging tip and surface. Several factors have been identified to contribute to the oxide size such as the electrical conditions of the interface, meniscus size and space charge buildup within the growing oxide. Three properties give local oxidation nanolithography a special character for the fabrication of nanometer-scale devices such as *in situ* characterization of the device, combination with pre-existing microelectronics methods and feature sizes in the sub-20 nm range.

Although some fundamental aspects of local oxidation are still not fully understood, the variety of semiconductor and metallic surfaces amenable for local oxidation experiments, the fabrication of diverse systems ranging from high density data storage memories to a variety of quantum devices and the extended use of atomic force microscopes will confer local oxidation a special status for investigating properties of nanometer-scale devices in the near future. On the other hand, technological applications for mass production of nanometer-size devices are not a far fetched possibility once routine operation of arrays of cantilever will become a reality.

ACKNOWLEDGMENTS

It is a pleasure to acknowledge many discussions with colleagues, among them Prof. Fernando Briones, Dr. John Dagata, Prof. Nicolás García, Dr. Francesc Pérez-Murano, Prof. Calvin Quate, Dr. Heinrich Rohrer, Dr. Juan José Sáenz and Dr. Eric Snow, as well as with close collaborators Montserrat Calleja, Dr. Fernando García and Marta Tello. Financial support has been provided by the European Commission (MONA-LISA, GRDI-2000-25592) and the Dirección General de Investigación Científica y Técnica of Spain (PB98-0471).

REFERENCES

1. G. Binnig and H. Rohrer, Rev. Mod. Phys. 71, S324 (1999).
2. D.W. Abraham, H.J. Mamin, E. Ganz and J. Clarke, IBM J. Res. Develop. 30, 492 (1986).
3. I.-W. Lyo and P. Avouris, Science 253, 173 (1991)).

4. D.M. Eigler and E.K. Schweizer, Nature 344, 524 (1990).
5. R. García, Appl. Phys. Lett. 64, 1162 (1994).
6. C.R.K. Marrian and E.S. Snow, Microel. Enginneering 32, 173 (1996).
7. R.M. Nyffenegger and R.M. Penner, Chem. Rev. 97, 1195 (1997).
8. R. Piner, J. Zhu, F. Xu, S. Hong, and C. Mirkin, Science 283, 661 (1998).
9. G. Binnig, M. Despont, U. Drechsler, W. Häberle, M. Lutwyche, P. Vettiger, H.J. Mamin, B.W. Chui, and T.W. Kenny, Appl. Phys. Lett. 74, 1329 (1999).
10. J.A. Dagata, J. Schneir, H.H. Harary, C.J. Evans, M.T. Postek, and J. Bennet, Appl. Phys. Lett. 56, 2001 (1990).
11. T. Thundat, L.A. Nagahara, P.I. Oden, S.M. Lindsay, M.A. George, and W.S. Glaunsinger, J. Vac. Sci. Technol. A 8, 3527 (1990).
12. H.C. Day and D.R. Alle, Appl. Phys. Lett. 62, 2691 (1993).
13. F. Pérez-Murano, G. Abadal, N. Barniol, X. Aymerich, J. Servat, P. Gorostiza, and F. Sanz, , J. Appl. Phys. 78, 6797 (1995).
14. R. García, M. Calleja and F. Pérez-Murano, Appl. Phys. Lett. 72, 2295 (1998).
15. R. García, M. Calleja and H. Rohrer, J. Appl. Phys. 86, 1898 (1999).
16. Y.-Y. Wei and G. Eres, Appl. Phys. Lett. 76, 194 (2000).
17. J.W. Lyding, T.C. Shen, J.S. Hubacek, J.R. Tucker, and C. Abeln, Appl. Phys. Lett. 64, 2010 (1994).
18. J.A. Dagata, W. Tseng, J. Bennet, J. Schneir, H.H. Harary, J. Appl. Phys. 70, 3661 (1991).
19. Y. Osaka, Y. Iuchi, M. Kawabe, J. Harris, J. Appl. Phys. 88, 1136 (2000).
20. H. Sugimura, T. Uchida, N. Kitamura, H. Masuhara, Appl. Phys. Lett. 63, 1288 (1993).
21. J. Shirakashi, M. Ishii, K. Matsumoto, N. Miura, M. Konagai, Jpn. J. Appl. Phys. 35, L1524 (1996).
22. F.S.-S. Chien, Y.C. Chou, T.T. Chen, W.-F. Hsieh, T.S. Chao and S. Gwo J. Appl. Phys. 89, 2465 (2001).
23. R. Maoz, E. Frydman, S. R. Cohen, and J. Sagiv, Adv. Mater. 12, 725 (2000).
24. Avramescu, A. Ueta, K. Uesugi, and I. Suemune, J. Appl. Phys. 88, 3158 (2000).
25. P. Avouris, T. Hertel, R. Martel, Appl. Phys. Lett. 71, 287 (1997).
26. E. Dubois and Jean-Luc Bubendorff, J. Appl. Phys. 87, 8148 (2000).
27. M. Calleja and R. García, Appl. Phys. Lett. 76, 3427 (2000).
28. J.A. Dagata, T. Inoue, J. Itoh, K. Matsumoto, H. Yokoyama, J. Appl. Phys. 84, 6891 (1998).
29. T. Teuschler, K. Mahr, S. Miyazaki, M. Hundhausen, L. Levy, Appl. Phys. Lett. 67, 3144 (1995).
30. K. Morimoto, F. Pérez-Murano, and J.A. Dagata, Appl. Surf. Sci. 158, 205 (2000).
31. H. Sugimura and N. Nakagiri, Jpn. J. Appl. Phys. 34, 3406 (1995).
32. F. Marchi, V. Bouchiat, H. Dallaporta, V. Safarov, D. Tonneau and P. Doppelt, J. Vac. Sci. Technol. B 16, 1 (1998).
33. K. Matsumoto, Y. Gotoh, T. Maeda, J. A. Dagata, J. S. Harris, Appl. Phys. Lett. 76, 239 (2000).
34. E.S. Snow, P.M. Campbell, F.A. Buot, D. Park, C.R.K. Marrian, R. Magno 72, Appl. Phys. Lett. 3071 (1998).
35. F.S. Chien, C.-L. Wu, Y.-C. Chou, T.T. Chen, S. Gwo, W.-F. Hsieh, Appl. Phys. Lett. 75, 2429 (1999).
36. R. García and A. San Paulo, Phys. Rev. B 60, 4961 (1999).
37. R. García and A. San Paulo, Phys. Rev. B 61, R13381 (2000).
38. J.N. Israelachvili, Intermolecular and surface forces. Academic Press, London, 1992.
39. M. Tello and R. García, Appl. Phys. Lett. 79, 424 (2001).
40. K. Morimoto, K. Araki, K. Yamashita, K. Morita and M. Niwa, Appl. Surf. Sci. 117, 652 (1997).
41. J.A. Dagata, F. Pérez-Murano, G. Abadal, K. Morimoto, T. Inoue, J. Itoh, and H. Yokoyama, Appl. Phys. Lett. 76, 2710 (2000).
42. F. Pérez-Murano, K. Birkelund, G. Abadal, K. Morimoto, and J. Dagata, Appl. Phys. Lett. 75, 199 (1999).
43. D. Stiévenard, P.A. Fontaine, and E. Dubois, Appl. Phys. Lett. 70, 3272 (1997).
44. Legrand and D. Stievenard, Appl. Phys. Lett. 76, 1018 (2000).
45. Y.R. Ma, C. Yu, Y.-D. Yao, Y. Liou, and S.-F. Lee, Phys. Rev. B 64, 195324 (2001).
46. H. Sugimura, T. Uchida, N. Kitamura and H. Masuahara, J. Vac. Sci. Technol. B12, 2884 (1994).
47. M. Lazzarino, S. Jun, B. Ressel, K.C. Prince, P. Pingue, and C. Ascoli, Appl. Phys. Lett. 81, 2842 (2002).
48. H.H. Uhlig, Acta Metall. 4, 541 (1956).
49. N. Cabrera and N.F. Mott, Rep. Prog. Phys. 12, 163 (1948).
50. T. Heinzel, R. held, S. Lüscher, K. Esslin, W. Wegscheider and M. Bichler, Physica E 9, 84 (2001).
51. K. Masumoto, M. Ishii, K. Segawa, and Y. Oka, Appl. Phys. Lett. 68, 34 (1996).

52. Irmer, M. Kehre, H. Lorenz and J.P. Kottaus, Appl. Phys. Lett. 71, 1733 (1997).
53. E.S. Snow, D. Park, and P.M. Campbell, Appl. Phys. Lett. 69, 269 (1996).
54. Wang, L. Tsau, K.L. Wang, and P. Chow, Appl. Phys. Lett. 67, 1295 (1995).
55. S. Gwo, C.L. Yeh, P.-F. Chen, Y.C. Chou, T.T. Chen, T.-S. Chao, S.F. Hu, and T.Y. Huang, Appl. Phys. Lett. 74, 1090 (1999).
56. F.S.S. Chien, Y.C. Chou, T.T. Chen, W.-F. Hsieh, T.-S. Chao, and S. Gwo, J. Appl. Phys. 89, 2465 (2001).
57. Song, B.M. Kim and G. Park, Appl. Phys. Lett. 76, 601 (2000).
58. Y. Gotoh, K. Masumoto, T. Maeda, E.B. Cooper, S.R. Manalis, H. Fang, S.C. Minne, T. Hunt, H. Dai, J. Harris and C.F. Quate, J. Vac. Sci. Technol. A 18, 1321 (2000).
59. E.B. Cooper, S.R. Manalis, H. Fang, H. Dai, K. Matsumoto, S.C. Minne, T. Hunt, and C.F. Quate, Appl. Phys. Lett. 75, 3566 (1999).
60. M. Tello, F. García and R. García, J. Appl. Phys. 92, 4075 (2002).
61. P.M. Campbell, E.S. Snow, and P.J. McMarr Solid-State Electronics 37, 583 (1994).
62. M. Calleja, J. Anguita, F. García and R. García (unpublished); M. Calleja, Doctoral Thesis, Universidad de Santiago de Compostela, April 2002 (Spain).
63. F.S. Chien, C.L. Wu, Y.-C. Chou, T.T. Chen, and S. Gwo, and W.F. Hsieh, Appl. Phys. Lett. 75, 2429 (1999).
64. T. Yasuda, S. Yamasaki and S. Gwo, Appl. Phys. Lett. 77, 3917 (2000).
65. Snow, P. M. Campbell, M. Twigg, and F.K. Perkins, Appl. Phys. Lett. 79, 1109 (2001).
66. L. Kouwenhoven and P. McEuen in Nanotechnology, G. Timp Ed., Springer-Verlag, New York 1999.
67. J. Shirikasi, K. Matsumoto, N. Miura, and M. Konagai, Jpn. J. Appl. Phys. 36, L1257 (1997).
68. K. Matsumoto, Y. Gotoh, T. Maeda, J. Dagata and J. Harris, Jpn. J. Appl. Phys. 38, 477 (1999).
69. S.C. Minne, J.D. Adams, G. Yaralioglu, S.R. Manalis, A. Atalar, C.F. Quate, Appl. Phys. Lett. 73, 1742 (1998).
70. V. Bouchiat, M. Faucher, C. Thirion, W. Wernsdorfer, T. Fournier and B. Pannetier, Appl. Phys. Lett. 79, 123 (2001).
71. U.F. Keyser, H.W. Schumacher, U. Zeitler, R.J. Haug and K. Eberl, Phys. Stat. Sol. 224, 681 (2001).
72. Dorn, M. Sigrist, A. Fuhrer, T. Ihn, T. Heinzel, K. Ensslin, W. Wegsscheider, and M. Bichler Appl. Phys. Lett. 80, 252 (2002).
73. R. Held, T. Vancura, T. Heinzel, K. Ensslin, M. Holland, and W. Wegscheider, Appl. Phys. Lett. 73, 262 (1998).
74. S.C. Minne, J.D. Adams, G. Yaraioglu, S.R. Manalis, A. Atalar and C.F. Quate, Appl. Phys. Lett. 73, 1742 (1998).
75. M. Lutwyche, C. Andreoli, G. Binnig, J. Brüger, U. Drechsler, W. Haeberle, H. Rohrer, H. Rothuzen, and P. Vetigger, Sensors and Actuators (1998).

12

Combined Approaches for Nanoelectronic Device Fabrication

Electron beam lithography and nanoimprint lithography

I. Martini, M. Kamp and A. Forchel
Lehrstuhl für Technische Physik, Bayerische Julius-Maximilians-Universität Würzburg, Germany

12.1. INTRODUCTION

Lithography plays a key roll in modern IC manufacturing industry. The increased performance of modern IC devices is strongly linked to an increase of lithographic resolution. For lithographic mask production and patterning of features down to a few nanometers, electron beam lithography (EBL) is a well established technique. Transferring the mask pattern into resist on an industrial high throughput level is mainly the domain of optical lithography. Using resolution enhancement techniques (RET), it possible to achieve dimensions of the order of 100 nm. According to the International Technology Roadmap for Semiconductors (ITRS) sub 100 nm patterning is a great demand for next generation lithography (NGL). Advanced technologies, such as extreme ultraviolet lithography (EUV), X-ray lithography (XRL), electron projection lithography (EPL), and ion projection lithography (IPL) are pushing towards into the domain of 35 nm.[1]

Nanoimprint lithography (NIL) has been proposed by Chou *et al.*[2] as an alternative process for pattern transfer down to the nanometer range. The method is based on embossing a mold into a thin polymer film. It has potentially high throughput due to parallel processing, and it is cost-efficient, since no sophisticated tools are required. Since then, various groups have concentrated on further developments of nanoimprint techniques.[3-8] For applications in device fabrication imprint lithography has to be integrated with conventional processing. Particularly issues like precise alignment and imprint on a prepatterned surface with significant height variations have to be addressed. First applications of NIL in electronic device fabrication have been reported by Chou *et al.*[9]

12.2. FABRICATION OF NANOELECTRONIC DEVICES

The mix-and-mach fabrication of nanoelectronic devices by imprint lithography was demonstrated by split-gate devices such as quantum point contacts (QPC) or quantum dots (QD). The patterns are defined by a combination of imprint lithography for the gate level with optical lithography for mesa and contact definition. A schematic diagram of the QPC is shown in Figure 1. The device is based on a modulation doped GaAs/AlGaAs heterostructure grown by molecular beam epitaxy (MBE). The two-dimensional electron gas (2DEG) with a mobility $\mu = 77$ m^2/Vs and a carrier concentration $n = 4\,4.06 \times 10^{11}$ cm^{-2} is located about 90 nm below the surface (see Figure 2). A negative gate voltage depletes the 2 DEG below the gates so that the carrier transport is limited to a small channel between the gate tips.

The QPC constitutes an ideal test device for the applicability of nanoimprint in nanoelectronic device fabrication. Although a QPC is a basic, well understood device, its fabrication involves challenges typically for device fabrication by imprint, e.g. alignment, uniform pressure, mold release.

The preparation of the basic transport structure consists of two major steps. First, the Hall bar is defined by wet etching with 0.5 NH_4OH + 0.5 H_2O_2+ 50 H_2O using photoresist patterned by optical lithography as etch mask. The second step is the deposition of the AuGe/ Ni/ Au for ohmic contacts. The metal of the contact layer is also used to form alignment marks for the imprint step. Prior to the metal evaporation, the sample is dipped into diluted HCl. After lift-off procedure, the contacts are annealed at 480 °C.

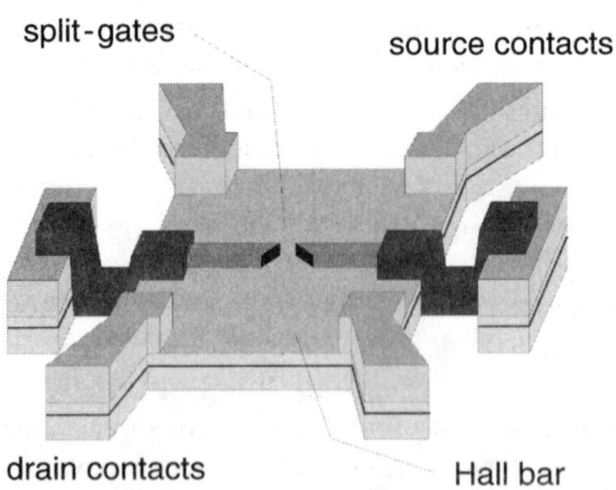

FIGURE 12.1. Schematic of the quantum point contact.

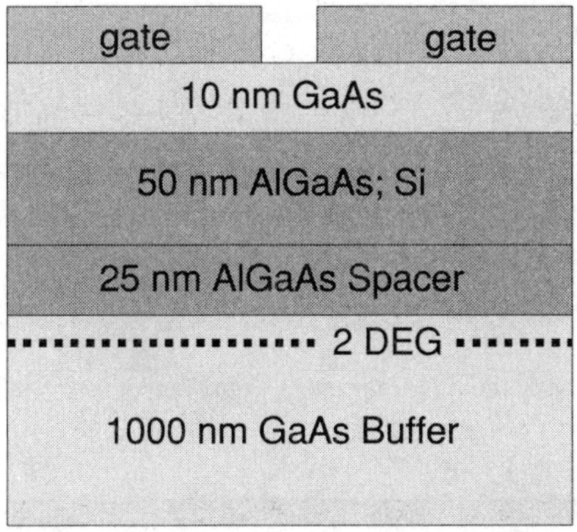

FIGURE 12.2. Cross-section of the GaAs/AlGaAs heterostructure.

12.3. MOLD FABRICATION

The molds are fabricated from a double side polished Si p<100> wafer with a thickness of 525 μm. Imprint lithography has been performed mainly on planar surfaces, but for integration in device fabrication, imprint on nonplanar surfaces is crucial. In our case, the contact pads on the mesa have a thickness of 400 nm which has to be compensated by a mesa on the mold. First, photoresist is spin-coated on the Si and the mesa is patterned with optical lithography. After Cr/Au evaporation and lift-off, the Si sample is dipped into HF 5% for 20s and then wet etched with KOH (30%)/ Isopropanol at 50 °C to a height corresponding to the thickness of the ohmic contacts. This mixture shows highly anisotropic etch rates in the Si <100> direction. Since the split-gates will be defined on this mesa a complete removal of the Cr/Au mask is essential for the subsequent process steps. Second, Cr alignment marks for the imprint step are fabricated by optical lithography. The third process step consists in the fabrication of the split-gate pattern. The stamp is spin-coated with PMMA 950K (100 nm thickness). The gate pattern is transferred by high resolution electron beam lithography followed by chromium evaporation (30 nm thick film). The chromium serves as etch mask for the subsequent ECR-RIE CCl_2F_2/Ar etching of the split-gates. Figure 3 shows an image of a split-gate mold.

A critical process step in imprint lithography is the mold release. If there is significant adhesion of the resist to the mold, high forces are needed for separation, which can lead to a partial destruction of the resist film. The best and most reproducible results yet were achieved by the technique of surface passivation using alcylsilanes. Organic silane compounds like alkyltrichlorosilanes are known to form very stable monolayers on silicon surfaces by formation of Si-O-Si bonds. The properties of a surface treated with these compounds will depend on the organic side-chain which can be tailored to the desired functionality. For our purposes we have chosen octadecyltrichlorosilane (OTS) $CH_3(CH_2)_{17}SiCl_3$ which will result in a hydrophobic surface because of its long alcylic chain. OTS is commercially available (e.g. by Sigma-Aldrich). It is applied simply by

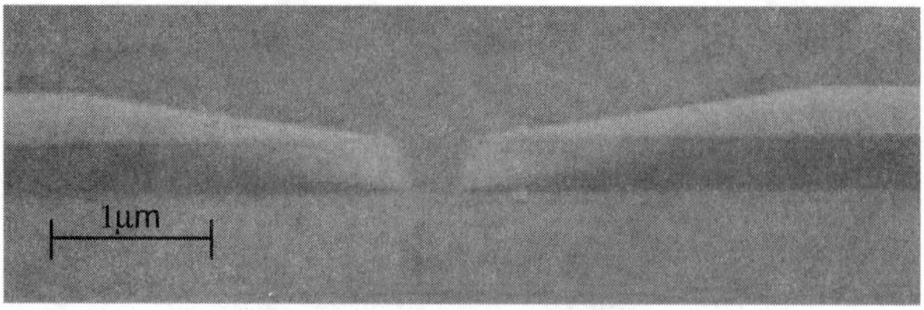

FIGURE 12.3. SEM image of a Si mold for imprinting a quantum point contact. The pattern height is 300 nm and the gate tip separation about 250 nm.

FIGURE 12.4. SEM image of a mold for a quantum dot.

immersion of the silicon mold in a 2% solution of OTS in toluene for about 5min, subsequently rinsing in toluene and blowing dry with nitrogen. Because water starts the crosslinking handling under nitrogen atmosphere is recommended, but working in a dry fume hood seems to be sufficient. No special treatment was performed on the silicon mold before immersion.

With the OTS layer adhesion of PMMA to the mold drastically reduced. Now in most cases the mold separates easily after imprint, while before large additional forces had to be applied. The anti-sticking treatment is stable up to a temperature of 170°C tried so far. One treatment can be used for several imprint cycles and it withstands cleaning in organic solvents like acetone or pyrrolidon.

To check the process stability lines and spaces patterns with a filling factor of about 1:1 were imprinted. Figure 5 shows a grating of lines and spaces with a period of 300 nm.

The structure was replicated from a mold with 100 nm deep etched grooves into 100 nm of PMMA 50K at a temperature of 160°C. Then 30nm of Chromium were evaporated and lift-off was performed. For better contrast the structure in the micrograph was etched about 50 nm deep. As can be seen the structure has been replicated in high detail. The same results could also be observed for patterns located at different positions of the mold.

FIGURE 12.5. A 1:1 lines and spaces structure with 300 nm period after lift-off.

12.4. ALIGNMENT

The metal marks of the contact level on the QPC sample and the Cr marks on the mold are aligned under backside IR-illumination in the optical mask aligner, as illustrated in Figure 6. Then, both sample and mold are transferred into a press and after pressure build up (about 100 bar) the press is heated up to 165 °C for 20 min. The split-gate pattern is embossed into the PMMA 50K (105 °C glass transition temperature). The lower molecular weight PMMA is chosen for its lower viscosity. The pressure is released after the press has reached room temperature. During the imprint process, the mold never touches the GaAs surface. The thin residual PMMA film is removed in a Ar/O_2 plasma.

The actual split-gates are fabricated by metal evaporation of Al or Ti/Au and lift-off. The last process step is contacting the split-gates to the contact pads with a 100 nm thick Al film. The complete process sequence is shown in Figure 7.

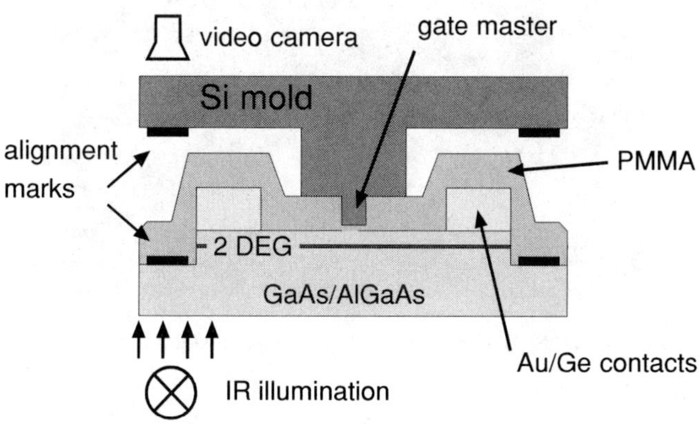

FIGURE 12.6. Schematic to illustrate the topology during alignment and imprint procedure.

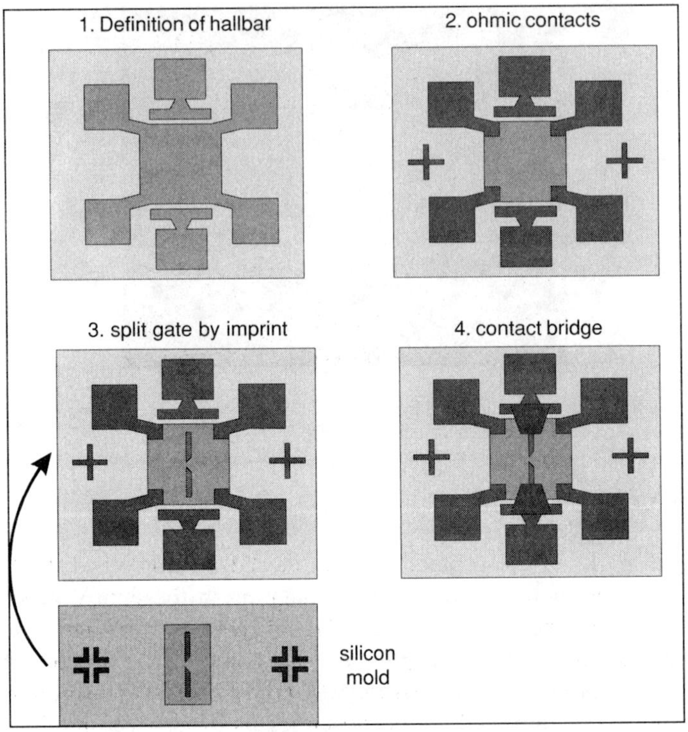

FIGURE 12.7. Process flow for the fabrication of quantum point contacts by imprint lithography.

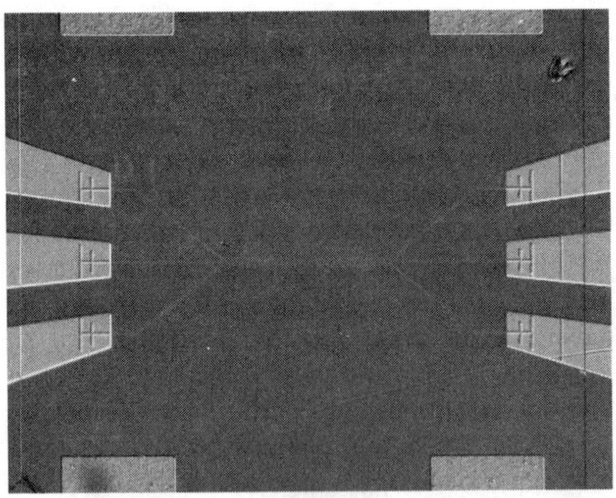

FIGURE 12.8. Optical micrograph of finished quantum dot device.

Figure 8 shows an optical micrograph of a finished quantum dot device. The thin wires are the Cr/Au gates defined by the imprint step. SEM micrographs of split gates are shown in Figure 8-10.

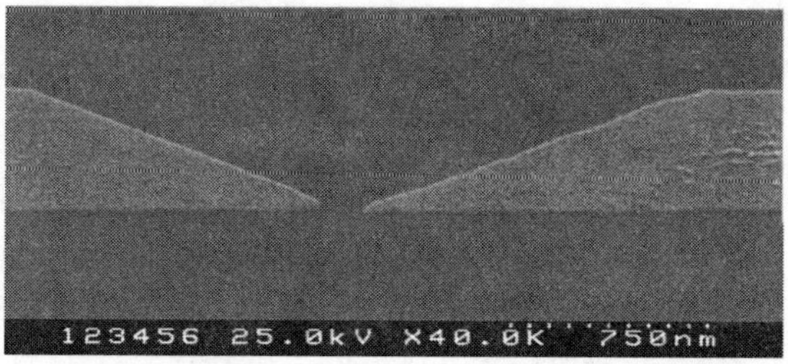

FIGURE 12.9. SEM micrograph of imprinted split gate after lift-off.

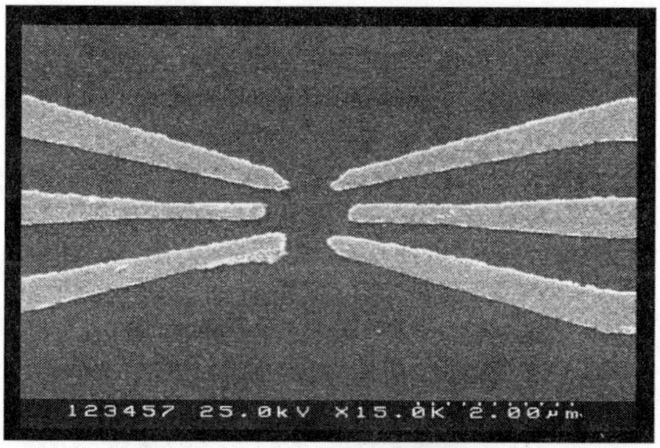

FIGURE 12.10. SEM mircograph of imprinted quantum dot after lift-off.

The alignment accuracy achieved with the process described above is around 1-2 μm, which is mainly determined by the mechanical tolerances of the optical contact aligner. Further and less important contributions to the alignment error are: Accuracy of mark detection, positioning errors of the sample stage, pattern distortion, etc. The pattern distortion is a critical issue in imprint lithography, because the high temperature and pressure during the imprint process can lead to a displacement of the pattern. This effect especially important when stamp and sample are made of different materials and have different thermal expansion coefficients.

We have tested the distortion of patterns fabricated by imprint lithography by aligning an E-beam written test pattern to a structure fabricated by imprint ligthography. For an easy assesment of the alignment acuracy under an optical microscop, a vernier pattern with a used. Figure 11 shows the schematic of the sample layout, which features alignment marks at a distance of 1 mm repeated in a square grid on the stamp. One scale of the vernier is placed in the center of each 1*1 mm square. The imprinted scale of the vernier has a period of 5 μm, the other uses a period of 5.04 μm. Alignment errors of 40 nm can easily detected under an optical microscope. The mold with one scale of the vernier lines (800 nm width, 5 μm period) was fabricated using our standard process for mold fabrication and etched to a depth of 350 nm.

The first part of the vernier scale was transferred by imprint into a 175 nm thick polymer layer on Si substrate, O_2 ashing of the residual polymer, metal evaporation (Cr)

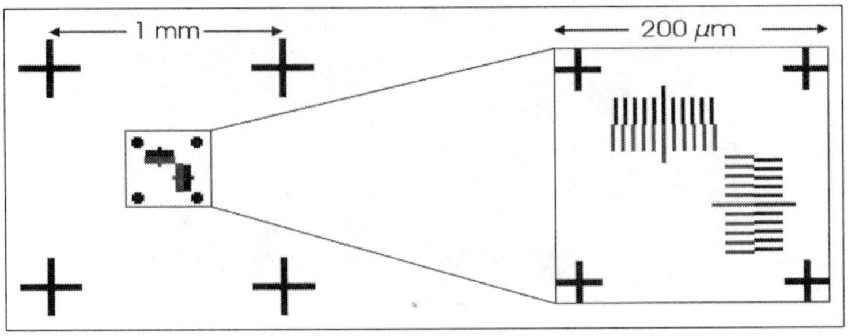

FIGURE 12.11. Layout of the aligment marks for the E-beam level (left). The enlarged section shows the Vernier pattern. The outer part of the vernier is patterned by imprint, the interior scale is exposed by electron beam lithography.

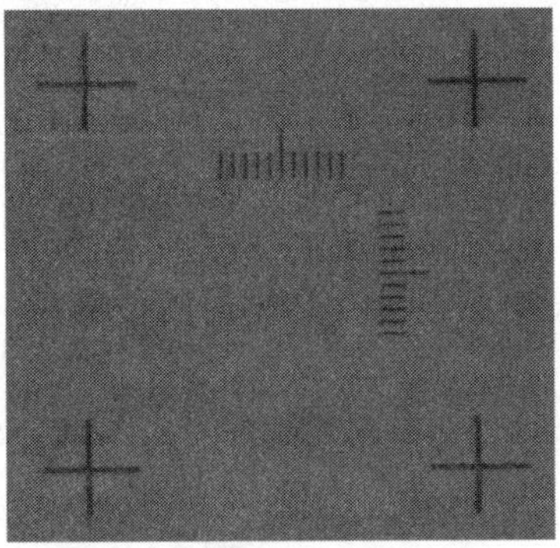

FIGURE 12.12. Optical micrograph of the mold with vernier pattern.

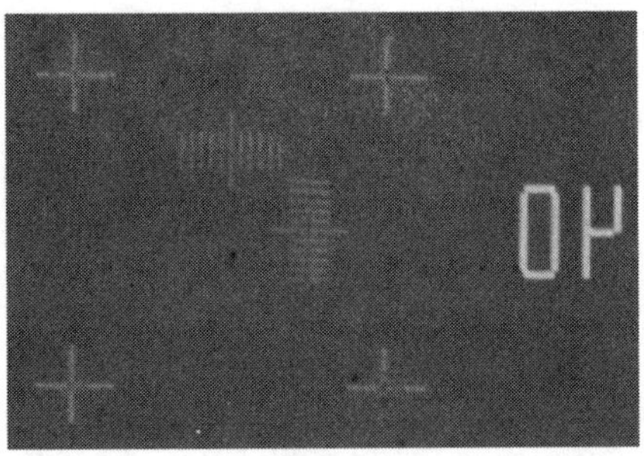

FIGURE 12.13. Optical image of vernier scales patterned by imprint (outer part) and electron beam lithography (interior part).

and lift-off. In order to improve the contrast of the marks, the imprinted features were dry etched in a RIE using CHF_3/Ar gas.

The second part of the vernier scale consists of the same line patterns than the imprinted ones, except that the lines have an additional offset (40 nm) starting from the center line. These lines are patterned by e-beam lithography, Cr/Au evaporation and lift-off. In Figure 13 an optical image with a horizontal and a vertical vernier scale is depicted.

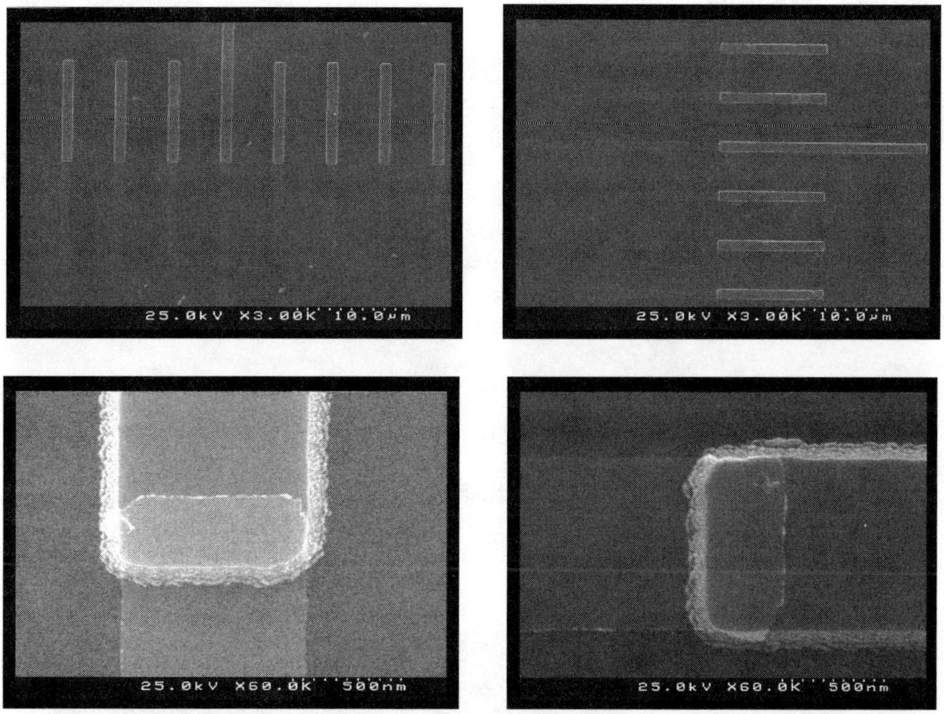

FIGURE 12.14. SEM images of a horizontal and the corresponding vertical vernier scale. The two picures below are zoomed on the center line where the e-beam patterned part matches with the imprinted part.

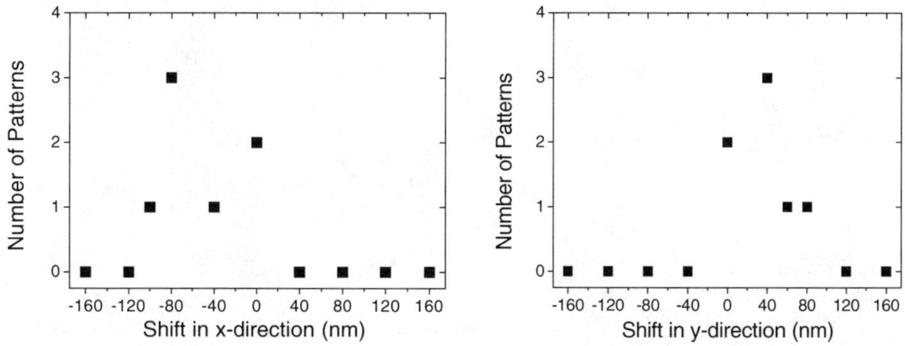

FIGURE 12.15. Alignment error of the of e-beam exposed patterns with respect to the imprinted patterns.

The alignment accuracy is tested by finding the matching line. Figure 14 shows SEM images of the horizontal and vertical vernier pattern.

The alignment error was read from a number of test patterns on the chip. Figure 15 shows the number of patterns versus the horizontal and vertical alignment error.

A systematic shift of around 40 nm is observed for both directions, the width of the distribution is around 80 nm.

The results indicate that the relative distances of patterns can be maintained during the imprint process. For a better overlay accuracy in mix and match applications, the most critical issue seems to be an improvement of the mechanical accuracy of the alignment procedure.

However, the design and fabrication of a tool which combines alignment and imprint is extremely challenging. Alignment accuracy better than 1 μm requires precise and sensitive optics and mechanics whereas the imprint step makes demands on the force and temperature stability of the equipment.

12.5. ALTERNATIVE POLYMERS

Besides using PMMA, imprint tests of quantum point contacts and quantum dots patterns were also performed in HYBRANE[10] polymer. The HYBRANE was spincoated on a Si substrate and baked for 2 min at 165° C on a hotplate. The thickness of the film ranges from 90 nm to 150 nm. We found that the imprint temperature has an important

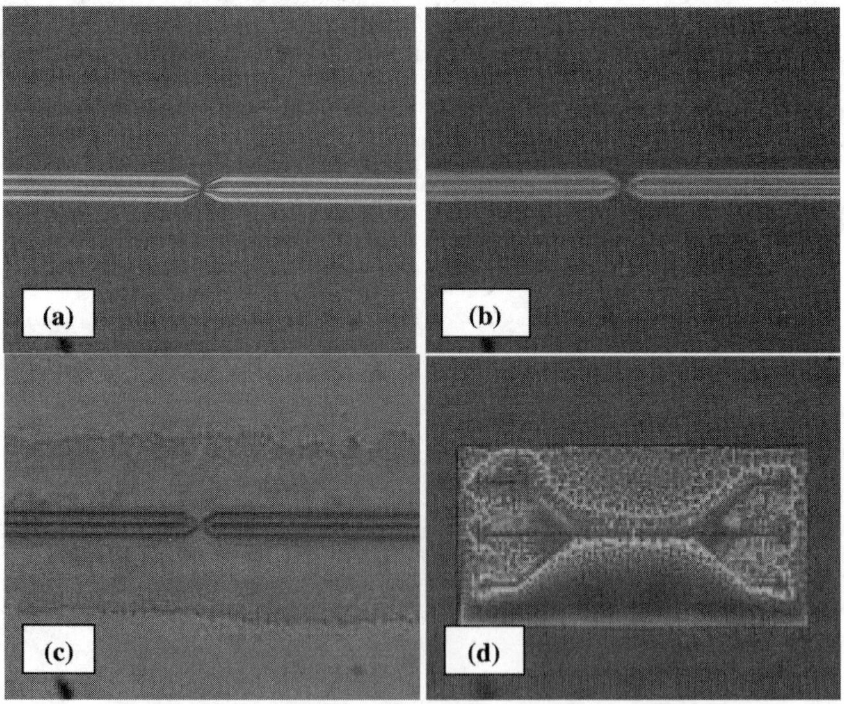

FIGURE 12.16. Optical micrographs of a quantum dot pattern after imprint in HYBRANE layer at different temperatures. a) 35 °C, b) 60 °C, c) 80 °C, d): 120 °C.

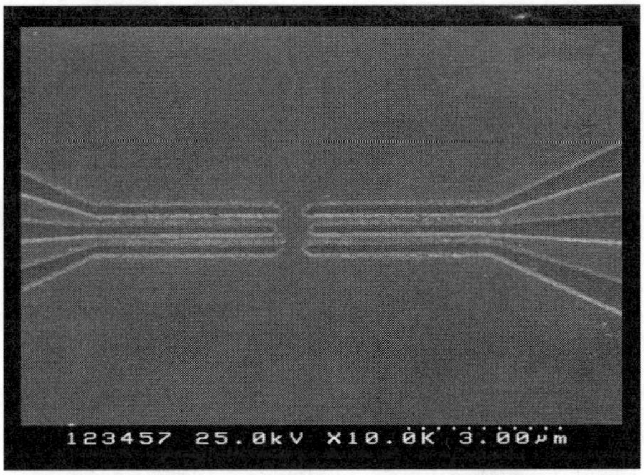

FIGURE 12.17. Imprinted Q-dot feature into HYBRANE film on Si (70 oC, 1.4 kN, release at RT).

influence on the pattern transfer. All tests were performed with a constant force of 1.2 kN and a stamp with the gate pattern of a quantum dot. Figure 16 shows optical micrographs of the pattern transferred into the HYBRANE film.

The imprint quality is good from RT up to 70°C. Above 70°C the formation of bubbles is observed. This may be due to residual solvent in the polymer film. The feasibility of a lift off process with a HYBRANE layer was also tested. HYBRANE is soluble in standard solvents, e.g. Toluene.

After mold release, the sample was ashed for 120 sec in a barrel reactor (65°C) in O_2 atmosphere (etch rate: 60 nm / min). A Ti/Au layer with a total thickness of 18 nm was evaporated. Figure 17 shows SEM micrograph of the pattern after metal deposition. Then lift off was performed in pyrrolidon followed by acetone rinse. The result is displayed in Figure 18.

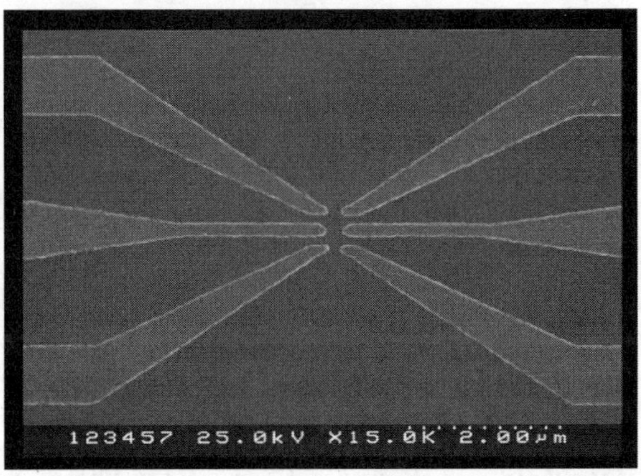

FIGURE 12.18. Different layout after Ti/Au evaporation and lift-off.

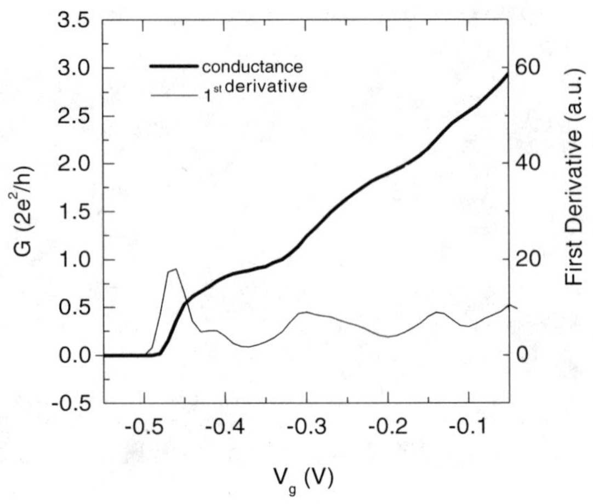

FIGURE 12.19. The conductance as a function of the gate voltage. The split-gate defined by imprint technology is a taper with a nominal tip separation of 280 nm.

12.6. CHARACTERIZATION

Transport measurements on the QPCs were performed at T = 4.2 K. As can be seen in Figure 13, the conductance G shows clearly observable modulation with varying gate voltage. Evaluating the derivative of the curve, plateaus at multiple values of $2e^2/h$ become visible, in agreement with standard literature.[11-12]

The conductance quantization in units of $2e^2/h$ in absence of a magnetic field can be understood in terms of ballistic electron transport between two electron reservoirs through N ($N = Int\,[k_F W/\pi]$) quasi-one-dimensional channels,[13]

$$G = \frac{2e^2}{h}N, \tag{1}$$

where k_F is the Fermi wave vector and W the width of the lateral confining potential. For a detailed theoretical description of the conductance quantization, especially the more general Landauer-Büttiker formalism, the reader is referred to van Houten *et al.*[13] This result demonstrates the integration of nanoimprint lithography with conventional processing for a basic nanoelectronic device structure.

Characterization of the quantum dot devices was performed in a He_3 cryo with at a temperature of 250 mK. A lock-in based measurement of the conductance with a small AC voltage of $U_s = 50\ \mu V$ (f = 13 Hz) was applied across the point contacts. The gate voltage of each was first swept from 0 to –2 V with the other gates grounded in order to confirm conductance quantization of the point contacts. Figure 20 shows a plot of the conductance of the three gate electrodes. Clear quantization of the conductance at multiples of $2e^2/h$ is observed for all three gates.

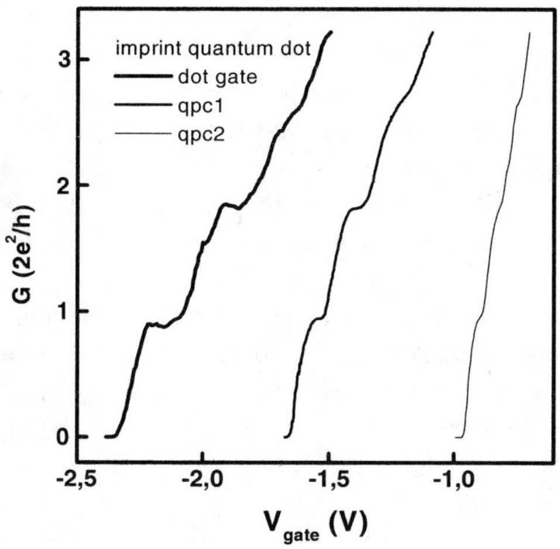

FIGURE 12.20. Conductance traces of an imprinted quantum dot for the three gates.

Figure 21 shows conductance traces for a variation of the dot gate voltage at four different bias conditions (0, -0.8 -1.0 and -1.1 V) of the QPC gates. The conductance decreases for increasing bias. For a voltage of - 1.1 V, the two QPCs operate as tunnel barriers and a dot is formed. The conductance now shows oscillations for a variation of the dot gate voltage, which are a clear indication of Coulomb blockade in the structure.

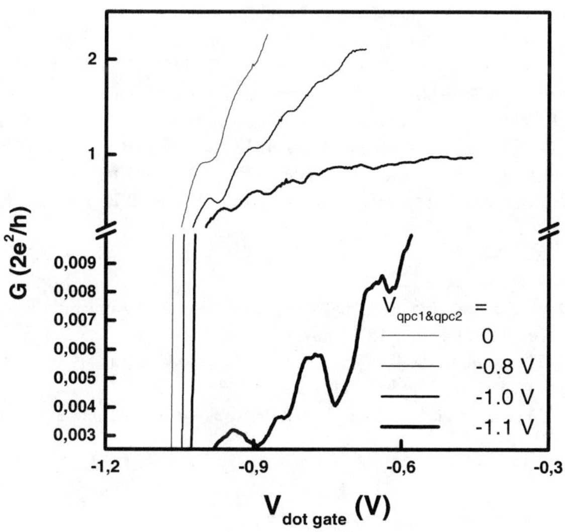

FIGURE 12.21. Indication of Coulomb blockade measured for varying dotgate voltage for QPC voltage of 1.1 V.

CONCLUSIONS

From today's point of view, optical lithography (including EUV) will remain the method of choice for high throughput patterning of structures down to the 100 nm regime. For patterning of structures down to the few nm range, both electron beam lithography and nanoimprint lithography are well suited. Therefore, the techniques are rather in competition than complement one another, whereby NIL requires EBL for the fabrication of molds.

Nanoelectronic devices, traditionally fabricated by a combination of optical lithography and EBL, can be fabricated by replacing EBL with NIL. Concerning the minimum feature size to be patterned by NIL, the limiting factor is the fabrication of the mold and therefore EBL. Compared to the costs of a average EBL system (5M USD), the equipment costs for NIL including alignment tools are of the order of 250k USD.

OUTLOOK

For laboratories with access to EBL, for the production of often varying patterns (in R&D environment) and optical masks, EBL will remain the preferred patterning method. NIL will be a suitable method for high throughput application for industrial processes, where high resolution at low costs is required, as e.g. in the biotechnology field.

REFERENCES

1. The International Technology Roadmap for Semiconductors http://www.itrs.net/1999_SIA_Roadmap/Home.htm.
2. S.Y. Chou, P.R. Krauss, P.J. Renstrom, Appl. Phys. Lett. **67** (21), 3114 (1995).
3. J. Haisma, M. Verheijen, K. van den Heuvel, J. Vac. Sci. Technol. B **14**(6), 4124 (1996).
4. H.-C. Scheer, H. Schulz, T. Hoffmann, C.M. Sotomayor Torres, J. Vac. Sci. Technol. B **16**(6), 3917 (1998).
5. R.W. Jaszewski, H. Schift, J. Gobrecht, P. Smith, Microelectron. Eng. **41/42** 575 (1998).
6. X. Sun, L. Zhung, W. Zhang, S.Y. Chou, J. Vac. Sci. Technol. B **16**(6), 3922 (1998).
7. D. Eisert, W. Braun, S. Kuhn, J. Koeth, A. Forchel, Microelectron. Eng. **46**, 179 (1999).
8. Lebib, Y. Chen, J. Bourneix, F. Carcenac, E. Cambril, L. Couraud, H. Launois, Microelectron. Eng. **46** 319 (1999).
9. S.Y. Chou, P.R. Krauss, W. Zhang, L. Guo, L. Zhuang, J. Vac. Sci. Technol. B. **15**(6), 2897 (1997).
10. DSM research, www.hybrane.com
11. B.J. van Wees, H. van Houten, C.W.J. Beenakker, J.G. Willian, L.P. Kouwenhoven, D. van der Marel, and C.T. Foxon, Phys. Rev. Lett. **60** (9), 848 (1988).
12. D.A. Wharam, T.J. Thornton, R. Newbury, M. Pepper, H. Ahmed, J.E.F. Forst, D.G. Hasko, D.C. Peacock, D.A. Ritchie, and G.A. Jones, J. Phys. C. **21**, L209 (1988).
13. H. van Houten, C.W.J. Beenakker and B.J. van Wees, Nanostructured Systems, Semiconductor and Semimetals **35**, 9 (1992).

13

Application of Nanoimprint Lithography in Magnetism

Y. Chen, M. Natali, S.P. Li and A. Lebib
Laboratoire de Photonique et Nanostructures, CNRS, Route de Nozay, 91460 Marcoussis, France

13.1. INTRODUCTION

Patterned magnetic structures are of particular importance for the development of new information technology. They can be used for ultra high-density recording,[1] for the fabrication of magnetic randomly accessible memories[2] as well as the emerging spin electronics such as spin transistors, resonant tunnel devices and non-volatile programmable logic.[3] Previously, high resolution magnetic structures were patterned using X-ray lithography [4], electron beam lithography,[5] interference lithography,[6] as well as ion irradiation induced mixing.[7,8] X-ray lithography is known to be an expensive technology which should not be considered for mass production of magnetic recording media. Electron beam lithography is limited by its writing speed for large area patterning.[9] While interference lithography is a simple technique, it is only suitable for the fabrication of regular pattern arrays. Finally, ion irradiation with a stencil mask provides an interesting alternative but still considerable efforts have to be made in order to show a performance of both high resolution and high throughput.

Comparing to the conventional techniques, nanoimprint lithography has shown a number of advantages.[10] It can be easily used to replicate high resolution patterns and it is a low cost technique with a high throughput manufacturing potentiality. For research purposes, nanoimprint lithography is quite flexible, versatile and robust which can be easily implemented in a laboratory environment.

In this chapter, we discuss the application of nanoimprint lithography for high resolution magnetic structure patterning. The outline of the chapter is as follows: In part 2 we review some of the fundamental concepts in micromagnetism and present the Landau Lifshitz equation. In part 3 we describe the process optimization of nanoimprint lithography for magnetic structure patterning. In part 4, we briefly describe some commonly used magnetic characterization methods and give a few examples of magnetic nanostructures fabricated by nanoimprint lithography.

13.2. PHYSICS OF PATTERNED MAGNETIC STRUCTURES

Most of patterned magnetic structures are made of ferromagnetic materials in which the macroscopic properties are described by the theories of Curie and Weiss.[11] Basically, the macroscopic behavior is described by averaging a very large number of microscopic entities. The magnetic configurations of these microscopic entities can be different from each other, but *atomic* magnetic moments inside each of them are coupled ferromagnetically.[12,13]

13.2.1. Micromagnetic theory

Microscopically, ferromagnetism of a material arises from the electrons spin magnetic moment, that according to the Pauli principle can align to minimize the electrostatic repulsion between the electrons. The energy associated with this spin alignment is the *exchange energy* and may be written phenomenologically as:

$$E_{ex} = -1/2 \sum_{i,j} m_{ij} \bar{M}_i \cdot \bar{M}_j \tag{1}$$

where $M_{i,j}$ denote either the magnetic moments of the electrons or in a finite element model, the magnetization density within cells i,j; m_{ij} being the exchange coupling constants. The integral form of equation (1) is:

$$E_{ex} = \int A_{ex}\,[(\nabla \cdot a_1)^2 + (\nabla \cdot a_2)^2 + (\nabla \cdot a_3)^2]\,dV \tag{2}$$

A_{ex} is the exchange coefficient and $a_i = \cos\alpha_i$ are the direction cosines of the magnetization in the Cartesian coordinate system.

In addition to the exchange energy, the *magnetostatic energy* is obtained by considering magnetic charges ρ and can be calculated as in electrostatics:

$$E_d = -\frac{1}{2} \sum_{i,j} \frac{\rho_i \cdot \rho_j}{r_{ij}} \tag{3}$$

where $\rho = -\nabla \cdot \bar{M}$ inside the body, $\rho = \bar{n} \cdot \bar{M}$ at the surface of the body. The demagnetization field H_d that is then given by:

$$H_d = -\frac{\partial E_d}{\partial M} = -\frac{1}{4\pi} \sum_i \frac{\rho_i \cdot \bar{r}_{ij}}{r_{ij}^3} \tag{4}$$

The third contribution comes from the *magnetocrystalline anisotropy* that determines easy and hard axes of magnetization. This term is originated from the spin-orbit coupling determines a connection between the electron spin orientation and the crystal structure. The magnetocrystalline energy for a crystal with cubic symmetry can be expressed in terms of the direction cosines of the magnetization with respect to the crystal axes:

$$E_c = K_1 V \left(a_2^2 a_3^2 + a_2^2 a_3^2 + a_2^2 a_3^2\right) + K_2 \left(a_2^2 a_3^2 a_2^2\right) + \cdots \tag{5}$$

where V is the crystal volume. In a crystal with uniaxial symmetry, indicating by θ the angle of M with respect to the symmetry axis:

$$E_c = K_1 \sin^2 \theta \tag{6}$$

Finally, if a magnetic field is applied all magnetic moments interact with the applied field, resulting in the *Zeeman energy*:

$$E_z = -\int \mu_0 \bar{M} \cdot \bar{H} \, dV \tag{7}$$

The time evolution of the magnetization is described by the Landau-Lifshitz equation:

$$\frac{dM}{dt} = -\gamma \, \bar{M} \times \bar{H}_{eff} - \frac{\gamma \, \alpha}{M_s} \bar{M} \times (\bar{M} \times \bar{H}_{eff}) \tag{8}$$

where γ is the electrons gyromagnetic ratio, α is a phenomenological damping parameter, and

$$\bar{H}_{eff} = -\frac{1}{\mu_0} \frac{\partial E_{tot}}{\partial \bar{M}} \tag{9}$$

where E_{tot} includes all above described energy terms:

$$E_{tot} = E_{ex} + E_d + E_c + E_{zee} \tag{10}$$

The first term on the right in equation (8) describes the torque on the magnetization that produce a precession of the magnetization around the effective magnetic field direction. The second term describes the dissipation or the origin of the slowing down of the precession.

13.2.2. Properties of low dimensional magnetic structures

The Landau-Lifshitz equation can be solved numerically by using a finite element approach once appropriate boundary conditions are specified. Several programs are available to perform this process both for 2D and 3D problems.[14] Using the Landau-Lifshitz equation the equilibrium magnetization configuration of a mesoscopic ferromagnetic element may be determined starting from some initial chosen configuration. The equilibrium state is assumed to be reached when the torque that determines the evolution of the magnetization falls below some chosen threshold value. To obtain meaningful simulations the size of the discretization cell in the calculation should be smaller than the exchange length, that is of the order of some nm. This clearly limits the size of the system that can be simulated with common available computing resources, typically to the submicron length scale. To obtain results that faithfully represent the real time scale of the dynamic phenomena the correct damping parameter must be chosen. Unfortunately this quantity, which depends strongly on material parameters, is known with sufficient accuracy only in a few cases.

Note that the exchange coupling is a short range interaction. The characteristic length scale of the exchange interaction is given by the exchange length

$$l_{ex} = \sqrt{A_{ex}/2\pi M^2}, \tag{11}$$

where M_s is the saturation magnetization density. In contrast, the magnetostatic interaction is long range. In a uniformly magnetized ellipsoid shaped body H_d can be described by the demagnetization tensor N,

$$H_d = -\mathbf{N} \cdot \bar{M}. \tag{12}$$

For a sphere $N_{xx} = N_{yy} = N_{zz} = 1/3$. In the case of a revolution ellipsoid with the longer axis along the z direction $N_{xx} = N_{yy} = N_\upsilon$ and $N_{zz} = N_\tau < N_\upsilon$. In other words, the demagnetization field and therefore the magnetostatic energy is smaller when the ellipsoid is magnetized along its long axis than in a direction perpendicular to it. Therefore, the alignment of the magnetization parallel to the long ellipsoid axis is energetically favorable. This leads to the concept of shape anisotropy, which introduces easy axes and hard axes of magnetization. Similar considerations are valid for other shapes. Demagnetization tensors N_{ij} can be found tabulated for different simple shaped bodies.[15,16]

As noted first by Weiss, a ferromagnetic material can have zero average magnetization while still having a non zero local magnetization. The reason for this is that typically in a "demagnetized" macroscopic ferromagnet there exist magnetic domains, i.e., regions that are magnetized in different directions. This situation in zero applied field for a large body is favorable since it minimizes the magnetostatic energy. However, as the size of the body decreases the energy cost associated with the domain walls that separate regions of different magnetization may become predominant and the body becomes a single-domain particle. A domain wall consists of a transition region of some finite width where the magnetization changes continuously from the direction of one domain to that of the adjacent domain. This variation in general corresponds to an exchange energy cost as well as to a magnetocrystalline and magnetostatic energy cost. Typical domain wall widths are of the order of ~10 nm. The size where the single-domain configuration (SD) becomes favored over a multi-domain state depends on the details of the competing multi-domain structure. For instance, for a 30 nm thick circular dot consisting of $Co_{0.5}Ni_{0.5}$/Pt multilayer, magnetized perpendicularly to the disk, a single-domain state is energetically favorable compared to a two-domain state for dot diameters below ~60 nm.[17]

Generally speaking, when a magnetic field is applied to a small ferromagnetic body complicated dynamical phenomena start to occur including domain nucleation and wall propagation, coherent rotation, buckling or curling processes. Various magnetisation configuration can occupy, in a remanent state of a magnet, such as multi or single domain state, leaf state, flower state and vortex state among others. The remanent configuration and reversal mechanism in a magnet are closely dominated by the patterned geometry, size, film quality, etc. The magnetization reversal is determined by either nucleation or domain wall propagation, depending on the quality of the film. Typical remanent states can be single domain, multi-domain, vortex, vortex with strong crystal anisotropy, etc. Both nucleation and domain wall propagation are thermally activated phenomena which strongly depend on the applied magnetic field. The nucleation probability P can have a gaussian distribution. The magnetic domains then expand by domain wall propagation at velocities strongly dependent on the applied field.

At micrometer length scale, the magnetization behavior is different. In particular, when the size of the object is comparable to a characteristic magnetic length scale, such as the exchange length (some nm), the domain wall width ($\sim$10 nm), the spin-diffusion length (10-200 nm) or the magnetic domain sizes (100 nm - 10 μm), many interesting phenomena are expected. For example, the magnetic anisotropy may result from shape anisotropy according to the pattern geometry. The magnetization reversal in dot arrays, for instance, is determined mainly by nucleation since domain wall propagation is prohibited between the dots. In general, the number, form and size of domains in a patterned structure are determined by the equilibrium of demagnetizing energy and the domain wall energy (including exchange energy and anisotropy energy). As the length scale of a small magnet is reduced to a critical size and thickness, the competition between exchange energy and magneto-static energy can result in a uniform magnetization distribution for which the entire magnet can be described by one magnetic moment.[18,19] Such single domain particles are basically required for the fabrication of magnetic memory chips, spin transistors and hard disk read heads.[1-3] Another important magnetic configuration is the flux-closure or vortex state which is also interesting because of the reduced magneto-static interaction between individual magnets.[20] To understand fully the magnetization reversal processes of both magnetization states, systematic studies have to be performed.

In Figure 1 we illustrate the shape anisotropy confined magnetization configurations for some magnets with simple geometry, e.g. pillar, disk and ring. Obviously, in a pillar

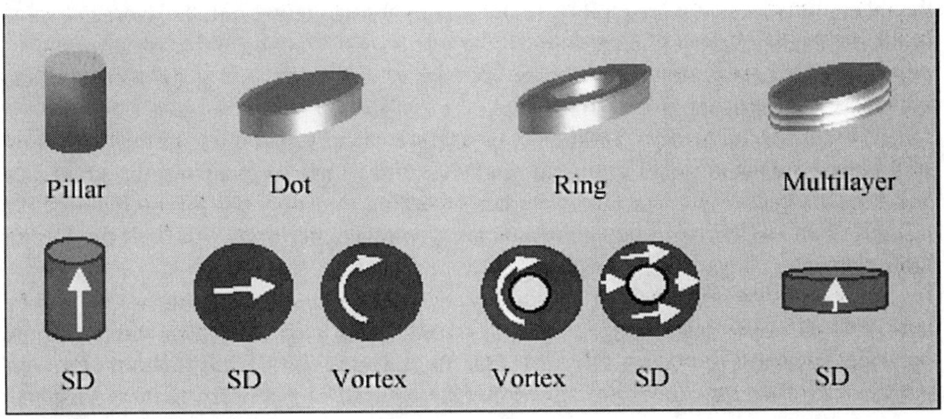

FIGURE 13.1. Stable magnetic configurations in some simple magnetic microstructures.

magnet the magnetic moments will tend to align in the direction of pillar axis to maintain minimum demagnetic energy. For a disk or ring-shaped magnet the minimization of exchange energy favors a uniform magnetized state (single domain) while minimization of magnetostatic energy required reduction of magnetic charge on the edges of magnets favors a flux-closure magnetization configuration (vortex). The dominated magnetization configuration in a small magnet depends on the result of competition between magnetostatic energy and exchange energy. For a submicron scale magnetic disk a smaller diameter is favorable for a single domain state, while larger diameter favors a vortex state. If the vortex core has been removed from the disk a dramatic decrease exchange energy causes the vortex state became always stable in ring magnets. However, a metastable single domain state may be trapped in large size ring magnets in magnetization reversal process. In Figure 1, we also include ultra thin multi-layer disks which present a magnetization perpendicular to the plane. Here, the perpendicular magnetization is basically caused by interface anisotropy between layers.

13.2.3. Super paramagnetic limit

For very small magnetic particles, the magnetization can become unstable because of thermal fluctuation. Let us consider a single domain magnetic particle of volume V with an uniaxial symmetry. The anisotropy energy of the particle is proportional to its volume, i.e.,

$$E_c = K_1 V \sin^2\theta. \tag{13}$$

This implies a potential barrier for the magnetization reversal along the easy axis. Then, the magnetization reversal rate becomes[21]:

$$\frac{1}{\tau} = Ae^{-KV/kT} \tag{14}$$

Here, the factor A depends on K_1, V, T and the saturation magnetization M_S, but its variation is not as important as the exponential term. With decreasing particle size (V), the relaxation rate increases rapidly. In the super paramagnetic limit, the averaged value of the magnetic moment of a particle will be zero which is not desirable for information storage. For bulk hcp cobalt, $K_1 = 5.3 \times 10^6$ erg/cm^3, kT = 26 meV at 300 K, the critical volume of Co particule is V$\sim$ 10 nm^3.

The current magnetic recording is based on statistically averaging the magnetization of a large number of small grains in magnetic thin films. By reducing the grain size and film thickness, one can obtain higher recording densities but the problem of the instability due to thermal fluctuations (super- paramagnetic limit) will limit the further improvement.

To avoid this, well defined and highly anisotropic magnetic nanoparticles can be used.[22] Each of the nanoparticles contains two discrete magnetic states with equal but opposite magnetic moments, allowing thus to store one bit of information. Two approaches are currently under investigation. One consists in synthesizing mono-disperse magnetic particles and then depositing them on a substrate by self-assembling. The first arrays suitable for magnetic recording were obtained by IBM using highly anisotropic FePt ferromagnetic nanoparticles.[23] Another approach is based on lithographic patterning. Both approaches are delicate and challenging for production at an industrial level. For example, for a practical use of patterned magnetic media, the required lithography resolution is of the order of 50 nm. For this reason nanoimprint lithography is attractive.

13.3. NANOIMPRINT LITHOGRAPHY PARAMETERS

The performance of the nanoimprint lithography depends on a number of factors: the quality of the mold, the characteristics of the polymer and the control parameters of the imprinting (temperature, pressure and duration) and the pattern transfer (reactive ion etching and lift-off).[24] For magnetic structure patterning, a number of specifications can be noted. In general, the required magnetic features are high density with a homogeneous distribution over a large surface area. For recording media, an accurate alignment may not be required but the edge roughness of each patterned elements needs to be minimized in order to narrow the switching field distribution.

13.3.1. Mold preparation

Mold patterns are generally defined by electron beam lithography. Patterns are then transferred into various substrates such as thermal oxidized silicon, Si, Si_3N_4, quartz or W by reactive ion etching or electroplating of nickel, followed by an anti-adhesion surface treatment. For high density recording, two types of recording configurations are considered. The first one consists of making a number of regular dot arrays that can be used for recording by proximity probe techniques.[25] The second one is to use a dedicated electron beam system to pattern circular tracks, in analogy to conventional compact disks. In both cases, the patterning speed should be high enough to allow covering at least a 2" surface within a reasonable time scale. Finally, the etching depth of the mold has to be correlated to the imprint conditions in order to increase the manufacturing speed.

It is known that the imprint speed depends not only on the viscosity parameter of the polymer but also on the geometry of the mold pattern.[28] In particular, the imprint speed can be considerably high for homogeneously distributed high density arrays. General speaking, the imprint speed at the beginning is relatively high and then decreases rapidly as soon as the mold cavity is filled.[26]

Figure 2 illustrates schematically three imprinting regimes. In regime I imprinting is incomplete and the mold cavities are not completely filled. In regime II the mold cavities are filled and imprinting is complete. In regime III the polymer is squeezed out from the mold area. To evaluate the imprint speed for simplicity, we consider a protruding disk of radius a located in the center of a wafer of radius R with an etch depth h_m. The imprint speed at the beginning (regime I) and at the end (regime III) of the printing process is

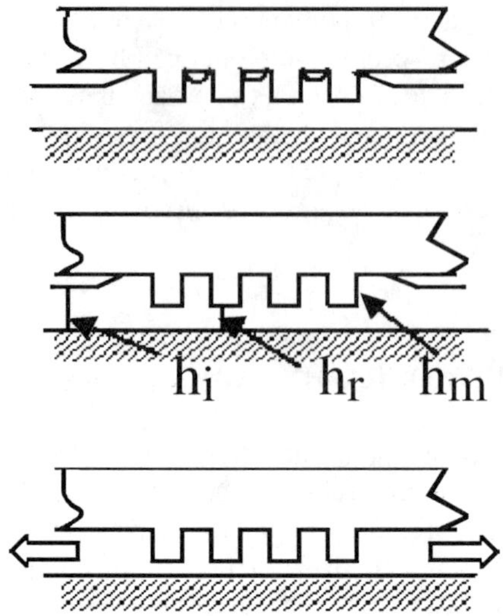

FIGURE 13.2. Schematic picture of mold filling regimes during imprinting.

given by:

$$u_a = \frac{4h^3}{3\pi\eta a^4}F \tag{15a}$$

$$u_R = \frac{4(h + h_m)^3}{3\pi\eta R^4}F \tag{15b}$$

respectively, where F is the applied force, η is the viscosity of the polymer, and h the thickness of the polymer under the disk. At the beginning of the imprint, the force is applied to a relatively small area (πa^2) and the polymer flow under the applied pressure is limited to the area under the disk. At the end of the imprinting, the force is charged over the entire mould surface (πR^2). In this case the mass transportation of the polymer is over a large distance and will be much more difficult than before. Obviously when $a << R$, $u_a >> u_R$. This would suggest that a fast imprinting can be obtained when polymer flow occurs locally on a short length scale. An important aspect in nanoimprint is the choice of the initial resist thickness h_i. Considering the conservation of the resist volume in regime II, the residual resist thickness, h_r which remains in the recessed area in the polymer after filling the mold cavity is closely correlated to the etching depth of the mold, h_m, and the initial thickness h_i by the relation:

$$h_i = h_r + f h_m \tag{16}$$

here $f = A_e/A_t$ is the ratio of the etched area (A_e) and the total area (A_t) of the mold.

TABLE 13.1. Initial polymer layer thickness required to fill completely the mold cavity leaving a residual thickness of 20 nm in the recessed polymer area. The mold etch depth is 150 nm, d is the period of the patterns.

Form	Filling factor	d	h_c
Lines (linewidth a)	$f = 1 - \frac{a}{d}$	2a	95 nm
Dots (radius a)	$f = 1 - \frac{\pi a^2}{d^2}$	4a	130 nm
Holes (radius a)	$f = \frac{\pi a^2}{d^2}$	4a	45 nm
Squares (size 2a×2a)	$f = 1 - \frac{4a^2}{d^2}$	4a	133nm

In Table 1, the factor f and the initial polymer layer thickness are given for several types of periodic patterns assuming a mold etch depth 150 nm and a desired residual polymer thickness of 20 nm.

In the case of a mold containing several periodic structures, the largest value of factor f should be used:

$$h_i = h + h_m \times \max\{f\}$$

For more complicated pattern geometries, a systematic investigation is needed to imprint all features. In some cases, additional etch areas can be introduced close to the large imprinting areas, which should be helpful to reduce the imprinting time.

13.3.2. Imprinting conditions

The imprinting condition changes with the polymer type and the pattern specification. In general, thermoplastic polymers are used and the imprinting is performed well above the glass transition temperature of the polymer. Polymers with a low molecular weight and a low glass transition temperature are preferable because of their low viscosity. Besides, the resistance of the polymer to reactive ion etching has to be reasonable for an easy process control. For uniformly distributed high density patterns, the imprint temperature can be substantially lower than that required for mixed small and large feature, because it is difficult to evacuate the polymer from large imprinting areas.

In Figure 3, the imprint temperature required for different polymers is plotted as a function of the feature size. Clearly, the larger the feature sizes, the higher the imprinting temperature. Compared to PMMA, hyper-branched polymers (hybranes) show more interesting properties. The glass transition temperature of the hyper-branched polymers can be tuned from 10°C to 70°C, depending on the molecules present in the branched groups [27-29]. Accordingly, nanoimprint can be done over a wide temperature range. Moreover, the viscosity of the hybranes decrease more rapidly with increasing temperature and they also have a relatively high etch resistance (the etch rate with O_2 plasma of these polymers is 1 nm per min., being half of the PMMA etch rate). Therefore, the

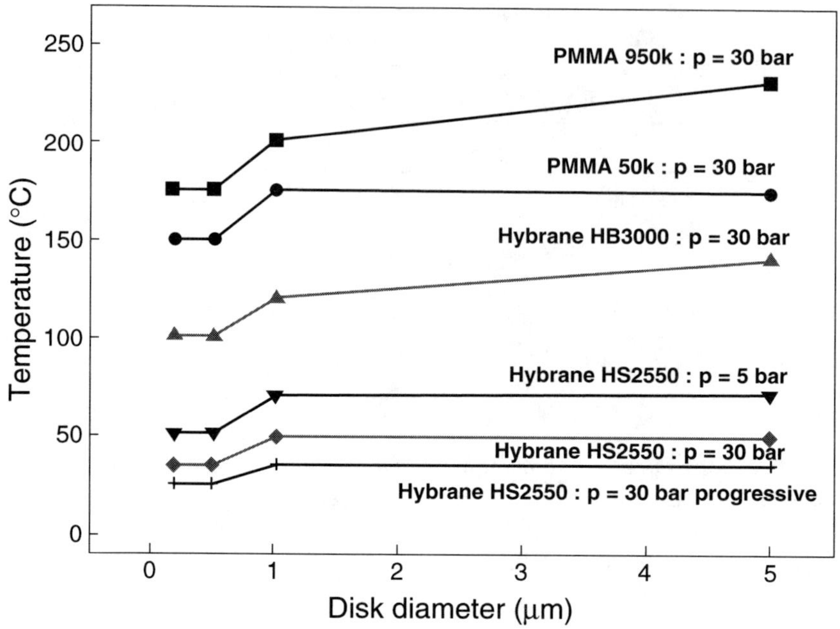

FIGURE 13.3. Required imprinting temperature versus feature size for different polymers.

process latitude of hyper-branched polymers is larger than that of PMMA, thus allowing a better size control. Another advantage is that some of the hybrane polymers can be used for a room temperature imprinting or a low pressure imprinting.[29] This is interesting for some particular cases (the magnetic properties for instance of ultra-thin Co/Pt multi-layers are stable only below at a temperature below 100°C). With hybrane 2550, 75 nm line and space gratings have been successfully replicated at 20°C and lift-off with a metal film. The same polymer has been used for low pressure imprinting as well as step and repeat experiences, showing comparable results.[29]

13.3.3. Pattern transfer

A number of pattern transfer techniques can be used to obtain high density magnetic structures, including lift-off, electroplating, dry etching as well as ion irradiation induced mixing. In addition, magnetic structures can also be obtained by deposition of a thin film on a pre-patterned substrate.

Lift-off is possible with nickel, cobalt, permalloy and other magnetic materials by electron beam assisted evaporation. The quality of lift-off magnetic structures depends on the deposition condition and also on the resist profile. The evaporation should cover the entire sample surface with a beam as perpendicular as possible to the substrate to avoid asymmetric deposition or thickness inhomogeneity. A good resist profile for lift-off should be slightly under-cut or vertical and the resist thickness contrast should be high enough (three times larger than the metal thickness). In order to obtain features of high aspect ratio, a tri-layer nanoimprint process can be used[27,30,31] which consists of

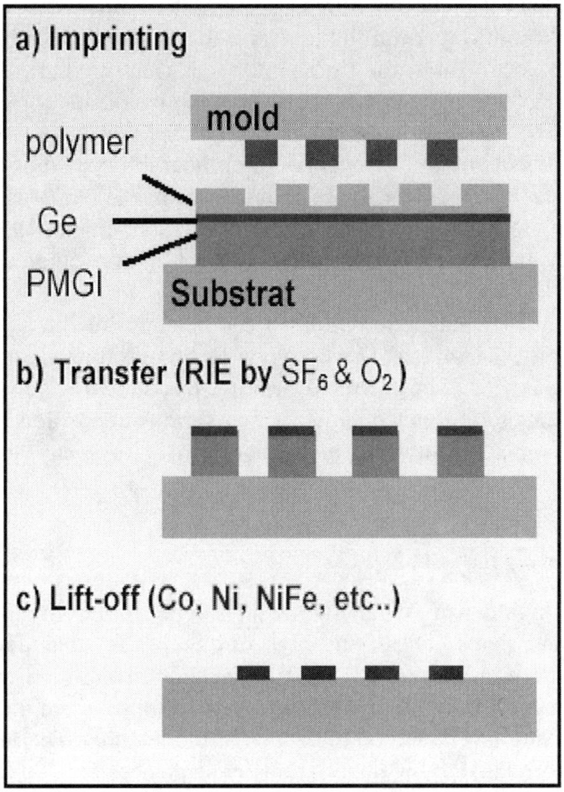

FIGURE 13.4. Schematic diagram of tri-layer based nanoimprint lithography

a thick bottom layer which is thermostable at the imprint temperature (PMGI), a thin intermediate layer (10 nm thick Ge) and a the top imaging layer (PMMA) (Figure 4).

After imprinting, the thickness contrast in the top layer is amplified by a sequential etching of the whole tri-layer system. This trilayer process is particularly useful in magnetic studies when different deposition thicknesses are required. Moreover, by over etching the bottom polymer layer, magnetic structures of different sizes can be obtained with the same mold.[32,33]

For larger aspect ratio features, electroplating is more appropriate.[34] Various magnetic materials can be used such as cobalt, nickel, permalloy, iron, etc. Here the height of the plated metal can be close to the resist thickness. Feature sizes as small as 20 nm can be deposited with gold and 50 nm with cobalt with an aspect ratio as high as 10.[35]

The third technique is ion milling. The mask for ion milling can be obtained by nanoimprint lithography and lift-off of a thin metallic film (usually Al or Ti). As a result of ion bombardment, atoms in the unmasked area are ejected from the magnetic thin films. This technique can be applied to a large range of magnetic materials but the problem of re-deposition of the etched material on the resist side-walls remains as its main drawback.[4]

To pattern planar magnetic structures, ion irradiation induced mixing is appropriate for ultra-thin Co/Pt multi-layer thin films. Here, the ion irradiation in unmasked areas induces atomic displacement at the Co/Pt interfaces (intermixing), which results in a remarkable change of the magnetic properties in the irradiated zone. Two techniques have been used for irradiation masking. The first one consists of using a patterned polymer[7] or dielectric material[36-38] obtained by lithography or lithography and etching methods. Feature sizes as small as 30 nm could be obtained. The second technique uses a stencil mask. High resolution could be obtained by irradiating the same sample several times with a slight mask displacement[8]. It is believed that nanoimprint lithography is useful for both cases.

Finally, high quality magnetic thin films can be deposited on a pre-patterned substrate such as silicon, GaAs, etc. The common technique for the substrate patterning is reactive ion etching[39,40] but chemical printing has shown a good performance as well.[41,42] The advantage of this technique is that various deposition techniques can be applied including typically sputtering, molecular beam epitaxy, etc., allowing high quality magnetic structures.

13.3.4. Manufacturing performances

Patterning of high density magnetic media requires a high lithography resolution, an accurate size and shape control and high throughput for manufacturing. Also the magnetization reversal in patterned structures depends strongly on the quality of pattern transfer. In practice, both lithography and pattern transfer can introduce defects or roughness in magnetic layers and on the edges of the features. The imprinting process has to be closely correlated to the pattern transfer sequences.

The resolution of the nanoimprint lithography seems to be limited only by the resolution of the electron beam lithography used in the mold fabrication. Previously, Chou *et al.* have demonstrated a minimum feature size as small as 6 nm with a pitch size down to 40 nm,[43] corresponding to a storage density of 400 Gbit/in^2. No comparable results have been reported from other groups, but the fabrication of feature sizes down to 50 nm (100 nm pitch) can be indeed achieved routinely using several types of polymers.[31]

Figure 5 shows a 60 nm pitch array, obtained by nanoimprint lithography and lift-off of a thin nickel film. Much effort has to be devoted to the mold fabrication in order to demonstrate the feasibility of terabit per inch manufacturing.

For 50 nm feature sizes, the reproducibility of the nanoimprint lithography is excellent. In particular, the above described tri-layer process showed a large process latitude and an accurate control of the critical dimensions. By changing the subsequent etch process, it is also possible to obtain an accurate size tailoring (Figure 6a). For example, the diameter of the Co dots fabricated with the same mold can be varied in the range from 100 nm to 175 nm (Figure 6b and c).

Another important issue is the long distance accuracy of the patterning replication. It has been shown that as soon as the same materials are used as the mold and sample substrates, thermal process induced pattern distortion is less than 30 nm (detected accuracy) over a distance of 24 mm. Large area fabrication has been shown to be possible with 4 inch[44,45] and 6 in wafers.[46]

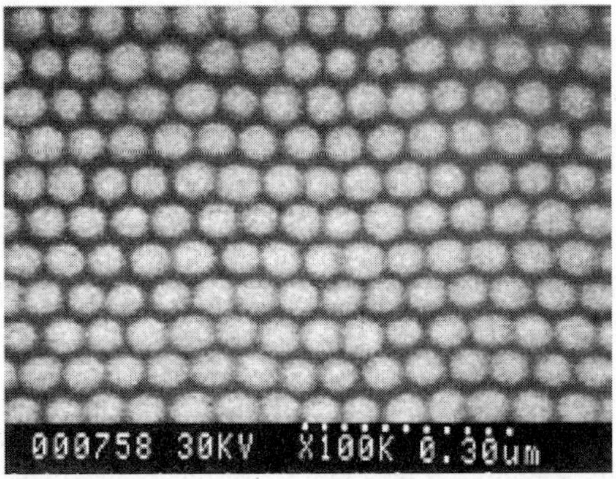

FIGURE 13.5. SEM image of a 60 nm period dot array obtained by nanoimprint and lift-off of 20 nm thick nickel.

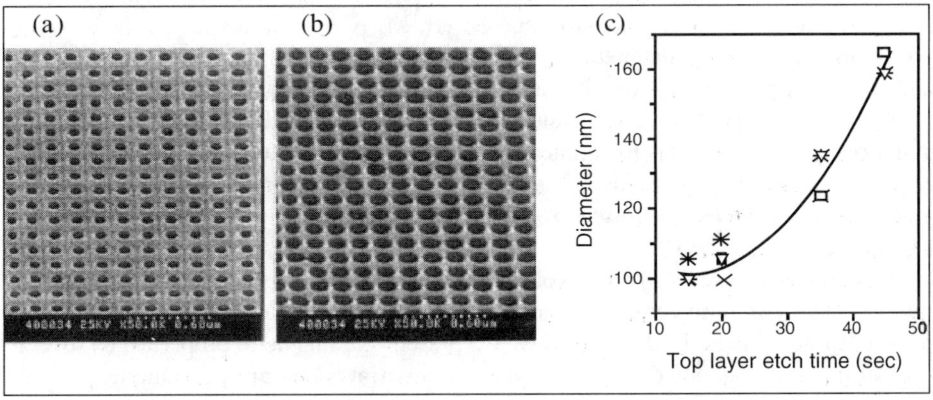

FIGURE 13.6. SEM images of PMGI resist patterns obtained from tri-layer nanoimprint: a) hybrane etching time 15 s, b) hybrane etching time 45 c, and c) diameter of PMGI holes as a function of hybrane etching time. Different symbols represent results obtained with the same imprint but different etch times.

Finally, the production rate of the nanoimprint lithography is defined by wafer loading, alignment, imprinting, separation and wafer unloading. Without accuracy alignment, the cycle time of nanoimprint lithography depends mostly on the imprinting (heating and molding) and the separation (cooling) speeds. Typically, each of them lasts one minute or so, indicating a throughput of 20 wafers per hour with advanced wafer loading and unloading facilities. Emerging approaches such as flash and repeat or room temperature imprint could improve considerably the throughput. Some of these approaches are discussed in Chapters 6 and 7. However, up to date, no automatic test has been performed and questions such as the mold duration and the total number of imprinting without further anti-stick mold treatment are still open.

13.4. PATTERNED MAGNETIC NANOSTRUCTURES

Various methods are used for the investigation of patterned magnetic structures. For example, a standard optical microscope with polarization analysis can be easily used for the observation of magnetic domains which are perpendicular to the plane (Co/Pt multilayers).[47] Since the resolution of the conventional optical techniques is limited by diffraction of the order of 1 μm, near field optical microscopy can be used for a high resolution magnetic domain imaging.[48] However, because of the important edge diffraction with respect to relatively weak magneto-optical contrast, etched magnetic samples are not well suited for near-field observation. It is shown that high resolution magnetic domains and their evolutions can be imaged in as grown[49] as well as ion irradiated samples.[7] Magnetic force microscopy (MFM) is the most powerful technique which allows imaging of both in plane and out of plane magnetization.[50-52] Besides, differential phase contrast electronic microscopy provides highest resolution but the magnetic sample has to be prepared on a thin membrane.[53]

For most applications, we are interested in the switching process, i.e., the process during which the average magnetization changes to its opposite direction. One way to describe the switching process is via the (B, M) hysteresis loop. In a quasi-static setting, the steady state magnetization distribution is reached at each fixed magnetic field. The average magnetization in a particular direction is measured and plotted against the external field. Typically, the resulting (H, M) curve depends on how H is changed, giving rise to the hysteresis phenomenon. Typically, micro-squids are used to measure hysteresis cycles of single particles[54] and surface sensitive techniques such as Magneto-optic Kerr effect (MOKE) are used to characterize ultra-thin magnetic layers or patterned magnetic structure arrays.

Previously, magnetic structures of different geometries have been studied, including squares, disks and more complex structures such as rectangles with flat or pointed ends as well as micro-rings. In the following, we present the magnetic properties of several types of them to illustrate the ability of the nanoimprint lithography technique.

13.4.1. Co dot arrays

Co dots of diameter varying in the range 100 nm to 500 nm and thickness 10 nm to 100 nm were studied by measuring both hysteresis loops of MOKE in the longitudinal geometry and MFM images. They give detailed information on the remanent state and magnetization reversal process as a function of Co dot thickness and size. Two types of hysteresis loops have been found, as typified in Figure 7. The Co dot arrays with the diameter/thickness combinations of 100 nm/30 nm (loop 7a) showed different behavior when compared to the behavior of the arrays with the diameter/thickness combinations of 100 nm/15 nm (loop 7b). Loop (b) presents a high remanence and a low switching field, which are typical of the single domain configuration. Loop (a) is characterized by the formation of a vortex state: As the applied field is reduced from minus saturation the maximum moment is retained until a critical field H_n is reached, at which the magnetization starts to be lost and nearly all magnetization is lost at the remanent state. Then, the magnetization progressively reappears as the field is increased from zero until the positive saturation is achieved. The sudden loss of magnetization to zero at remanence

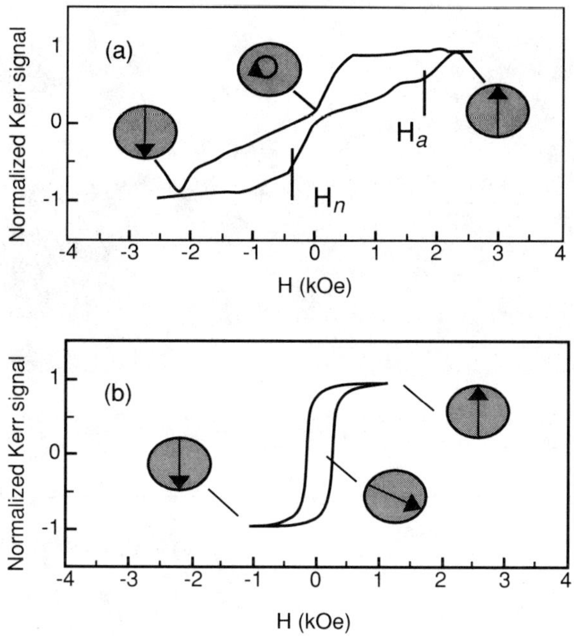

FIGURE 13.7. MOKE hysteresis loops for circular Co dots of diameter D=100nm and different thickness (a) t=30nm, (b) t=10nm.

is a characteristic of the formation of a flux closure configuration (FC) in which the magnetization vectors remains parallel to the edge of the dot and forms a vortex centered around the dot center. This state lowers the system energy by minimizing magnetostatic energy. Increasing the field then deforms the FC configuration by pushing the vortex center towards the half part of the dot where the magnetization vectors are anti-parallel to the external field, until all magnetic moments are eventually reversed, and becomes parallel to the field at the vortex annihilation field H_a.[32]

Both vortex and single-domain states can be visualized by MFM. Figure 8a shows a MFM image of the vortex state at remanence for a dot with 30nm thickness and 500nm diameter. The black contrast at the dot center corresponds to the out of plane magnetization of the vortex core. A typical MFM image for a single-domain dot is shown in Figure 8b for a dot with 500nm diameter and 15nm thickness. In this case, magnetization reversal is controlled by coherent rotation. To support such a magnetization configuration an anisotropy is required. Due to the circular symmetry of the Co dots, the only remaining anisotropy is then the weak in-plane anisotropy intrinsic to the polycrystalline cobalt. The transition between the two states can be understood as a competition between a reduction of exchange energy in the near single domain state versus a reduction in magneto-static energy in the vortex state.

In order to study the inter-dot coupling, cobalt arrays of the same period (200nm) and different diameters (100-150 nm) and thickness 30nm have been studied.

In Figure 9 MOKE hysteresis loops are shown for arrays of Co dots with D = 100nm (a) and D = 150nm (b). A shift of the vortex nucleation and annihilation fields to lower

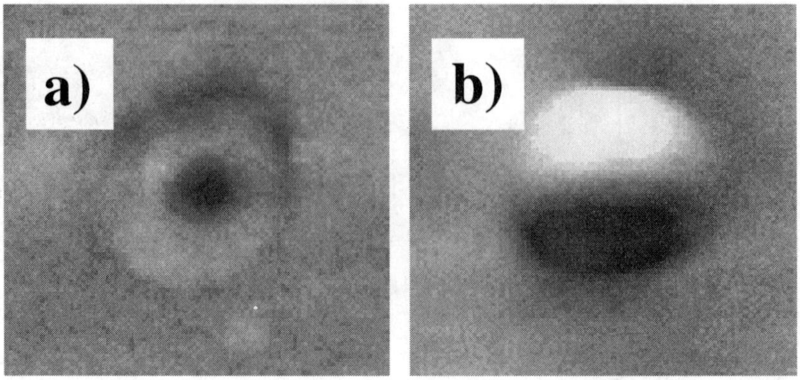

FIGURE 13.8. MFM images of 500 nm Co dots with a thickness of (a) 30 nm showing a vortex state and (b) 15nm thickness showing a monodomain state.

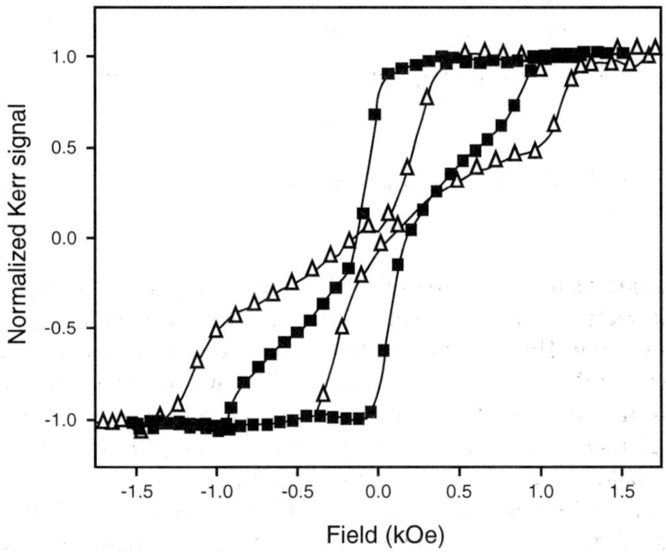

FIGURE 13.9. Effect of inter-dot coupling on the hysteresis loops of Co dots with fixed period P=200nm and two different dot diameter : D=100nm (open triangles) and D=150nm (solid squares).

values are observed for the more closely packed dots, that is a consequence of the stabilization of the single-domain state by the dipolar coupling between dots. Similarly for dots with a fixed diameter and increasing separation the vortex nucleation and annihilation fields depend on inter-dot distance and can be accounted for by the action of the dipole interaction field between dots.[33]

A further illustration of interaction effects is shown in Figure 10 where we observe in MFM images the spatial distribution of vortex and SD states for a closely spaced dot array D = 500 nm / S = 600 nm (a-c) and for a widely spaced dot array D = 550 nm / S = 1000 nm (d). The samples were first saturated in a field of –1.5 kOe and then images

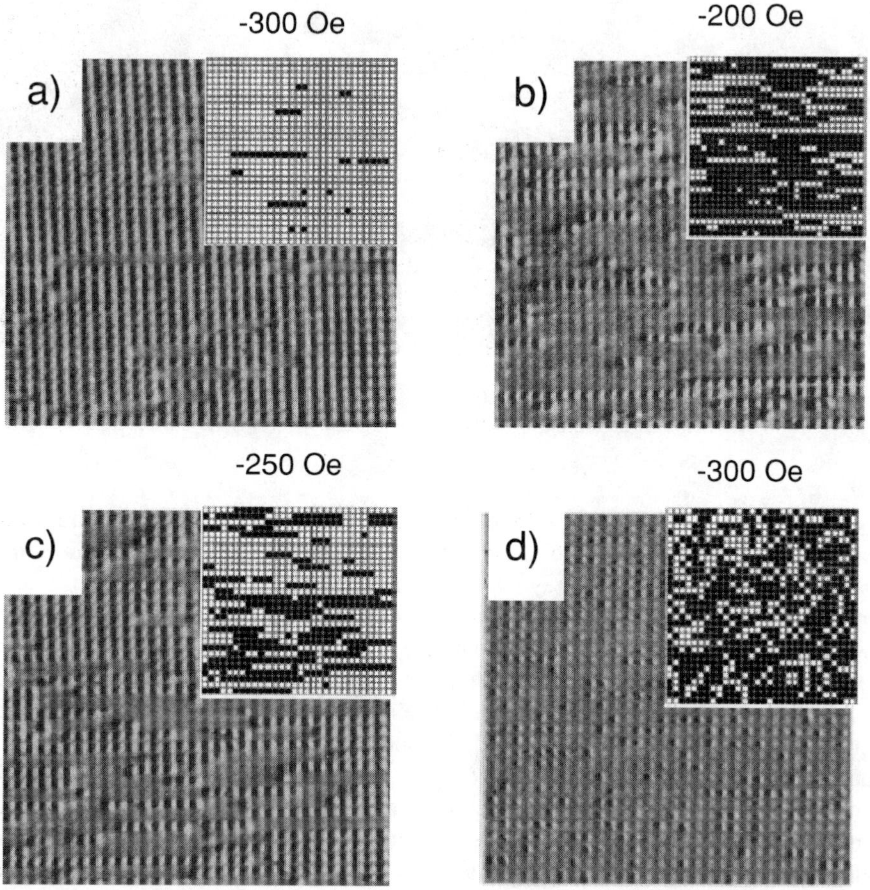

FIGURE 13.10. Distribution of SD and vortex states (a-c) in arrays of closely packed dots with diameter D=550nm and period P=600nm; thickness 30nm, (d) in an array of widely spaced dots D=550nm, P=1000nm.

were taken at several lower field values. While SD dots are characterized by a dipolar contrast vortices appear as dots with a weak magnetic contrast. The insets show graphic elaborations of the MFM images and represent vortices as black dots and SD states as white dots. In the closely spaced dot array vortices from chains parallel to the applied field direction (Figure 10a-c), while they nucleate randomly in the widely spaced dot array (Figure 10d). The chain formation is closely related to the interaction between adjacent dots and results from a nucleation cascade.[55]

13.4.2. *Micro-rings*

Magnetic micro-rings have attracted a particular attention because of the possibility of constructing the so-called vertical magneto-resistance random access memory (VMRAM).[20] Compared to dot geometries, the ring shaped elements can have stable flux

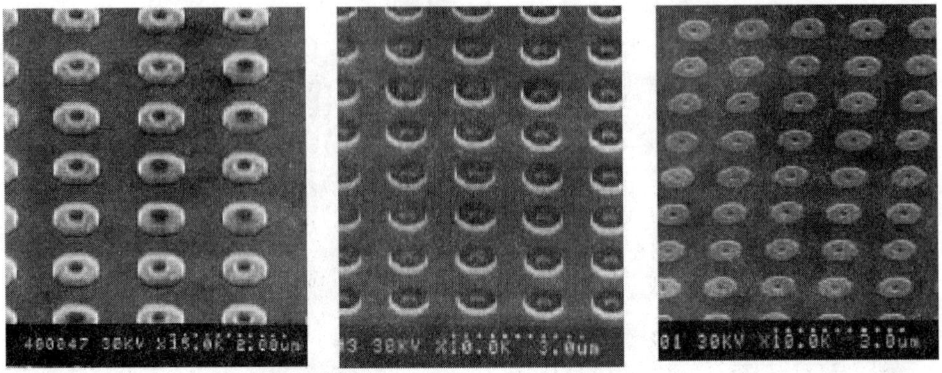

FIGURE 13.11. SEM images of a ring array in the mold (a), in the resist after imprinting (b) and lift-off (c) of 50 nm thick Cobalt rings. External diameter 700 nm and internal diameter 300 nm.

close (FC) magnetization states without the formation of the central vortex core. Consequently, undesired interactions between adjacent magnets of the device are minimized because of the absence of the stray field in the FC states.

Submicron rings were obtained by nanoimprint and lift-off of Co films of thicknesses varying in the range of 10 to 100 nm.

Figure 11 shows SEM images of a typical micro-ring array obtained before (a) and after (b) lift-off of a 50 nm thick Co film. The inner and outer diameters of these rings are respectively 0.3 and 0.7 μm. Instead of a perfect circular ring, an octagonal shape is adopted to facilitate the MFM imaging of the FC configuration and to reduce the electron-beam patterning time of the mold. According to a computation study, there should be no significant difference between the results obtained with octagonal and circular rings. Both MFM imaging and numerical simulation based on micromagnetic theory[56] have confirmed the formation of stable FC states in micro-rings (Figure 12).

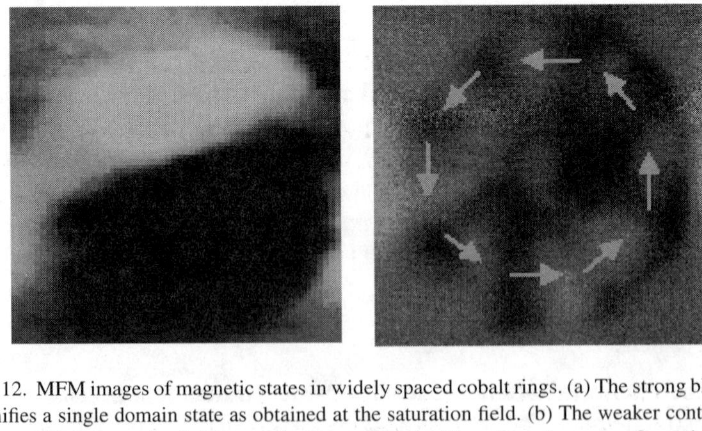

FIGURE 13.12. MFM images of magnetic states in widely spaced cobalt rings. (a) The strong black and white contrast signifies a single domain state as obtained at the saturation field. (b) The weaker contrast compared to the SD state, with 8 alternating bright and dark segments signifies a flux closure configuration.

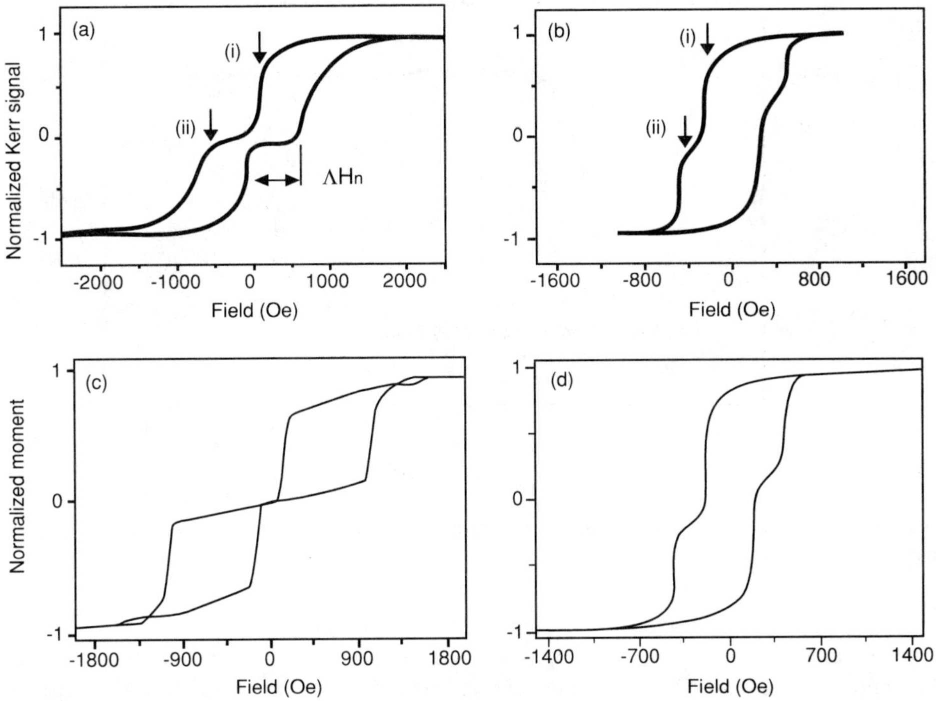

FIGURE 13.13. (a), (b) Typical hysteresis loops measured on rings with the same external/internal diameter of 700 nm / 300 nm respectively but for different thickness of 50 nm (a) and 20 nm (b). (c), (d) Calculated hysteresis loops for the corresponding Co rings.

MOKE hysteresis loop measurements in longitudinal geometry were also performed on Co rings of different thickness and sizes. Two types of behaviors have been observed: one involving the FC state and another the near-single domain state.

Figure 13 shows two examples of hysteresis loops obtained with the same Co ring thickness (20 nm) but different diameters (300 and 400 nm). The observed size and thickness dependencies are comparable to those obtained with Co rings made by electron-beam lithography, from which a phase diagram in ring diameter and thickness could be established.[57,58]

13.4.3. Co/PtNi multilayer structures.

High quality Co/Pt and CoNi/Pt multi-layers can be obtained by r.f. sputtering. The multilayer configuration allows easy and precise adjustment of magnetic properties through the layer thicknesses and number of periods. To illustrate the effect of confinement, patterning has been done by nanoimprint and ion milling on a thin film of Pt(3.4nm)/{CoNi(0.4)/Pt(0.3)}. Both as-grown thin films and patterned structures have perpendicular magnetization.

Figure 14 shows MOKE loops under polar configuration. Clearly, the magnetic reversal in sub-micrometer dots strongly depends on the dot size. With decreasing the

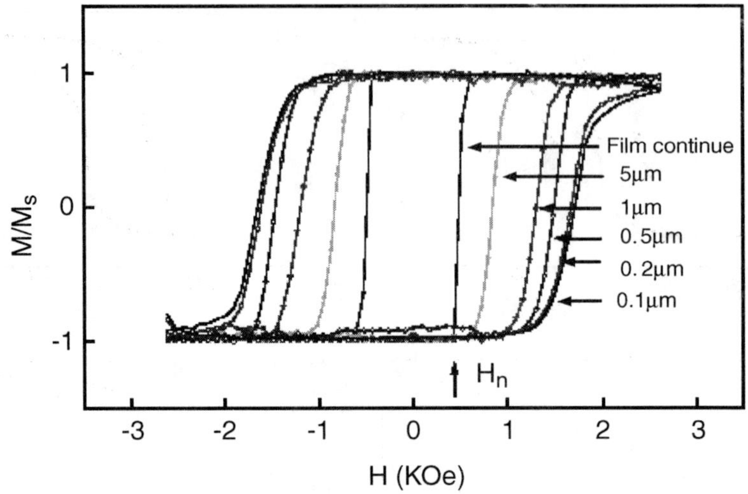

FIGURE 13.14. Magnetic hysteresis loops of the CoNi/Pt multiplayer thin film and etched dots of different diameters.

dot diameter, the saturation field (H_s), the coercitive field (H_c) and the hysteresis loops become more and more rounded. The magnetization reversal begins in the dots with the same field (H_n) indicates that H_n is partially intrinsic. Similar results were obtained with patterned ultra-thin films of Au/Co/Au(111)[4]and multi-layer of CoNi/Pt.[17] It has been shown that the wall propagation can be blocked at the dot edges and the magnetization reversal generally passes through a nucleation process followed by the wall propagation. The reversal behavior depends on the dot size through the spatial distribution of the nucleation centers. Therefore, the hysteresis loops become more rounded with a larger coercitive filed as the diameter decreases.

13.5. CONCLUSION

Nanoimprint lithography is one of the most promising techniques for high density magnetic structure patterning because of its ultra-high resolution, parallel processing and low cost. It is particularly suitable for patterning of magnetic recording media because only one-step replication is required without precise alignment. For more fundamental studies, nanoimprint lithography is sufficiently flexible, efficient and compatible to most of the pattern transfer techniques. This allows a fast processing which is helpful for understanding how the geometry of nanostructures produced by nanofabrication processes (size dispersion, deformation, and edge effects) affects the stability and the dispersion of the magnetization reversal properties in magnetic nanostructure arrays. Finally, application of nanoimprint lithography can also be extended to the fabrication of magnetic memories and new spin electronic devices.

ACKNOWLEDGMENTS

The authors would like to thank Drs. I.L. Prejbeanu, L.D. Buda, U. Ebels and K. Ounadjela (IPCMS Strasbourg) for important contributions in the MFM imaging. This work was partially supported by the European Commission through project contracts "SPINUP" and "SUBMAGDEV".

REFERENCES

1. S. Y. Chou, Proc. IEEE 85, (1997), 652.
2. G.A. Prinz, K. Hathaway, Phys. Today 48, (1995), 24 MRAM.
3. G. A. Prinz, Science **282,** (1998), 1660.
4. F. Rousseaux, D. Decanini, F. Carcenac, E. Cambril, M.F. Ravet, C. Chappert, N. Bardou, B. Bartenlian, and F. Veillet, J. Vac. Sci. & Technol. B 13, (1995), 2787 .
5. C. Haginoya, S. Heike, M. Ishibashi, K. Nakamura, and K. Koike, J. Appl. Phys. 85, (1999), 8327.
6. C.A. Ross, H. I. Smith,T. Savas, M. Schattenburg, M; Farhoud, M. Hwang, M. Walsh, M.C. Abraham, R.J. Ram, J. Vac. Sci. & Technol. B17, (1999), 3168.
7. C.Chappert, H.Bernas, J.Ferré, V.Kottler, J.PJamet, Y.Chen, E.Cambril, T.Devolder, F.Rousseaux, V.Mathet, H.Launois, *Science* **280**, (1998), 1920.
8. B. Terris, L.Folks, D. Weller, J.E.E. Baglin, A.J. Kellock, H. Rothuizen, and P. Vettiger, *Appl. Phys. Lett.* 75, (1999), 403.
9. For a general review of conventional and non conventional nanofabrication methods, see Y. Chen and A. Pépin, Electrophoresis, (2001), 122, 187.
10. S. Y. Chou, P.R. Krauss, and J. Renstrom, Appl. Phys. Lett. **67**, (1995), 3114.
11. C. Kittle, Introduction to Solid State Physics, 7th edition (John Wiley & Sons, 1996).
12. Hubert, R. Schäfer, Magnetic domains, Springer, Berlin (1988).
13. J.A.C. Bland and B. Heinrich (eds), *Ultrathin Magnetic Structures I & II*, Springer-Verlag, Berlin (1994).
14. J. Miltat, in Applied Magnetism, edited by R. Gerber, C. D. Wright, and G. Asti, NATO ASI Series (Kluwer, Dordrecht, 1994).
15. C.J. Hegedus, G. Kadar, and E. Dalla Torre, J. Inst. Math. Appl., (1979), 24279.
16. R. Moskowitz, E. Dalla Torre and R.M.M. Chen, Proc. IEEE, (1966), 54, 1211.
17. M.A.M. Haast, Ph. D. thesis, University of Twente, The Netherlands (1999).
18. W. F. Brown, Jr., J. Appl. Phys., (1968), **39,** 993.
19. R. P. Cowburn and M. E. Welland, Appl. Phys. Lett., (1998), **72**, 2041.
20. J. Zhu and Y. Zheng, and G. Prinz, J. Appl. Phys., (2000), **87**, 6668.
21. J.L Dormann, Revue Phys. Appl., 1981, 16, 275, SUPERMAGN.
22. L. Krusin-Elbaum, T. Shibauchi, B. Agyle, L. Gignac, D. Weller, Nature (2001), 410, 444.
23. S.Sun, C.B. Murray, D. Weller, L. Folks, and A. Moser, Science, 2000, 287, 1989.
24. Nanoimprint techniques, H.C. Scheer, H. Schulz, T. Hoffmann, C. M. Sotomayor Torres, in *The Handbook of Thin Films*, H.S. Nalwar (Ed.) Academic Press (2001).
25. P. Vettiger, M. Despont, U. Drechsler, U. Dürig, W. Häberle, M. I. Lutwyche, H. E. Rothuizen, R. Stutz, R. Widmer, and G. K. Binnig, IBM J. of R & D 44, 323 (2000).
26. M. Otto, M. Bender, B. Hadam, B. Spanggenberg, and H. Kurz, Microelectro. Eng., (2000), 57/58, 361.
27. Lebib, M. Natali, S.P. Li, E. Cambril, L. Manin, Y.Chen, H. M. Janssen, R.P. Sijbesma and E.W. Meijier, Microelectron. Eng. (2001), 57/58, 411.
28. Lebib, Ph. D. Thesis, Univeristé Paris 7, France (2001).
29. Lebib, Y. Chen, E. Cambril, P. Youinou, V. Studer, M. Natali and A. Pépin H. M. Janssen and R.P. Sijbesma, *Microelectron. Eng.* (2002) in press.
30. Y.Chen, A.Lebib, S. Li, D. Peyrade, M. Natali, A.Pepin and E. Cambril, Eur. Phys. J. Appl. Phys. (2000), 12, 223.

31. Lebib, Y.Chen, F.Carcenac, E.Cambril, L. Manin, L.Couraud and H.Launois, Microelectron. Eng., (2000), 53, 175.
32. Lebib, S. P. Li, M. Natali and Y. Chen, J. Appl. Phys. (2001), 89, 3892.
33. M. Natali, A. Lebib, S.P. Li, E. Cambril, Y.Chen, J. Vac. Sci. & Technol. (2001), B19, 2779.
34. W. Wu, B. Cui, X. Sun, W. Zhang, L. Zhuang, L. Kong, and S. Y. Chou, J. Vac. Sci. & Techno., (1998), B16, 3825.
35. G. Simon, Ph. D. Thesis, Univeristé Paris Sud, France (1997).
36. T.Devolder, Y.Chen, H.Bernas, C.Chappert, J.PJamet, and E.Cambril, Appl. Phys. Lett., (1999), 74, 3383.
37. T.Devolder, C. Chappert, V. Mathet, H. Bernas, Y. Chen, J. P. Jamet, and J. Ferré, J. Appl. Phys, (2000), 87, 8671.
38. T.Devolder, C. Chappert, Y.Chen, E. Cambril, H.Launois, H. Bernas, J. P. Jamet, and J. Ferré, J. Vac. Sci. & Technol., (1999), B17, 3177.
39. Y. B. Xu, , A. Hirohata, S.M. Gardiner, M. Tselepi, J. Rothman, J. Klauim, L. Lopez-Diaz, J. A. C. Bland , Y. Chen, E. Cambri, F. Rousseaux., IEEE Trans Magn. 37, 2055 (2001).
40. J. Rothman, M. Kläui,, L. Lopez-Diaz, C.A.F. Vaz, A. Bleoch, J. A. C. Bland, Z. Cui, and R. Speeks, Phys. Rev. Lett., (2001), 86, 1098.
41. S. P. Li, W. S. Lew, J. A. C. Bland, L. Lopez-Diaz, M. Natali, C. A. F. Vaz, Y. Chen, *Nature,* (2002), 415, 600.
42. S. P. Li, W. S. Lew, J. A. C. Bland, L. Lopez-Diaz, C. A. F. Vaz,, M. Natali, Y. Chen. *Phys. Rev. Lett.,* (2002), *88*, 87202.
43. S.Y. Chou, P.R. Krauss and PJ. Renstrom, Science, (1998), 272, 85.
44. A.Lebib, Y.Chen, J.Bourneix, F.Carcenac, E.Cambril, L.Couraud and H.Launois, Microelectron. Eng., (1999), 46 , 319.
45. W. Zhang and S.Y. Chou, results presented in the EIPBN 2000 conference (USA).
46. Heidari, I. Maximov. And L. Montelius, J. Vac. Sci. & Technolo., (2000), 18, 3557.
47. Y.Chen, F.Carcenac, V.Kottler, C.Chappert, N.Essaidi, H.Launois, *J. Vac. Sci. Technol.,* (1998), B**16**, 3830.
48. E. Betzig, and J.K. Trautman, Science, (1992), 257, 189.
49. V.Kottler, C.Chappert, N.Essaidi and Y.Chen, IEEE *Trans. Mag*. **34**, 2012 (1998).
50. G.A. Gibson, S.Shultz, J. Appl. Phys., (1993), 73, 4516.
51. M. Hehn, K. Ounadjela, J. Bucher, F. Rousseaux, D. Decanini, B. Bartenlian, and C. Chappert, Science, (1996), **272 ,** 1782.
52. Dieny, A. Lebib, Y. Chen; JMMM (2001).
53. J.N. Chapman, M.R. Scheinfein, J. Mag. Magn. Mater., (1999), 200, 729.
54. M. Jamet, W. Wernsdorfer, C. Thirion,, D. Mailly, V. Dupuis, P. Mélinon, Phys. Rev. Lett., (2001), 86, 4676.
55. M. Natali, A.Lebib, Y.Chen, L. Prejbeanu, U.Ebels, L. Buda, K. Ounadjela; *Phys. Rev. Lett.* in press.
56. A simulation code to calculate the magnetization configuration and its field evolution for 2D flat elements is described on http://math.nist.gov/oommf.
57. S. P. Li, D. Peyrade, M. Natali, A. Lebib, Y. Chen, U. Ebels, L. D. Buda and K. Ounadjela, Phys. Rev. Lett., (2001), 86, 1102.
58. Y. Chen, A. Lebib, S. P. Li, M. Natali, D. Peyrade, and E. Cambril, Microelectron. Eng., (2001), 57/58, 405.

14

Application of Microcontact Printing and Nanoimprint Lithography

Towards device fabrication

Georg Schmidt, Tatjana Borzenko, Massimo Tormen, Volkmar Hock, Laurens W. Molenkamp
Physikalisches Institut, Universität Würzburg, Germany

14.1. INTRODUCTION

Microcontact printing (μCP) and Nanoimprint lithography (NIL) have both proven to be high resolution lithography processes suitable for large area parallel lithography.[1-7] This property makes them both promising for application in device fabrication, where high throughput is an issue. Especially NIL is used by numerous groups due to its high resolution and potential for applications.[8-10]

Besides resolution issues and the ability to cover large areas in a single step, two additional points have to be addressed when a lithography process is to be used in device fabrication. The first issue is the suitability of the mask which is created during lithography for various further processes from ion implantation to etching on various materials, the second one is the overlay accuracy and the possibility of alignment in subsequent lithography steps or mix and match procedures.[11-13] In the following these two points will be addressed in detail focusing on the application of μCP in a full-blown fabrication process. Possible solutions developed in our group at Würzburg University will be presented.

The basic microcontact printing process remarkably resembles a simple printing process using paper, a stamp and an ink. A master is fabricated by molding an elastomer in such a way that the patterns which are to be printed persist as protrusions at the surface. This master is inked with a suitable liquid (e.g. Eicosanethiole). Normally the liquid is left to dry on the surface leaving a chemically active surface layer. The elastomer stamp is then brought in contact with the sample onto which the lithography step is to be performed. On a suitable surface (a match for eicosanethiole is e.g. a gold layer) the ink layer is transferred in the regions of contact, forming a closed self assembled monolayer (SAM), while the regions in which no contact took place remain uncovered. The advantage of inks forming SAM's compared to other materials is the fact that small holes in the layer are closed by self organization and thus a perfect protection against a variety of wet etchants is given. Typically further processing can consist of wet etching of the gold layer, completing a full pattern transfer into a thin metal layer.[14]

The description given above already makes clear that the chemistry involved and the ability of the ink to form an SAM are crucial for the performance of the process.[3,15] Alkane-thiols can form SAM's on gold due to their sulfur bonds, however they will not perform similarly on silicon, oxides or metals with oxidized surfaces. One might imagine that for patterning different materials, the use of different suitable inks could be a possible means for application of μCP in a wide variety of situations, however this is a cumbersome road to go, because it seems very unlikely that for any process a suitable combination a suitable combination of ink and substrate can be found.

It is much more desirable to have a μCP process which does not rely on different inks, and which is nonetheless suitable for all materials that have to be patterned. The surface protection obtained by the process also has to be stable in different methods of pattern transfer for which normally polymer resists are used like in optical or e-beam lithography. A self assembled monolayer can be highly stable towards wet etchants, however, it will hardly withstand a reactive ion etching process or UHV conditions and will be almost non existent for a high energy particle beam in an ion implantation process.

Besides the problem of the material and process compatibility the printing also faces further problems arising from the surface topology of the samples. In planar

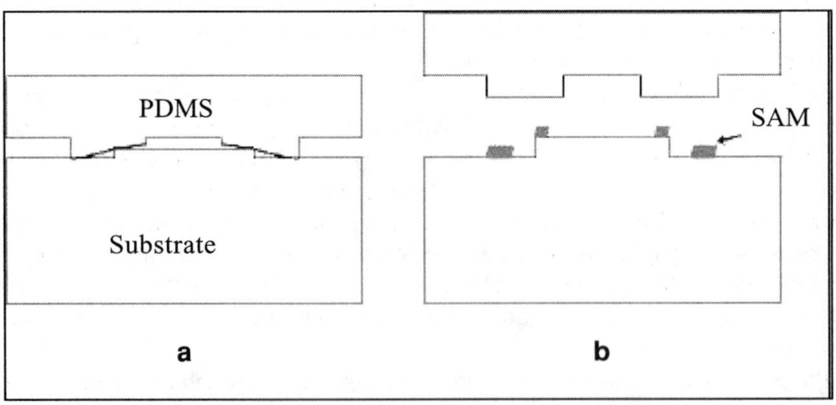

FIGURE 14.1. Microcontact printing on a sample with a non-planar surface. At the edges, the stamp is deformed (a) leading to an incomplete protection of the surface (b).

processes, printing can well compete with irradiation based lithography tools, also printing on spherical or cylindrical surfaces has been demonstrated successfully. But when non-planar structures, e.g., mesas for transistors or trenches for storage capacitors are present on the sample, only the 'hills' of the sample make contact to the stamp whereas part of the 'valleys' stay separated from the printing tool (Figure 1).

14.2. PROCESS

The process presented in the following combines the high resolution and performance of the printing with self assembled thiol-monolayers on gold with the versatility of optical or e-beam resists, which can be used for almost every step in semiconductor processing, including ion implantation, reactive ion etching, metallization and lift-off. At the same time the process provides a planarization process to make the soft imprint lithography suitable for highly non planar substrates.

Up to now, the most successful experiments with soft imprint lithography have been carried out with alkanethiols on thin gold films.[3,16,17] In order to introduce soft imprint into standard silicon technology it is obvious to make use of the expertise that already has been acquired in this area. However, gold is neither compatible with silicon technology, nor can a thin gold layer be used as a mask for lift off, trench etching or ion implantation. A logical approach is to use an additional layer between the gold and the silicon, which avoids any contact between metal and semiconductor and may at the same time stabilize the system for further processing. For many years, ultra high resolution electron beam lithography makes use of a process which utilizes a thin titanium layer to pattern a thick polymer layer by oxygen plasma, in order to achieve a resist which is suitable for lift-off or RIE processes.[18] We have now modified this process to make it suitable for soft imprint lithography. The main resist consists of a layer of PMMA on which a thin gold layer is deposited (Figure 2a). The gold layer is patterned by soft imprint and wet chemical etching (Figure 2b, 2c). The pattern is transferred into the PMMA by oxygen plasma etching (Figure 2d). The patterned PMMA then serves as a standard resist for any kind of pattern transfer (Figure 2e, 2f).

The stamps are fabricated from polydimethylsiloxane (PDMS, Sylgard 184 from Dow Corning) using a master and standard casting technique.[1] The master was prepared by photolithography with 1 micro thick AZ resist on silicon wafers. After a 30 min degassing at room temperature, the elastomer was cured at 60 °C for 2 hours and then carefully peeled from the master. Finally, the stamp was placed on a glass substrate for easy handling.

The substrates used were silicon (100) wafers with a low p-type doping. Either PMMA 600K or 950K were spun on and dried at a temperature of 160 °C for 90 minutes. The corresponding layer thicknesses were 270 nm (600K) and 120 nm (950K). On top of the PMMA, 20 nm of gold were evaporated by thermal evaporation in a high vacuum system with a base pressure of 1.4×10^{-4}Pa. We used a solution of Hexadecanthiole ($C_{16}H_{33}SH$) in Ethanol and a PDMS stamp to perform the soft imprint process. The stamp was wetted with the thiole solution for 45 s and then dried in a flow of dry nitrogen. The stamp was carefully brought in contact with the sample for 10 s. After the removal of the stamp, the sample was etched in aqueous solution of $K_3Fe(CN)_6$,

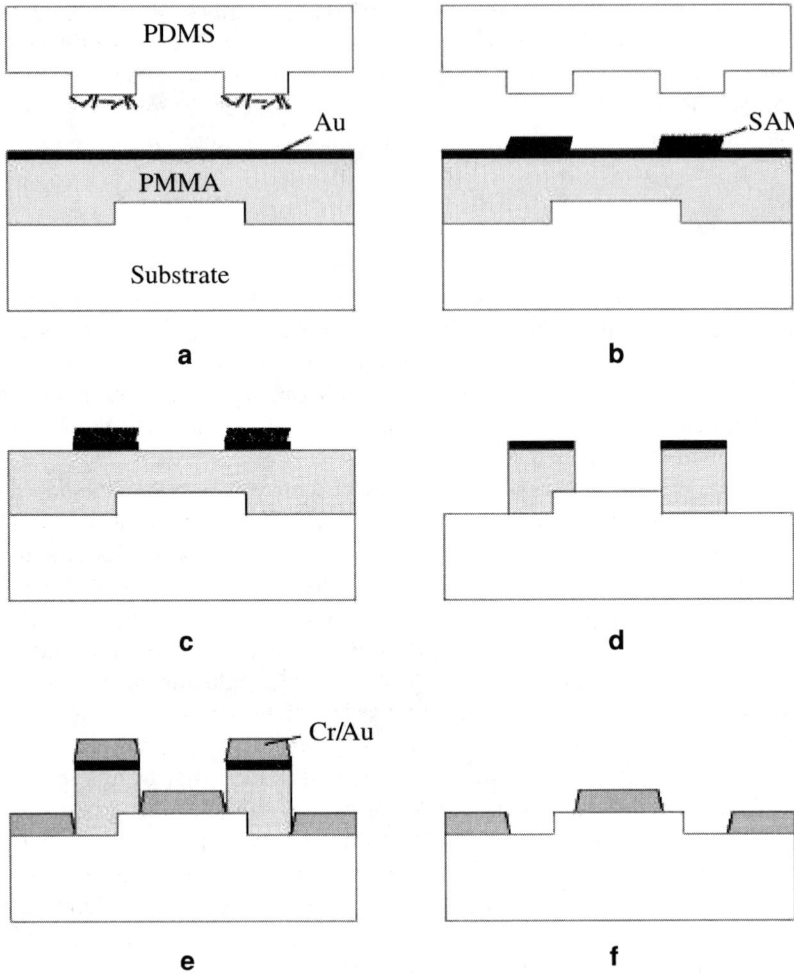

FIGURE 14.2. The printing process on a multilayer resist. The sample is planarized using PMMA and covered with gold (a). A self assembled monolayer is then printed on the sample (b). The gold is etched by wet etching but remains intact underneath the SAM (c). The polymer is patterned by oxygen plasma (d). Metal is evaporated (e) and lift-off is performed (f).

$K_4Fe(CN)_6$, $Na_2S_2O_3$ and KOH. This solution was already described by Whitesides et al.[19]who used $K_2S_2O_3$ instead of $Na_2S_2O_3$. The etching was done for 6 minutes until the parts of the gold which were not protected by thiols were completely removed. After drying the sample with nitrogen, it was placed in a reactive ion etching system where an oxygen plasma etching was performed at a pressure of 0.65 Pa and a power of 10 W for three minutes. At this stage the sample already resembles a standard e-beam resist system which has been exposed and developed. Due to the imperfect anisotropy of the RIE step at the operation conditions a considerable undercut under the gold occurs, which may

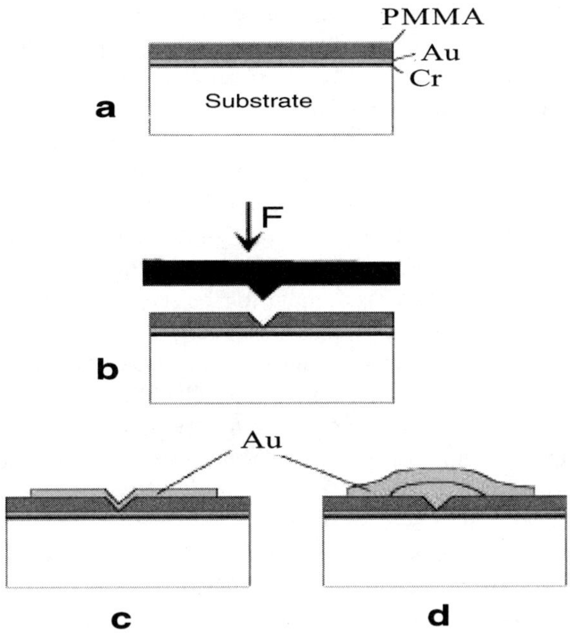

FIGURE 14.3. Cr/Au-structures fabricated by microcontact printing, wet and dry etching, metal evaporation, and lift-off.

be used for metal evaporation and lift-off. A double layer of 10 nm of Cr and 20 nm of Au were then evaporated and a lift-off was done in boiling acetone. The final result is shown in Figure 3.

During all process steps no considerable change in geometry could be observed. Even the smallest patterns of the stamp were transferred with a high fidelity.

With the process described above, MCP is available for all different varieties of semiconductor processing. However, device fabrication typically includes more than one level of lithography. Each of those levels has to be aligned on top of the previous ones, which is normally done by optical means in UV lithography. For this alignment several requirements have to be met. On one hand, it should be possible to match several marks on the lithographic master (typically a mask) with marks on the sample that have been defined in the previous levels of lithography. On the other hand the overlay accuracy has to be higher than the placement accuracy needed in the process, which requires a high mechanical stability without any lateral distortion. Both of these conditions are not fulfilled by the elastomer stamps which are typically used in MCP.

A stamp consisting only of an elastomer suffers strong distortion by a) its own weight, b) forces induced by the handling[20,21] and c) by solvent induced swelling and thermal expansion.[22,23] All these effects can easily lead to lateral displacement of structures in the range of several micrometers over the range of a few mm, making mix and match processes virtually impossible.

Both issues can be addressed by introducing a thin elastomer stamp which is supported by a transparent solid carrier. This carrier not only supplies the necessary lateral stability which is needed in order to obtain the overlay accuracy, but also allows for easy handling and optical alignment through the master which now consists of a glass carrier and a homogeneous thin (a few mum) transparent elastomer film. Moreover it solves other stability problems which can be present in thick elastomer stamps.

The fabrication of a thin stamps is more complex than the fabrication of thick stamps due to the necessity to provide a planar surface. Thick stamps level out a slight bending of the surface due to their high elasticity. For thin stamps the planarity has to be inherent to the stamp itself. Due to this fact, the mold for the stamp fabrication in our case was fabricated on highly planar and highly stable glass disks, normally used in optics which exhibit almost no bending under pressure and have a surface planarity of less than 30 nm. On such a disk, PMMA 950k was spun and baked. The film thickness obtained was 600 nm. The film was covered by 20 nm of gold and then exposed by a 30 kV electron beam. After removal of the gold film, the PMMA was developed in a mixture of Toluene and isopropanole. Feature sizes between 100 nm and 40 μm were defined over a 12×12 mm^2 area with an overall filling factor of 50%.

Using this mold various stamps with different elastomer layer thicknesses and different carrier materials were fabricated. The elastomer for the stamp was a polymerized mixture of Vinyl terminated PDMS and MethylHydrosiloxane – Dimethylsiloxane (DMS-V05 and HMS-301). The amount of catalyst which was used for the polymerization was chosen in order to allow for a curing after approx. one hour. Silicon and Quartz substrates were used as carriers for the stamp. Both were pretreated for enhanced adhesion of the PDMS to the carrier material.

In two cases a force of 1.5 kN was applied during the molding, resulting in stamps with a residual thickness of a few 100 nanometers only (stamp a on a quartz master and stamp b on silicon). For comparison, a thicker stamp was also fabricated without using additional pressure during molding. This stamp had a thickness of 200 μm. The height of the protruding features was 600 nm in all cases corresponding to the thickness of the initial PMMA layer.

To investigate the suitability of the stamps for printing, the stamps were immersed in a 0.2 mM solution of eicosanthiol in ethanol for one minute and blown dry with nitrogen. The printing experiments were performed on silicon wafers on which 5 nm of Cr and 20 nm of Au were thermally evaporated.

For the thin stamps (a and b) full contact was only obtained when an external pressure was applied. Pressure was between 2.5 and 4MPa and was applied using a pneumatic press. For the third stamp, the pressure was approx. 1kPa, three orders of magnitude less than the pressure needed to obtain good results for the thin stamps. Finally, the SAM's pattern was transferred into Au by wet chemical etching [1 M KOH, 0.1 M $Na_2S_2O_3$, 0.01 M $K_3Fe(CN)_6$, 0.001 M $K_4Fe(CN)_6$] of the unprotected metal areas.[14]

The results clearly indicate the advantages of the thin stamps on rigid substrates. For stamp a the reproduction of the pattern is perfect over the whole stamp area (Figure 4a). In the same printing process, large features as well as lines of 1um width and 100 nm spacing could be reproduced. For thicker stamps especially small grooves in the elastomer tend to collapse, making a reproduction impossible. For stamp b which was

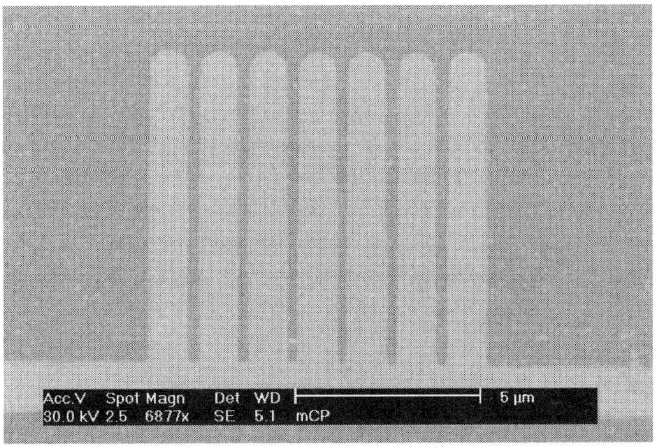

FIGURE 14.4. Results from microcontact printing with flat stamps on rigid carriers. For stamp a (a) the structures are nicely reproduced.

fabricated on a slightly less rigid substrate (silicon wafer instead of thick glass master) the shape of the features is still nicely reproduced, however, the uniformity of the printing is worse (Figure 4b).

For the thicker stamp, the sagging effect becomes clearly visible leading to a strong distortion of the patterns (Figure 4c).

In addition, FET gates were printed in between optically fabricated source drain structures (Figure 5). In these experiments, the glass master was used and the structures were aligned looking through the stamp using a xyz stage and an optical microscope. Alignment could be done to a precision of less than one micron and the overlay accuracy was good over an area of 5 × 5 mm. These results clearly show that MCP is suitable for the fabrication of devices in a multilevel lithography process when thin stamps on rigid carriers are used.

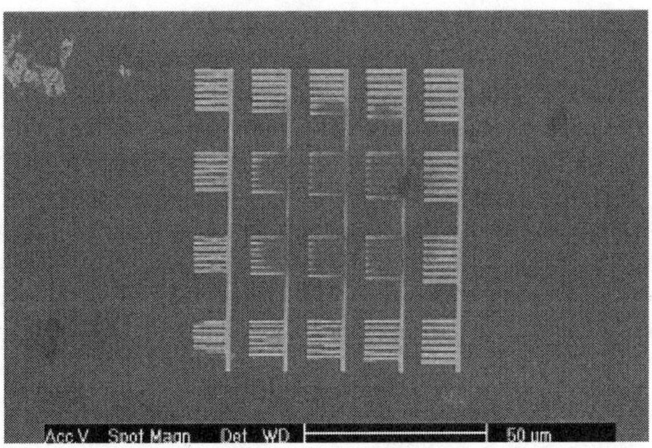

FIGURE 14.5. The result for stamp b (b) inhomogeneous protection is visible.

14.3. NANOIMPRINTING OF POINT CONTACTS

Although a real multilevel process for NIL allowing for the realization of complex electronic devices has not been developed yet, it is nonetheless possible to use NIL for the fabrication of quantum devices for which only a few fabrication steps are necessary. Although quite simple in form, these devices can require extremely high resolution lithography and their performance can serve as a tool for the assessment of NIL in fabrication of devices. In the following, a technological scheme for the fabrication of metallic point contacts is presented. Nanoimprint lithography (NIL) plays a key role in the process determining the critical dimension of the point contact.

14.3.1. Introduction

Electron transport through point contacts with a diameter of 10 nm or less is an interesting topic in transport physics.

One of the most reliable processes for the fabrication of these devices has been the lithographic formation of holes in thin freestanding membranes e. g. Si_3N_4. The holes are fabricated using high resolution electron beam lithography and dry etching processes, followed by metal deposition on both sides of the membrane. Although quite simple, this method requires expensive equipment and the supporting membranes are sensitive to mechanical stress.[24,25] The latter makes the handling during electrical measurement more difficult than for other devices.

Our fabrication scheme uses NIL to fabricate point contacts in polymer membranes located on a solid substrate. The resulting structures were investigated at low temperatures.

The NIL process optimized for point contacts is described in detail below. NIL is based on deformation of a thermoplastic polymer under high pressure and at a temperature above the glass transition temperature. A rigid stamp with a surface relief is pressed to the polymer film deposited on a substrate, the system is subject to high temperature and high pressure and, after cooling down, the stamp is separated from the film. The remaining relief of the polymer film reproduces the negative of the stamp pattern.

14.3.2. Process

The technological process of the point contact fabrication includes several steps which are shown schematically in Figure 6.

To define the small dimension of the point contact, stamps with sharp protruding tips have been prepared (Figure 7). The stamp size is 1×1 cm2. The stamp pattern has a specific geometry with areas containing single tips and areas with arrays of tips.

The arrays are intended for microscopical inspections of the stamp and the imprint results, while the areas with single tips are used for electrical measurements of single point contacts. The stamps have been fabricated from a 500 um thick Si(100) wafer. Holes of 200 nm in diameter are defined in a 350nm thick PMMA 950K resist film by electron beam lithography (EBL). 100 nm thick Ti dots of 0.2 um diameter, prepared on the Si surface after the EBL by metal deposition and subsequent lift-off, serve as a mask

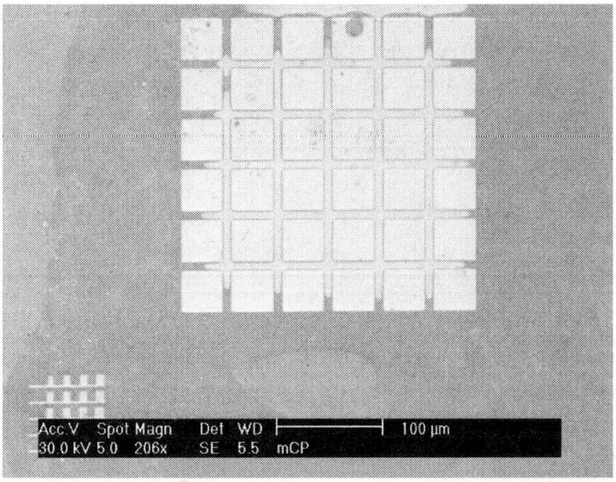

FIGURE 14.6. For stamp c (c) the sagging effect make the pattern transfer impossible.

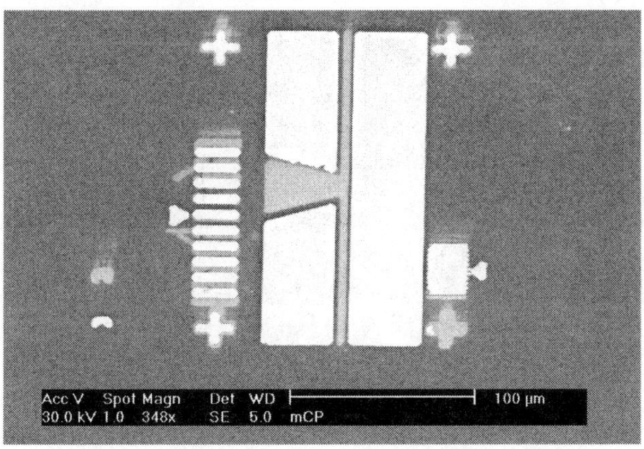

FIGURE 14.7. FET structure. The gate was fabricated using MCP and wet etching. The source/drain contacts were fabricated by optical lithography and lift-off. The gate structure was aligned optically through the stamp.

for ion beam etching of the Si. The ion beam etching of the materials has been done in a MRC 822 sputtering system with 1500eV Ar+ ions. The current density has been varied between 0.2 and 1.6 mA/cm2. At a current density of 1 mA/cm2, the etching rate of Si was about 450 Å/min. Because of the presence of a small amount of oxygen in the discharge chamber, the selectivity in etching between Si and Ti was more than 3:1. Due to lateral etching of the mask and facet formation on the mask edges, the diameter of the Ti dot diminishes during etching, and at an etch-depth about 200 nm, the Ti is completely removed from the top of the structures. Mask facetation and material redeposition on the side walls of the etched structures finally result in sharp silicon tips with height from 200 to 250 nm and an approximately 40° vertex angle is formed. From SEM images we infer that the radius of the tip curvature can be as small as 5 nm. Inspection of the tip arrays reveals a high uniformity in shape, which results from the high precision patterning of the mask by electron beam lithography. We estimate that the height of the tips on one sample (more than 10000 tips) varies by less than 10%. Even in the worst case the lateral size of the top of the tips was not larger than 20 nm. This small dimension should define the smallest point contact that can be obtained by our process. The areas with arrays and single tips are bounded by crosses, which serve for alignment during further processing.

Step 2 consists in the preparation of the stamp and the sample for the imprint process. The functional layers are deposited on a Si substrate with a high ohmic resistance in order to avoid any influence of the substrate during electrical measurements. First, the Si substrate is treated for improved adhesion. It is cleaned following a reduced standard RCA cleaning procedure followed by oxygen plasma treatment. Subsequently, a sticking layer of 10 nm Cr and a 100 nm Au layer are deposited by electron gun evaporation. Three subsequent layers of 110 nm of PMMA 50K are spun onto the sample and baked. The multiple layer coating helps to reduce the number of defects which may be formed during spin-coating. As will be shown and discussed below, one of the layers (the bottom one) must be processed to decrease the probability of pin-hole creation during imprinting. With the help of low-rate oxygen reactive ion etching, the obtained resist thickness is adjusted to the height of the stamp structures (we fabricated five different stamps with slightly different tip heights). Then, the stamp is soaked in a 2% solution of dodecyltrichlorosilane in toluene for one hour in order to provide good release after imprint. This treatment is carried out in a glove-box in dry nitrogen atmosphere.

Imprinting (Figure 6b) is performed using a pneumatic press which delivers a maximum force of 1500 N. A special insert has been used to provide a optimized parallelism of the sample and the stamp. The glass transition temperature of the polymer that we used was 115°C. The imprinting temperature was varied from 200° to 210°C, and the maximum applied pressure was 100 bar. The stamp and the sample were kept in contact under high pressure and high temperature for 30 minutes. After cooling, the stamp was separated from the substrate.

For creating the point contact, the pits embossed in the polymer must now be filled by metal (Figure 6c, 6d). After imprinting, a thin polymer layer remains on the bottom of the engraved pits, preventing direct electrical contact. Using a low power oxygen plasma the remaining polymer on the bottom of the pits was removed to guarantee electrical contact to the Au layer.

There are two obvious ways of metallization of the pit – (i) metal evaporation through a shadow mask with a hole geometry corresponding to the stamp topology, or (ii) electroplating through the small openings in the resist mask. The electroplating is more convenient as it can be carried out after short etching. In case of insufficient electrical contact, no gold is deposited and the process can be repeated after an additional etching step. The first method is easier to be realized and it was used for our first experiments. The diameter of the top metal contact is 500 um and bonding can easily be realized using thin wires and conductive glue. An additional window (200x300 μm^2) to the bottom Au layer was opened by exposure to a high electron dose ($\sim 1 \times 10^{-3}$ C/cm^2) and developing in isopropanol, which is known not to harm PMMA. Through this window the second contact for the transport measurements is made to the bottom gold layer.

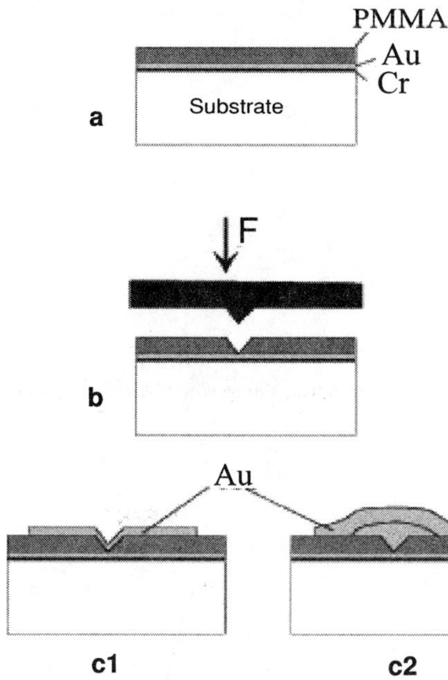

FIGURE 14.8. Main steps for the fabrication of metallic point contacts. A multilayer which consists of two metal layers and a polymer is deposited on the substrate (a). A sharp tip is embossed into the polymer leaving a conically shaped hole (b). After removing the polymer which remains at the bottom of the hole, metal is introduced either by evaporation (c1) or by electrodeposition (and subsequent evaporation) (c2) leaving a metallic point contact between top and bottom metallization.

14.3.3. Results

We have tested the entire scheme of the point contact fabrication and the contacts have also been characterized electrically (Figure 8).

The measured results are typical for contacts with a diameter larger than 10 nm, however, it is probable that by some optimization contact widths of ten nanometers or less will be achieved.

The main goal of the present work, however, was to study the imprinting step, namely, the shape of the engraved pits in order to understand the ability of NIL to produce small contact dimensions. Moreover, we have studied the masking properties of the PMMA layer that serves not only as a medium for patterning but as an insulating element of the electrical scheme as well.

The stamps with sharp tips and the imprinted surfaces have been studied in a scanning electron microscope. The clear and uniform patterns of the pits are presented in Figure 9.

However, it is impossible to estimate the pit bottom size with the obtained images. It could be estimated from the cross-section of the cleaved plane across the array if the structures are not deformed during the cleavage. We have used some additional methods for inspection of the pits. In particular, the engraved pits have been studied by contact mode AFM (DME Rasterscope 4000) (Figure 10).

FIGURE 14.9. SEM micrograph of an imprint stamp which was used for point contact fabrication.

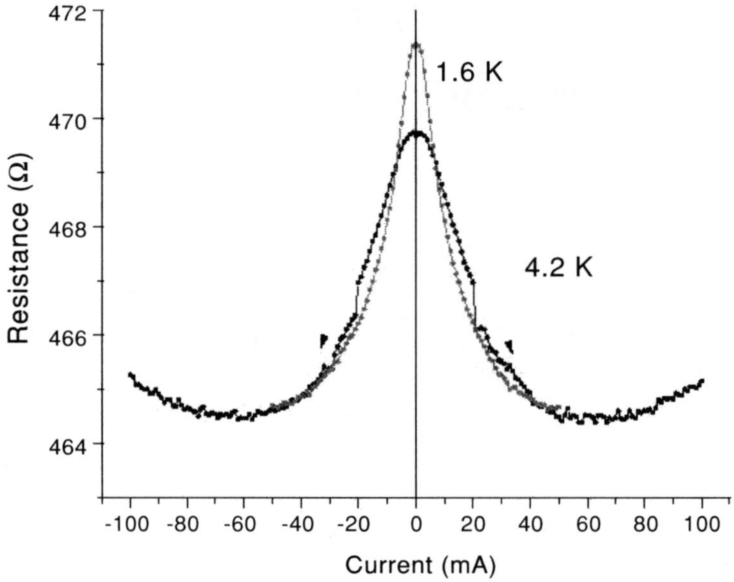

FIGURE 14.10. Resistance plotted versus current of a Ni/Au point contact fabricated by NIL and electrodeposition.

The obtained images give the exact value of the top hole diameter but do not show the real shape of the pits because of the finite size of the used cantilevers and the specific structure of the pattern. For a better understanding we have created replicas from the engraved polymer and studied them by SEM.

A thin layer of Ti/Au has been deposited onto the polymer surface after imprinting, which serves as a plating base for Ni electroplating in a Ni-sulfamate solution. The Ni layer was grown to a thickness of about 1nm and then a thick layer of Cu was grown up to 0.5-1mm thick in a copper sulfate solution. The PMMA film can be completely dissolved in acetone and a metal replica is obtained. We have studied thick metal replicas

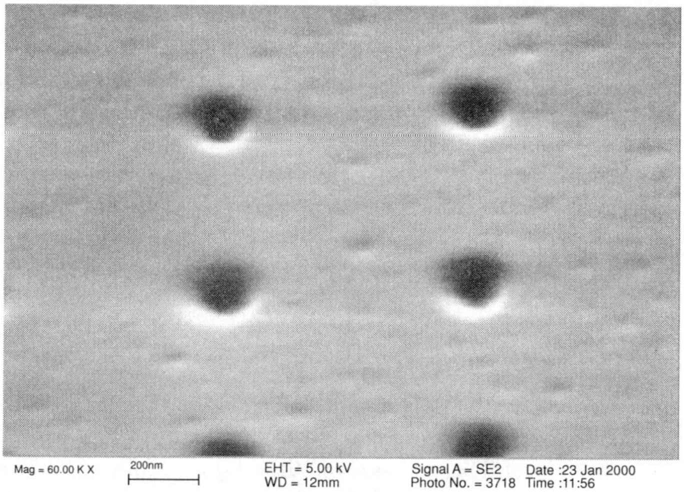

FIGURE 14.11. SEM micrograph of pits embossed into PMMA using the stamp with sharp tips.

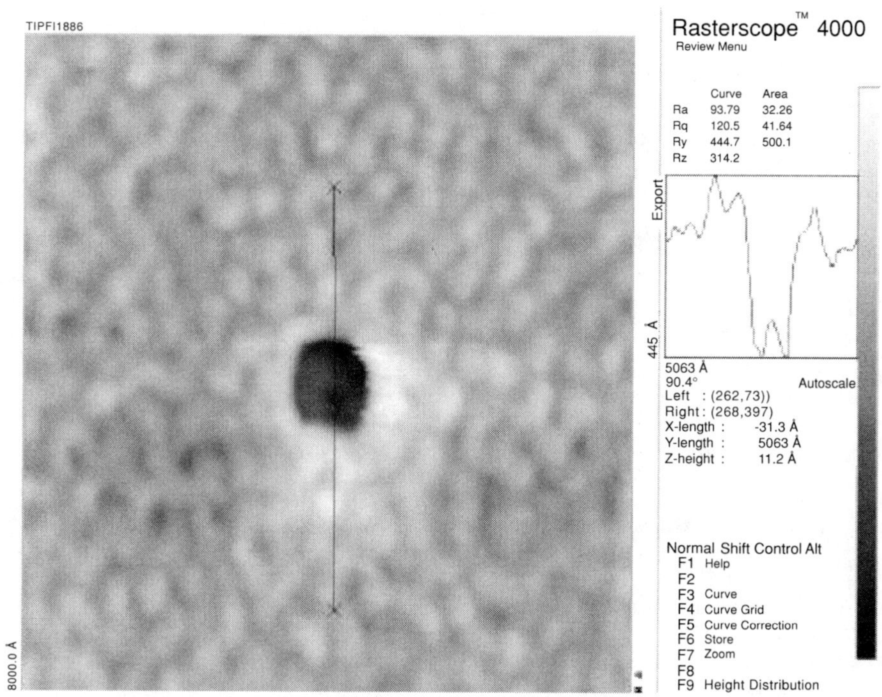

FIGURE 14.12. AFM picture of a single pit in PMMA and crossection (right).

that have self-separated from the polymer surface due to stress using a Philips XL30 SEM (Figure 11). The tips of the replica are not as sharp as the tips of the stamp. This can be explained by imperfect imprint, destruction of the top of the tip during separation, or partial filling of small pits with metal. However, the top size of the replica structures is estimated to be about 40 nm. We consider that some improvements in the accuracy of the imprint process and inspection techniques in the nearest future will help us confirm the suitability of the technique to obtain pits which are small enough for the fabrication of the point contacts.

CONCLUSIONS

The fabrication of metallic point contacts has shown that at least for special processes nanoimprint lithography can replace electron beam lithography in device fabrication. However for more complex processes multilevel lithography and alignment are still issues which restricts the application of NIL. Nevertheless, for microcontact printing the application of multilayer processes and the use of thin stamps on solid carriers can create a versatile tool even for several subsequent lithography steps with different subsequent processing steps including sub-micron alignment.

ACKNOWLEDGMENTS

The work was supported by the ESPRIT project No. 29097 (SPINUP). The authors thank Mrs. M. Emmerling (Department of Technical Physics of the Physical Institute of the Wuerzburg University) for help in the fabrication of the stamps for imprinting.

REFERENCES

1. A. Kumar and G. Whitesides, Appl. Phys. Lett., 1993, 63, 2002.
2. J.L. Wilbur, A. Kumar, E. Kim, G. M. Whitesides, Adv. Mater., (1994), 6, 600.
3. Y. Xia and G. M. Whitesides, Angew. Chem. Int. Ed. Eng., (1998), 37, 550.
4. L. Libiouville, A. Bietsh, H. Schmid, B. Michel, and E. Delamarche, Langmuir, (1999), 15, 300.
5. L. B. Goetting, T. Deng, and G. M. Whitesides, Langmuir, (1999), 15, 1182.
6. X. M. Yang, D. A. Tryk, K. Hasimoto, and A. Fujishima, Appl. Phys. Lett., (1996), 69, 4021.
7. L. Yan, X.M. Zhao, and G. M. Whitesides, J. Am. Chem. Soc. , (1998), 120, 6179.
8. S.Y. Chou, P.R. Krauss, and P.J. Renstrom, Nanoimprint lithography, J.Vac.Sci.Technol. B14, (1996), (6) 4129-4133.
9. S.Y. Chou, P.R. Krauss, and P.J. Renstrom, Imprint lithography with 25-nm resolution, Science, (1996), 272, 85-87.
10. S.Y. Chou, P.R. Krauss, and P.J. Renstrom, Imprint of sub-25 nm vias and trenches in polymers, Appl.Phys.Lett. , (1995), 67(21) 3114-3116.
11. L. Guo, P.R. Krauss, and S.Y. Chou, Nanoscale silicon field effect transistors fabricated using imprint lithography, Appl.Phys.Lett. , (1997), 71(13) 1881-1883.
12. P.R. Krauss and S.Y. Chou, Nano-compact disks with 400 Gbits/in2 storage density fabricated using nanoimprint lithography and read with proximal probe, Appl.Phys.Lett. , (1997), 71(21) 3174-3176.

13. I. Martini, D. Eisert, S. Kuhn, M. Kamp, L. Worschech, J. Koeth, and A. Forchel, Fabrication of quantum point contacts by imprint lithography and transport studies, Proc. EIPBN 2000, to be published in J.Vac.Sci. Technol.B.
14. X. M. Zhao, J. L. Wilbur, and G. M. Whitesides, Langmuir, (1996), 12, 3257.
15. D. Qin, Y. Xia, A. J. Black, G. M. Whitesides, J. Vac. Sci. Technol., (1998), B **16**, 98.
16. H. A. Biebuyck, N. B. Larsen, E. Delamarche, B. Michel, IBM J. Res. Develo.**41**, (1997), 159.
17. J. J. Wilbur, A. Kumar, H. A Biebuyck, E.Kim, and G. M. Whitesides, Nanotechology **7**, (1996), 452.
18. W. Langheinrich, H. Beneking, Microelectronics Engineering (1991), **83**, 225.
19. Y. Xia, X. M. Zhao, E. Kim, G. M. Whitesides, Chem. Mater., (1995), **7**, 2332.
20. A. Bietsch, B. Michel J. Appl. Phys. (2000), 88, 4310.
21. E. Delamarche, H. Schmid, B. Michel, H. Biebuyck, Adv. Material, (1997), 9, 741.
22. B. A. Grzybowski, S. T. Brittain, and G. Whitesides, Rev. Sci. Instrum., (1999), 70, 2031.
23. H. Schmid, B. Michel Macromolecules, (2000), 33, 3042.
24. N.N. Gribov, S.J.C.H. Theeuwen, J. Caro, and S. Radelaar, A new fabrication process for metallic point contacts, Microelectronic Engineering, (1997), 35(1-4) 317-320.
25. N.N. Gribov, S.J.C.H. Theeuwen, J. Caro, E.van der Drift, F.D.Tichelaar, T.R. de Kruijff, and B.J.Hickey, Fabrication of metallic point contacts: a new approach for devices with a multilayer or a heterointerface, J.Vac.Sci.Technol., (1998), B16(6) 3943-3947.

15

Optical Applications of Nanoimprint Lithography

J. Seekamp

University of Wuppertal, Institute of Materials Science and Department of Electrical and Information Engineering, Germany

15.1. INTRODUCTION

Optical circuits and devices are constantly growing in relevance in every day live. This refers not only to obvious applications like cameras and eyeglasses but includes modern measurement and control systems in, for example, cars. Optical fibres have taken over information technology with their bandwidth not remotely rivalled by terrestrial or satellite radio connections. The classical optical materials were glasses and transparent crystals. The high refractive indexes that most of these materials posses in comparison to polymers are bought for high weight and more extreme process parameters.

The weight and fragility argument won the case for plastic eyeglasses. The ease of processing in mould and extrusion tools opened the way for cameras with plastic optics. The CD is a polymer optical unit, which, despite it being a high precision device, enjoys only the appreciation of a book or an advert brochure.

Thus printing, pressing and moulding techniques have been widely used to produce optical devices for a long time. However, these applications are based on single devices subsequently assembled into optical systems. The typical size of these systems ranges from cm to m. All of them are based on devices rather than being integrated systems. The term integrated has a double meaning in the field of optics. It describes mainly miniaturised optical benches, individually made and aligned. The second meaning refers, analogous to the electronic world, to "monolithical" systems produced in huge quantities by parallel processing on maximised areas. The second meaning is the one used in this chapter. It is within this second definition that the positive contribution of nanoimprint lithography (NIL) will be maximised.

There exist basically two ways to introduce the main advantages of NIL, its very high resolution (<10 nm lateral size), parallel character and moderate to high throughput

(>1060 cm^2/hour per 150 mm wafer press), to the research on and production of optical systems.

The first is the use of NIL as a mask patterning technique. This uses the option to pattern silicon oxide or metal films as etch masks if not using the printed polymer itself. The second way heads for complete optical devices and systems defined in a single printing step. This concept has been realised for integrated optical spectrometers[1] already, using the coarser sibling of NIL, LIGA (*LIthographie* (lithography), *Galvanik* (galvanic), *Abformung* (moulding)), which was developed in the early 1980s.[2,3] NIL widens the perspective opened then to subwavelength devices in the visible and near infrared spectral range, i.e., including the optical information technology windows around 1.3 μm and 1.5 μm for medium to large-scale production.

The focus of this chapter is therefore on photonic devices allowing to shrink systems, which are of cm scale today to mm-sized monolithical devices. For integrated optical systems NIL makes higher level integration on the basis of photonic systems in medium to large numbers feasible for the first time.

15.2. CANDIDATES FOR AND EXAMPLES OF PRINTED OPTICAL DEVICES

Replicated gratings were replicated by imprinting techniques in the 1940s already. CDs, addressing a mass market in video technology were firstly made in 1971 and published from 1973.[4] These two early examples of printed optical devices fell short only one order of magnitude in terms of the lateral sizes published in the early NIL papers by S.Y. Chou.[5,6] However, the need for ultra small optical devices and subwavelength devices in a more confined sense only arose around that time. This section introduces a set of sub wavelength optical devices, which showed remarkable performance and unsustainable demands to the fabrication technology before. Their fabrication technology mainly is electron beam lithography, a serial, expensive tool with insufficient throughput. After a list of candidates, examples for the successful transition from the serial world of single beam writing to the parallel one of NIL are given.

15.2.1. Candidates

Information technology strongly demands for devices to add or drop information channels to and from optical backbones. This translates into add and drop filters inspired by the technology of optical fibres with Bragg gratings. An example of such a device structured by a combination of X-ray- and UV-lithography was given by Lim et al. in 1999.[7] The grating period was 244 nm and the completed add-drop system was made in InGaAs/InP.

The second structure is a subwavelength transmission grating to control the polarisation of light emitted by a vertical cavity surface emitting laser.[8] The third is a detector with improved performance. It is a metal-semiconductor-metal (MSM) photodetector with metal structures patterned by electron beam lithography.[9] The electrodes are 25 nm wide separated 25 nm. The minimum response time achieved with these structures was

0.87 ps for low temperature grown GaAs and 10.7 ps for crystalline silicon as the semiconductor material.

15.2.2. Optical devices made by NIL

Of the candidates mentioned above, the MSM detector was made by NIL in 1999.[10] The minimum electrode spacing was 200 nm. The main purpose of this work was to investigate NIL induced changes in the dark current of these detectors. Any change of the dark current would indicate a change in the electronic material properties, most likely arising from damage. The dark currents of the electron beam lithography patterned devices of reference[9] were of the order of 10 nA - 100 nA. The dark current for the NIL produced sensors was in the same range. More extensive characterisation of detectors with electrode spacing between 2 μm and 200 nm, comparing devices structured by UV- and electron beam lithography with NIL made devices, revealed unchanged dark currents for pressures of 2 MPa and 4 MPa. Increased dark currents were found for a pressure of 6MPa. Addressing the issue of NIL induced damage with respect to optical structures like the laser polarisers mentioned before; investigations of quantum well stacks were performed on $GaAs/Al_{0.3}Ga_{0.7}As$ and $Ga_{0.47}In_{0.53}As/InP$ to assess the change of emission properties due to NIL.[11] The stack consisted of three quantum wells with different emission wavelengths and in different depths of the substrate. Exposure to temperatures up to 190°C and pressures up to 20 MPa during the printing process induced no significant change of the photoluminescence and photoluminescence excitation spectra. Only a pressure of 60 MPa at 190°C led to changes in the quantum well placed at a maximum depth of 3 nm. Thus NIL is suitable as a mask patterning technique for optically active devices. An integrated waveguide polariser structurally similar to the device of reference[7] was presented by Wang et al.[12] Two types of waveguides incorporating gratings were realised. The first model is patterned in SiN on silicon, defining the mask for the 190 nm period grating by NIL, metal lift off and dry etching prior to a metal deposition on top of the structure and the fabrication of a superimposed rib waveguide by UV-Lithography and dry etching. This results in a two-level metal grating on top of the waveguide. The second type consists of a grating in silicon oxide patterned using NIL, metal lift off and dry etching. On this 190 nm period grating aluminium is deposited as for the first species and a rib waveguide in PMMA is defined using NIL again.

Directly printed active optical devices were presented in 1997.[13] To achieve distributed feedback lasers benzocylobutene (BCB) was imprinted with a grating and subsequently coated with die-doped hydroxyquinoline aluminium (Alq). The threshold power to observe laser action was about 12 kW/cm^2 at a wavelength of 657 nm. Structures printed on a flexible substrate exhibited tunability, although at the non-neglegible excitation power of 400 kW/cm^2. Feedback structures, i.e., organic materials with an imprinted surface corrugation were the subject of several studies thereafter and will be discussed in more detail in section 3.[14,15] The sensitivity of light emitting organic materials to chemical treatment or radiation, motivated studies on the influence NIL would have on the luminescence properties of these materials. For die-doped Alq, e.g., no decrease in photoluminescence intensity was found.[16] Yu et al. presented a passive optical device using the high resolution of NIL for a polariser consisting of metal gratings directly evaporated onto a printed polymer pattern.[17] The advantage of this structure lies in the double grating achieved this way, i.e., two levels of parallel metal stripes reflect the light and increase the reflectivity of the device as well as its polarising efficiency. The ratio between intensities in two orthogonal polarisations is about 200.

Most of the optical devices which have been realised using NIL use periodic structures, often with subwavelength periods. Structures of this sort have applications beyond polarising or distributed feedback structures. Polymer waveguides with periodic surface corrugations could serve as optical systems and they could be made in a single printing step using NIL.

15.3. NANOIMPRINT LITHOGRAPHY OF PHOTONIC DEVICES

Periodic changes of the refractive index, e.g., due to corrugations of the surface in the scale of the wavelength, are often referred to as photonic crystals because their geometry provides a photon density of states remotely similar to the electronic density of states in solid state materials.[18,19] Natural examples of these structures are the butterfly wing or opals, both of which were identified as potentially interesting and worth to copy.[20–22] In this section a view will be given on photonic structures printed or potentially printable using NIL. Firstly the term photonic crystal should be defined for the scope of this chapter. Here a photonic crystal is a structure with a periodically varying refractive index. This variation of the refractive index induces a dispersion relation within the structure owed not to the material properties alone but to the geometry of the device as well. A comparatively simple example for such a periodic structure is a grating used as a waveguide rather than as a diffractive element (Figure 1). The most prominent phenomenon of the altered dispersion relation of this structure is the Bragg reflex.

However, this class of simple structures offers a broader variety of optical phenomena suitable to add functionality to a printed device. To describe this variety a short excursion is undertaken next.

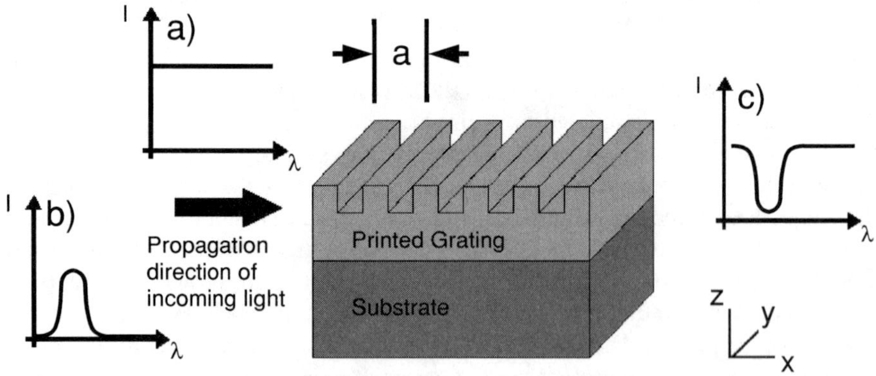

FIGURE 15.1. Sketch of a "grating waveguide" with the expected reflectance (b) and transmission (c) spectra for a given incident spectrum (a).

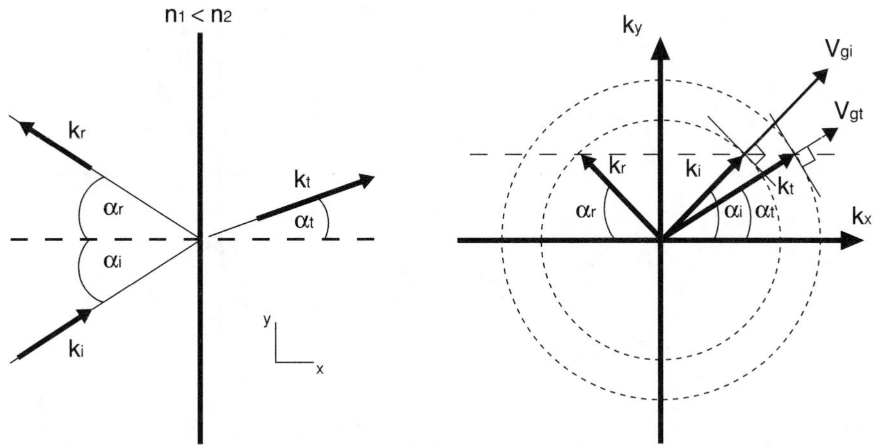

FIGURE 15.2. Reflection and refraction of a beam on the planar interface between two dielectrics. Propagation directions (a) and wave vector diagram (b) for the interface processes. The dash dotted lines in the wave vector diagram are the iso frequency circles in the media 1 and 2. The group velocity for all waves is parallel with the respective wave vector.

Given a simple interface between two dielectric materials and the reflection/refraction of a beam at this interface the well known schematic given in Figure 2 (left) describes the reflection and refraction in terms of beam optics. The tangential component of the wave vector $\mathbf{k_y}$ stays constant and the perpendicular component changes. The right part of Figure 2 shows the wave vector diagram for this case. This representation of wave vectors, uncommon for this simple example, is chosen as it provides easy access to processes occurring in media with a periodic variation of the refractive index.

If the refractive index change is periodic, the description of the system utilising the Floquet theorem yields a periodic continuation of the wave vector diagram with integer multiples of the lattice vector as the period (Fig. 3). The periodicity leads to a split in the dispersion if the change of the refractive index is sufficiently high (Fig. 3 thick line). The dash-dotted Lines in Figure 3 represent the case of a vanishing modulation of the refractive index. The group velocity determines the propagation direction of a wave:

$$\mathbf{v}_g(\mathbf{k}) = \nabla_k \omega \tag{1}$$

Hence the group velocity is orthogonal to the iso frequency surface. On the left hand side of Figure 3 the continuos line indicates the iso frequency line for the first order Bragg reflection for light impinging perpendicular to the ribs of the grating. Light that propagates parallel to $\mathbf{k_2}$ and $\mathbf{k_3}$ will propagate as for the equivalent unstructured slab waveguide. The wave vector and the group velocity vector of the wave are parallel. The coordinates and descriptors in Figure 3 are chosen in agreement with Figures 1 and 2. Light with wave vector $\mathbf{k_1}$ will not propagate in the grating waveguide and is reflected. On the right hand side an iso frequency line for a higher frequency is sketched. A wave with the wavevector $\mathbf{k_3}$ will still propagate as before. Its wave vector and group velocity

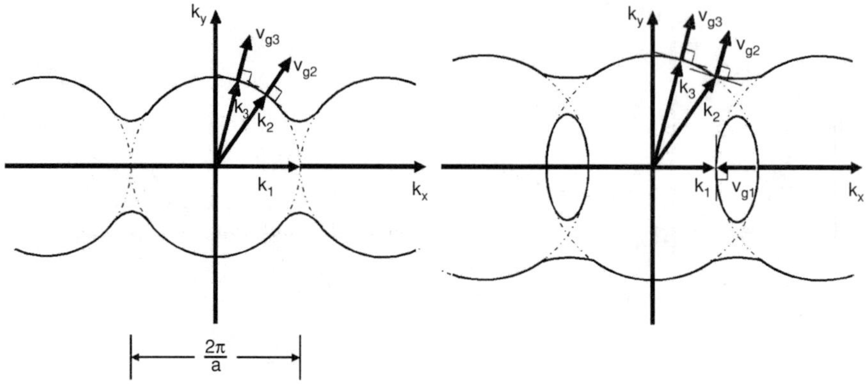

FIGURE 15.3. Wave vector diagram for a 1D periodic grating. The dash dotted lines indicate the iso frequency circles for a vanishing modulation of the refractive index. The solid line indicates the iso frequency line of the periodic structure. The drawing on the left shows the case of the lowest frequency suitable to observe a Bragg diffraction. The sketch on the right depicts the case of a slightly higher frequency. Wave vector and group velocity of the wave are not necessarily parallel anymore.

remain parallel. A wave with the wave vector $\mathbf{k_2}$ propagates parallel to light with the wave vector $\mathbf{k_3}$. Its wave vector and group velocity are not parallel. Waves along $\mathbf{k_2}$ and $\mathbf{k_3}$ are collimated. A wave with $\mathbf{k_1}$ will propagate in the opposite direction compared to an equivalent unstructured waveguide as $\mathbf{k_1}$ and $\mathbf{v_{g1}}$ are anti parallel. Yet it is propagating in the waveguide unlike in the first case.

A more detailed description for this example of a 1D periodic structure could be found in reference.[23] An experimental study of these phenomena in dry etched rather than printed anorganic gratings is available in reference.[24] In conclusion, a 1D photonic structure could already provide a function. The example given in Figure 3 demonstrates the collimation of light. Based on Ta_2O_5 as the waveguiding material a beam splitter was realised with a dry etched grating as the photonic structure.[24] For this experiment light was impinging on a grating dry etched into Ta_2O_5 at an angle. The light of wavelengths 672.8 nm and 676.7 nm passed this grating and was deflected with angles of 17° and 33.5°, respectively, achieving a suppression of the respective other channel of at least 12 dB with a damping below 5 dB by the grating structure. The relative deviations from the Bragg peak were Ω =-0.0187 andgΩ =-0.0128 respectively.

The first step to characterise such gratings made by nanoimprinting was to investigate the Bragg diffraction of a grating printed in polystyrene on silicon oxide.[25] The period of the grating was about 1 μm and the sixth and seventh order Bragg diffraction of it were observed. An atomic force microscope image of the stamp and an imprint of this structure are shown in Figure 4. The characteristic of this grating was measured using butt coupling with a 200 μm core diameter fibre to couple into and out of the waveguide structure. To proceed to comparative and more quantitative measurements a grating of a similar geometry was printed into a die-loaded polymer. The setup shown in Figure 5 permitted the pathlength of the photoluminescence to be varied to the edge of the structure with a higher accuracy and flexibility than accessible by cutback measurements.[25] For that purpose a polymer specifically developed for NIL, MRI 8000 of mrt GmbH,

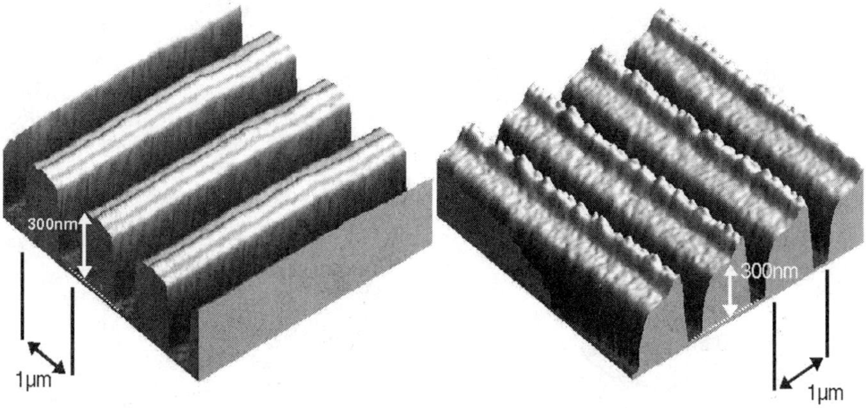

FIGURE 15.4. Tapping mode atomic force microscopy images of a silicon stamp of lines dry etched into silicon with a period of 1 μm and a depth of 300 nm (left) and a print of this stamp into a PS film on silicon oxide. The geometry transfer is very good. The interactions of the scanning probe tip and the sample lead to a narrowing of grooves on the stamp and on the sample image. While the transfer of details and the grating period are almost perfect, the freestanding edges of the polystyrene round. This is attributed to anelastic properties of the polymer.

Berlin, was loaded with a perylene die covalently bonding it to the polymer backbone.[26] This polymer spun on silicon oxide was imprinted to a depth of 300 nm leaving a residual layer thickness of 350 nm. The printed grating had a period of 800 nm. In Figure 6 the edge emission of a 4 μm wide waveguide printed to the same depth on the same sample is compared to the edge emission of a grating for light travelling about 100

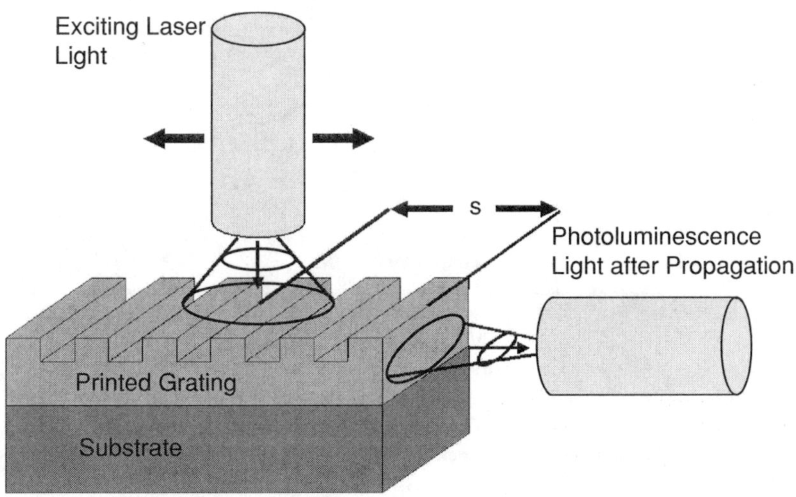

FIGURE 15.5. Transmission measurement setup for a 1D PC structure. The transmission is characterised using the photoluminescence of a dye bound to the polymer backbone. The exciting laser light is impinging from a monomode waveguide with 9 μm mode diameter. A multimode fibre with 200 μm core diameter collects the emitted light.

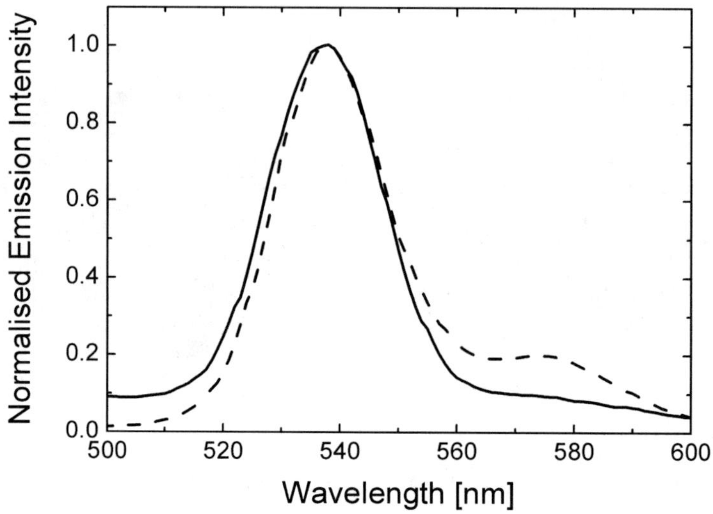

FIGURE 15.6. Edge emission of a 4 μm wide rib waveguide. The waveguide was printed to a depth of 300 nm into the dye-loaded MRI8000, allowing for a 350 nm residual layer. The excitation was by 458 nm laser light (dashed line). The solid line corresponds to the edge emission of a 800 nm period grating printed to a depth of 300 nm on the same sample after passing 100 periods of the grating (solid line).

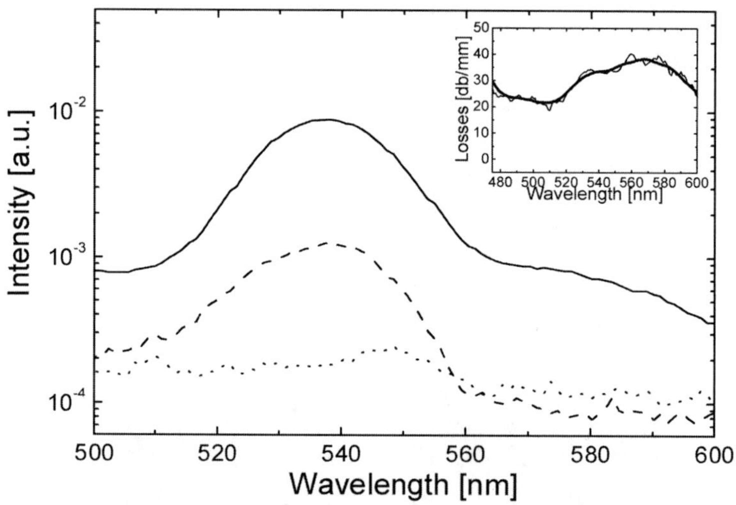

FIGURE 15.7. Edge emission of the grating from Figure 6 after light paths of about 100 grating periods (~80μm, solid line), 600 grating periods (~460μm, dashed line) and 900 grating periods (~720 μm, dotted line). The inset shows the damping in the structure assuming the Beer-Lambert law still applies for this ultrarefractive structure.

grating periods. The second emission peak of the dye at 580 nm is almost completely suppressed.

The modulation of the surface need not to be very deep to observe an effect on guided light.[24] In fact, the modulation has to be as shallow as possible to avoid unnecessary losses. For the example given in Figure 6 absorption measurements were made by moving the position of the laser beam spot away from the edge. Figure 7 shows the emission for light propagating for about 100 grating periods together with spectra for about 600 grating periods and 900 grating periods. After 900 periods the luminescence peak of the dye at 530 nm is completely damped and becomes indistinguishable to the noise level. The inset of Figure 7 shows the absorption of the grating under the daring assumption that the Beer-Lambert law still applies. For the 4 μm wide rib waveguide, absorption losses, measured and analysed in the same way were found to be at least two orders of magnitude lower than for a propagation perpendicular to the 800 nm period grating.

15.4. OUTLOOK AND CONCLUSION

While this chapter is written research on NIL of optical devices continues in interesting and promising directions. One representative, the topic of two dimensional photonic structures in polymer slab waveguides, shall be mentioned here.[25,27] These two dimensional photonic structures are currently tested for organic laser application.[28]

The first transfers from more conventional techniques for nanostructuring to NIL proved the capability of this approach. The transfer of very small features was used to realise polarising devices with sub wavelength period gratings. Lasing in organic materials was achieved using printed grating structures. And add drop filter concepts were shown that could be transferred to the NIL process.

The area of photonic devices provides room for the next step. Examining the light propagation in periodic structures more closely a way was shown to integrate optical systems with complex functionality on small areas (Figure 8). Summarising this approach the new nanostructuring by NIL and the still unconventional optics chosen in the field of photonic devices may provide a combination for high performance low cost and small size integrated optical systems of tomorrow.

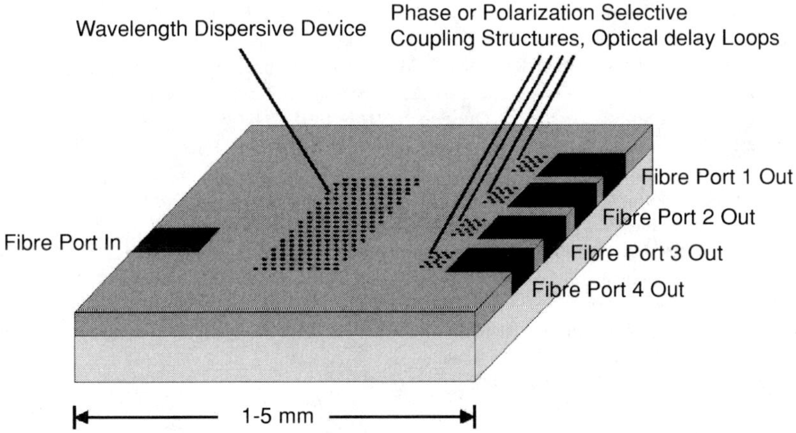

FIGURE 15.8. Schematic of an integrated optical system based on ultrarefractive elements.

ACKNOWLEDGEMENTS

The author would like to acknowledge the financial support by the European Union grants CHANIL (IST 1999-13415), APPTECH (IST 2000-29321) and MONALISA (GRD-CT2000-00349), the VW foundation, the German ministry of research and education (BMBF) grant (01BK253) and the German research council (DFG) grant (SCHE 580/3-1).

REFERENCES

1. D. Sander, J.Müller, Proceedings ECIO97 **1997** 248-251
2. D. Muschmeyer, W.Ehrfeld, SPIE Proc. 803 **1987** 72-79
3. E.W. Becker, W. Ehrfeld, D. Munchmeyer, H. Betz, A. Heuberger, S. Pongratz, W. Glashauser, H.J. Michel, R. von Siemens, Die Naturwissenschaften 69 (11) **1982** 520-523
4. P. Kramer, K.Compaan, Proc. IEEE international convention and exposition Vol. II **1973** 39.2-4
5. S.Y. Chou, P.R. Kraus, P.J. Renstrom, J.Vac.Sci.Technol. B 14(6) **1996** 4129-4133
6. S.Y. Chou, P.R. Kraus, P.J. Renstrom, Appl.Phys.Lett. 67(21) **1995** 3114-3116
7. M.H. Lim, T.E. Murphy, J.Ferrera, J.N.Domask, H.J. Smith, J.Vac.Sci.Technol. B 17(6) **1999** 3208-3211
8. L. Zhuang, S. Schablitzky, R.C. Shi, S.Y. Chou, J.Vac.Sci.Technol. B 14(6) **1996** 4055-4057
9. S.Y. Chou, M.Y.Liu, IEEE J.Quant.Elec. 28(10) **1992** 2358-2368
10. Z. Yu, S.J. Schablitzky, S.Y. Chou, Appl.Phys.Lett. 74(16) **1999** 2381-2383
11. S. Zankovych, J.Seekamp, S.Romanov, C.M. Sotomayor Torres, I. Maximov, M. Beck, I. Shonubalko, L. Montelius, D. Reuter, P. Schafmeister, A.Wieck, J. Ahopelto, MNE 2002, to be published in Microelec. Eng. **2002**
12. J. Wang, S. Schablitzky, Z. Yu, W. Wu, S.Y. Chou, J.Vac.Sci.Technol. B 17(6) **1999** 2957-2960
13. M. Beggren, A. Dodabalapur, R.E. Slusher, A. Timko, O.Nalamasa, Appl.Phys.Lett. 72(4) **1998** 410-411
14. J.M. Lapton, B.J. Matterson, I.D.W. Samuel, M.J. Jory, W.L. Barnes, Appl.Phys.Lett. 77(21) **2000** 3340-3342
15. J.R. Lawrence, P. Andrew, W.L. Barnes, M. Buck, G.A. Turnbull, I.D.W. Samuel, Appl.Phys.Lett. 81(11) **2002** 1955-1957
16. J. Wang, X. Sun, L. Chen, S.Y. Chou, Appl.Phys.Lett. 75(18) **1999** 2767-2769
17. Z. Yu, P. Deshpande, W. Wu, J. Wang, S.Y. Chou, Appl.Phys.Lett. 77(7) **2000** 927-929
18. J.D. Johannopoulos, R.D. Meade, J.N. Winn, *Photonic Crystals . Molding the Flow of Light*, Princeton University Press **1995**
19. S.G. Johnson, *Photonic crystals: From Theory Practice*, PhD thesis, MIT **2001**
20. V.N. Astratov, V.N. Bogamolov, A.A. Koplyanskii, A.V. Prokofiev, L.A. Samoilovich, S.M. Samoilovich, Yu.A. Vlasov, Il Nuovo Ciemento 17D **1995** 1349-1354
21. M. Müller, R. Zentel, T. Maka, S.G. Romanov, C.M. Sotomayor Torres, Adv.Mater. 12(20) **2000** 1499-1503
22. K.M. Ho, T.C. Chan, C.M. Soukoulis, R. Biswas, M. Sigalas, Solid State Comm. 89(5) **1994** 413-416
23. D.N. Chigrin, C.M. Sotomayor Torres, Opt.Spectr. 91(3) **2001** 484-489
24. R. Zengerle, *Lichtausbreitung in ebenen periodischen Wellenleitern*, PhD Thesis, Universität Stuttgart **1979**
25. J. Seekamp, S. Zankovich, A.H. Helfer, P. Maury, C.M. Sotomayor Torres, G.Böttger, C. Liguda, M.Eich, B.Heidari, L.Montelius, J.Ahopelto, Nanotechnology 13(5) **2002** 581-585
26. Ch. Finder, M. Beck, J. Seekamp, K. Pfeiffer, P. Carlberg, I. Maximov, F. Reuther, E.L. Sarwe, S. Zankovich, J. Ahopelto, L. Montelius, C. Mayer, C.M. Sotomayor Torres, MNE 2002, to be published in Microelec. Eng. **2002**
27. C. Liguda, G. Böttger, A. Kuligk, R. Blum, M. Eich, H. Roth, J. Kuhnert, W. Morgenroth, H.Elsner, App.Phys.Lett. 78(3) **2001** 2434-2436
28. http://optics.org/articles/news/8/14/1 **2002**

16

Biotechnology Applications of NIL

Lars Montelius and Babak Heidari
Lund University, Div of Solid State Physics & The Nanometer Consortium, Lund University, Sweden

16.1. INTRODUCTION

Nanoimprint lithography (NIL) is a very interesting technique for biotechnological applications due to the fact that nanostructures can reproducible be formed at a low cost. The real massive exploration of nanotechnology in biotechnological applications has not been hindered by lack of ideas of what to do but more about how to get it done at a sufficiently low cost. Biotechnology has for many years taken up technologies from microelectronics when their life time in the microelectronics arena has been completed. However, with the introduction of nanoimprint lithography a new scenario is established and it is possible for the biotechnological community to produce various kinds of advanced nanostructured surfaces without the need to build up a complete nanotechnology process line and without the need to take over used equipment from the microelectronics labs. Nevertheless, the NIL technique has not yet been widely adapted by the biotechnology community. The reason for this is first of all being due to small exposure of this technique to the bio-community. This is however rapidly changing right now thanks to the worldwide boost of nanotechnology that presently is taking place, partly due to the US Nanotechnology Initiative that in turn made Asian and European governments and research organizations to launch similar kind of large programs devoted to nanotechnology and -science.

16.2. INTRODUCTION TO NIL

In nanoimprint lithography[1] a stamp with protrusions or extrusions are employed. The stamp is placed into contact with a spin-coated polymer layer on top of a substrate held at a temperature above the glass transition temperature of the polymer and then when a pressure is applied, the stamp is pressed into the polymer layer that deforms. The substrate is cooled quickly and finally the pressure is reduced and finally the stamp and substrate are released from each other. Hence a replication of the stamp features is produced in the polymeric layer. Then a variety of subsequent additive or subtractive processing can be performed e.g. i) a metal layer can be deposited by evaporation and lift-off, ii) the pattern can be transferred into the substrate by etching using the resist as an etch-mask. Alternatively, if the employed polymer is a conductive one, the patterned conductive polymer in itself could e.g. be the electrode. Thus, for batch fabrication of devices where patterning is required, NIL is a promising and ideal method. Therefore it is of interest to explore the possibilities of such methods for fabrication of various biosensor and bio-chip structures, i.e. polymeric structures /electrodes as well as metal electrodes, maybe also in combination with each other. Obviously a large class of devices that ideally can be made with NIL is patterns in conductive polymers for electronic applications.[2,3] However, that kind of structures for molecular electronics is a whole research field of its own and has to be reported elsewhere.

16.3. BIOTECHNICAL APPLICATION AREAS FOR NIL

At first sight NIL seems just to another nanolithography method that offer no certain advantages to the biotech community. However, this is not the case since the technique for the first time allow parallel processing capability and maybe even more important the technique does not use any kind of high energy beams of photons, electrons, X-rays or ions to produce the fine structures in the resist materials employed. The technique only pushes a stamp into a moderately heated polymer. This will in most of the cases not effect the fragile biological surface which means that new patterning avenues are opened up whereby structures may be made in various resists put on top of the biological surfaces in contrast to other methods where the biology has to be added after producing the structures in a surface. However, if one prefers it is of course possible to make the biological structuring both ways.

Furthermore, the technique opens up the possibilities to produce patterns in multilayers of polymers having different kinds of functionalities. For instance, one could produce trenches in a hydrophobic material put on top of a hydrophilic material that would make it possible to make nanofluidic channels for various kinds of bio-devices such as e.g. μTAS-systems.[4,5] This means that the technique opens up endless possibilities for applications in the biotechnologies. One way trying to structure these application areas is to classify them according to:

- Production of structured biological active surfaces.
- Production of electrode systems for bioanalytical or sensor purposes.
- Production of channels for handling of fluids.
- Production of complete labs similar to μTAS but in the nanoscale, "nTAS".
- Production of nanotextured surfaces for investigations of e.g. protein adsorption.
- Production of surfaces and channels allowing separation of molecules.
- Production of surfaces and channels for single molecule detection.
- Production of surfaces and structures that change their properties as a function of added mass, e.g. similar to QCM.

The low cost associated with the lithography process also makes it an ideal candidate for industrial biotechnological projects. In the following will be described various projects that NIL have been utilized and where it has had a clear role. There exists many more biotechnological applications where NIL not yet have been used, i.e. where up to now the nanolithography have been made with another method (most often e-beam lithography), but that is beyond the scope of this article.

16.4. EXAMPLES

16.4.1. Fabrication of nano-bio sensors

One of the most employed structures for biosensor applications is the interdigitated electrodes resulting in an effective electrode area magnification at a given footprint area and hence increased capacitive and impedometric signals. Also the surface sensitivity is greatly enlarged for nanostructured interdigitated electrodes since the electric field lines probe not only the media in between the electrodes but also above them. Thus, by making the electrodes closer and closer, the more surface sensitive the sensor will become. We have previously shown that when the electrode sizes and interelectrode distances are becoming increasingly small, the detection capability will increase substantially.[6] By applying a voltage of e.g. 100 mV, and using a structure with a spacing of 100 nm between the electrodes, the electrical field created is roughly equal to 1 MV/m, which is large enough to induce non-linear dipole effects of the species present between the electrodes. This non-linear response can be used as a fingerprint for sensor applications, and may e.g. be explored in electronic nose concepts. Furthermore, by employing interdigitated electrodes with inter-electrode distances in the nanoscale, unwanted electrochemical reactions will be avoided since only a small voltage is needed to achieve such a large electric field. Also due to the small interelectrode distances one might be able to study redox-cycling in a real miniaturized system which, to the extent of the authors knowledge, not yet has been possible to pursue.

The key issue for interdigitated nanobiosensors is however, not the actual linewidth or pitch, but mere the possibility to fabricate a lot of identical sensors, needed both for making multi-sensor arrays as well as for enabling correct interpretation and validation of the measured data. This requires the possibility to manufacture identically scaled sensor structures. For multisensor arrays, each sensor should be predominant sensitive to one chemical or biological component of the substance under investigation. This can be achieved by utilizing electrical potential driven adsorption/desorption immobilization procedures. Thus, by using sensor arrays and employing some kind of multivariant analysis of the obtained signal pattern, it should be possible to tell the composition of an unknown substance. Furthermore, biosensors for use in the medical and food areas are often required to be disposable, often due to the fact that enzymes used for enzymatic detection of the analytes are difficult to stabilize making their lifetime short. It is thus clear that there exists a demand to fabricate smaller and smaller detection systems in a cost-efficient way. For instance, in the biomedical sector there is an urge for the possibility to perform analysis in a cost-effective and fast way in order to have analytical answers already at the first visit to a doctor. This implies the use of arrays in a disposable material such as polymer biochips. Another large area is the pharmaceutical industry requiring quality assured test protocols for studying the effects various formulations of medical aids will have. One such method is the study of how a cell is affected by the medicines used. This can indeed be performed using e.g. patch clamp techniques. These methods are today very labour intensive but by making an array of disposable planar electrodes with incorporated fluidic channels, a boost for such measurements can be envisaged.

A complete nanobiosensor structure consisting of a 200 μm•200 μm large area containing 100 nm sized interdigitated nanoelectrodes with varied interelectrode distances has been fabricated using nano*i*mprint lithography in combination with UV-lithography.[7] Stamps were made by e-beam lithography in 950K PMMA spun onto a thermally oxidized 300 nm thick SiO2 layer on top of a 2" Si-wafer, followed by oxygen RIE employing a 30 nm thick lift-off defined Cr-layer as etchmask. The nanostructures defined on the 2" Si/SiO2wafer were defined at 5 different evenly distributed locations on the wafer surface. At each location a 200μmx200μm large area of patterns were exposed. Each area consisted of interdigitated lines and spaces with 50, 75, 100 and 200 nm dimensions, respectively. Printing was made onto n-type doped SiO_2/Si wafers having a 100 nm thick spun layer of 950K PMMA on their top faces. The metal nanopatterned structures were then aligned with a metal contacting layer using UV-litography and subsequent lift-off. The contacting layer consisted of bondpads as well as connecting lines to the respective sides of the finger patterns. In the composite Figure 1 the resulted aligned structure of nano- and micropatterns at various magnifications is displayed. As can be seen from the magnified part shown in Figure 2, the linewidths of the fabricated metal lines are very uniform without any considerable edge roughness.

The complete structure has been characterized with admittance spectroscopy using a HP Precision LCR meter in the frequency range 20 Hz - 1 MHz and relevant electrical parameters were calculated from the spectra. The resistance between the finger was found to be 1 GΩ over the entire frequency region showing that there was no leak current on the surface. The capacitance showed little variation over the frequency range and was 17.0 pF at 1 kHz and 15.6pF at 1 MHz.

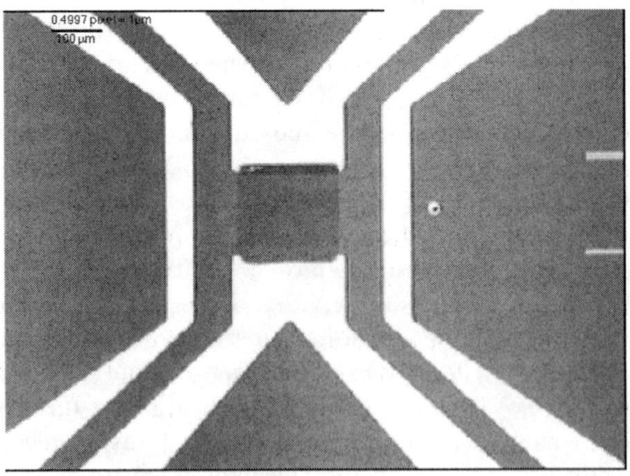

FIGURE 16.1. The complete contacted structure using NIL in combination with UV-lithography. The size of the figure is ≈ 1,3 mm × 1 mm.

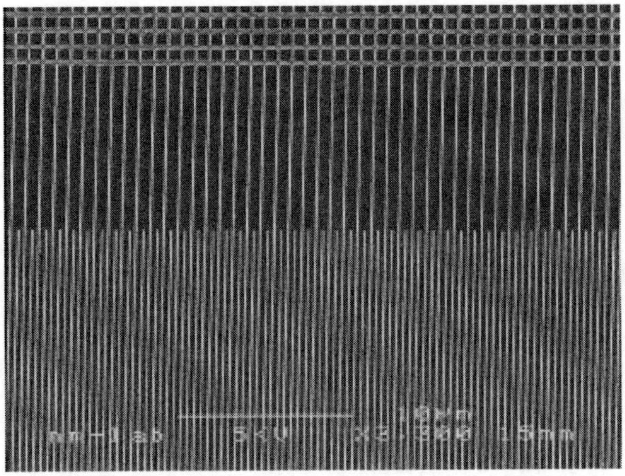

FIGURE 16.2. SEM micrographs showing part of the 200x200 μm^2 large interdigitated metal array defined by NIL and Cr lift-off.

16.4.2. Fabrication of nanoscaled electrodes for molecular devices

In a recent publication Austin and Chou discuss and show how easily nanocontacts can be fabricated using NIL in combination with electromigration.[8] First a relatively long and thin (20 nm) and shallow (15 nm) metallic lead is formed between two larger metal contacts. The whole device is made in a single NIL processing step together with a lift-off process. Then current is established between the two contacts and through the nanowire until it breaks due to electromigration. In this way reproducible and stable lateral sets of Au-electrodes having a gap of down to 8 nm were produced. This method is interesting since many contacts are made in parallel but a gap of 8 nm is still too large to allow single molecules to be attached in the gap for detailed studies.

16.4.3. Fabrication of channels for nanofluidics

In a recent paper by Cao et al.[9] the fabrication of 10 nm enclosed nanofluidic channels using NIL in combination with a deposition of amorphous SiO_2 was presented. They were able to create millions of enclosed nanofluidic channels smaller than 10 nm on a 100 mm wafer. The process started by defining a stamp with holography and printing resist tracks on a Si/SiO2 substrate followed by a etching step to make 50 nm sized trenches. Then the trenches were covered with a layer of SiO_2 deposited on top of the trenches. In this process the trenches were further reduced in size and become close to 10 nm in size. This finished the fabrication and since the SiO_2 layer was transparent it was possible to image using CCD videomicroscopy flouroscently stained long genomic DNA molecules that were effectively stretched in the nanochannels (see Figure 3).[10]

In the lab of Prof Yong Chen at LPN/CNRS, Marcoussi, France work is in progress to make electrophoresis chips by NIL based fabrication of channels with incorporated nanosized pillars that will act as an artificial separation gel in order to improve DNA separation. This work is reviewed in Chapter 17. Also electrokinetic fluidic pumps are

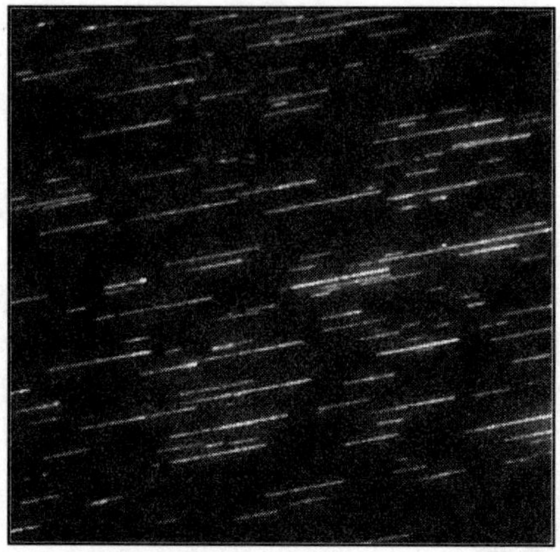

FIGURE 16.3. CCD image of lambda-phage DNA concatemers stretched in nanofluidic channels. The scale bar is 30 μm. (from reference 10, courtesy J O Tegenfeldt)

investigated in which the channels need to have asymmetric electrodes included in order to have an asymmetric electric field that will drive the fluid.

CONCLUSIONS

Nanoimprint lithography has the potential to be a favourable method for making nanoscaled structures for various bio-applications. The reason is the ease of using NIL in combination with metal lift-off processes. Also, the method is generic since it allows fabrication of structures ranging in size from micrometers to nanometers. Furthermore, it can be used to fabricate structures in polymer layers such as fluidic channels in combination with electrodes or contact points allowing separation of various kinds of molecules in combination with detection. When considering the possibilities offered there is an overwhelming potential in biotechnology for nanoimprint lithography.

ACKNOWLEDGEMENTS

The author is grateful to all people at the divison of Solid State Physics and The Nanometer Consortium that has contributed to and supported the work presented in this chapter. Especially I would like to put forward the following individuals that in various capacities have been deeply engaged in the Nanoimprint project: Tord Stjernholm at the workshop, Eva-Lena Sarwe, Mariusz Grazcyk and Ivan Maximov for help with processing issues and Babak Heidari, my former student who made all the ground-breaking work and the present students in the NIL group Marc Beck and Patrick Carlberg for their invaluable efforts and work with the development of the NIL technology and last but not least my dear collegue and friend Torbjörn Ling who has been invaluable in helping and introducing me to the interesting field of biotechnology. Finally, I would also like to thank all the members of the EU-IST MEL-ARI/NID projects NANOTECH and CHANIL for helpful discussions and support, i.e. all the people in the labs of Prof's Scheer and Sotomayor-Torres in University of Wuppertal, Prof Ahopelto at VTT, Prof Cardinaud at CNRS/Univ of Nantes, Prof Mayer at Univ of Duisburg and all the employees at MicroResistTechnologies GmbH, Berlin headed by Mrs Gabi Gruetzner. Finally, I would like to acknowledge the financial support provided by the Swedish Strategic Research Council and the European Commission.

REFERENCES

1. S. Y. Chou, P. R. Krauss, and P. J. Renstorm, Appl. Phys. Lett., 1995, **67**, 3114., Science, (1996), **272**, 85.
2. T. Mäkelä, T. Haatainen, J. Ahopelto, H. Isotalo, Synthetic Metals, (2001), **121,** 1309.
3. T. Mäkelä, T. Haatainen, J. Ahopelto, and H. Isotalo, J. Vac. Sci. Technol., (2001), **B 19**, 487.
4. B. H. Weigl, and P. Yager, Science (1999), **283**, 346.
5. J. T. Santini, M. J. Cima, R. Langer, Nature, (1999), **397**, 335.
6. L. Montelius, J. O. Tegenfeldt and T. G. I. Ling, J. Vac. Sci. Techn., (1995), A **13**, 1755. 7.
7. L. Montelius, B. Heidari, M. Graczyk, I. Maximov, E-L. Sarwe and T. G. I. Ling, Microelectronic engineering, (2000), 53, 521-524.
8. M. Austin and S. Y. Chou, J. Vac. Sci. Technol., (2002), **B20**, 665.
9. H. Cao, Z. Yu, J, Wang, J. O. Tegenfeldt, R. H. Austin, E. Chen, W. Wu, S.Y. Chou, Appl. Phys. Lett. (2002), **81**, 174.
10. J. O. Tegenfeldt, H. Cao, R. H. Austin, E. C. Cox, S. Y. Chou, *Nanofluidic Channels for DNA analysis*, Invited talk at the Medicon Vallec Academy conf, October 8, 2002, Malmö, Sweden

17

Soft Lithography and Imprint-Based Techniques for Microfluidics and Biological Analysis

Anne Pépin and Yong Chen
Laboratoire de Photonique et de Nanostructures (LPN), Centre National de la Recherche Scientifique (CNRS), Marcoussis, France

17.1. INTRODUCTION

In the past decade, microfabrication has become increasingly important to biology and biomedical applications. The fabrication techniques initially developed for the microelectronics industry, e.g. photolithography and associated material processing, along with the silicon micromachining methods later developed for micro-electromechanical systems (MEMS), are now being applied to the miniaturization and integration of biological and chemical analysis systems.

The first application to attract wide attention was the development of gene chips, i.e. DNA microarray chips for genetic sequence analysis.[1,2] By coupling photolithography and biotechnology, two-dimensional arrays of reaction cells of the order of 100 μm × 100 μm, each containing a different set of immobilized single-stranded DNA sequences, can be microfabricated on appropriate substrates. These microarrays are exposed to samples containing possible complementary targets, and the degree of hybridization is measured. DNA chips are now widely used in areas such as genetic disease diagnosis, drug design or toxicology and a number of companies are commercializing DNA and gene expression microarrays (Affymetrix, Nanogen, Motorola, etc.).[3] The fabrication of arrays of surface-bound capture molecules is of high interest for a variety of biological applications. In particular, ligand-binding assays based on protein microarrays hold great promises for biosensing, investigating proteomics and protein-protein or enzyme-substrate interactions, studying cell adhesion and motility, or building immunoassays based on antigen-antibody binding. However, due to the fact that protein structures are more complex compared to nucleic acids, protein microarrays have only recently started to emerge.[4]

In parallel with these planar devices the sole function of which is to enable detection of a binding reaction by a measurement of its intensity (via fluorescence, chemiluminescence, radioactivity, electrochemistry, etc.), the importance of three-dimensional and multifunctional microfluidic analytical devices - so-called lab-on-a-chip devices or micro-total analysis systems (μTAS) - has been quickly expanding over the past decade.[5-8]

The possibility of manipulating minute volumes of fluids, typically in the nanoliter to picoliter range, in microfabricated channels offers many advantages: reduction in reagent and sample volumes, reduced waste, automation, massive parallel processing, all leading to drastic improvements in performance, speed, throughput, reliability and overall cost. However, functional lab-on-a-chip devices which integrate complete biochemical processes require diverse building blocks such as fluid injectors, filters, pumps, valves, mixers, separation elements, detectors, etc. often including optical, electrical and mechanical elements. Such devices therefore present significant challenges for micro- and nano-engineering. For example, a large variety of fluid pumping schemes have been investigated, including pressure-driven, pneumatic, piezoelectric, electrostatic, electromagnetic, surface tension, and electrokinetic effects.[5-7]

Ideally, a lab-on-a-chip should be able to carry out a complex laboratory process, for instance going from the sampling of a cell of interest from a biological fluid to the determination of its complete protein expression. As of today, no such system exists, however microfluidic devices performing simpler tasks, like cell sorting, gene amplification through polymerase chain reaction (PCR), separation of biomolecules, DNA sequencing, etc. are quickly emerging.[5]

Although still in their infancy, microfluidic devices hold great promises for clinical diagnostics, environmental analysis, high throughput screening and drug discovery as well as therapeutic drug delivery. High market potentials for the pharmaceutical industry and healthcare business are leading to vigourous technological development worldwide. In particular, the use of microfluidic circuitry for precise sample delivery to microarrays, and even for combinatorial chemical synthesis of probe molecules, is gaining significant interest. A large number of research laboratories around the world are currently developing lab-on-a-chip activities and an increasing amount of companies are involved in commercial exploration (Caliper Technologies and Agilent Technologies, ACLARA Biosciences, Fluidigm, Gyros AB, Cepheid, BioMicroSystems, Micralyne, Micronics, Nanostream, Tecan, STEAG microParts, etc.).[9] Microfluidic devices also find applications in many other fields, such as optical microsystems, ink jet printers, energy conversion or integrated circuit cooling. We will concentrate in this chapter on biological analysis and biochemical applications.

The first effort on μTAS was launched a decade ago[10] and addressed the transfer of separation methods, mainly capillary electrophoresis (CE), into microfabricated glass microchannels. Since then, the development of chip-integrated CE has received increasing interest.[5] Conventional electrophoresis in fused silica capillaries could easily translate to glass, quartz or oxidized silicon chips because operating fundamentals remain the same.

Fluid flow in microchannels is commonly generated either by pressure or, for aqueous solutions and polar solvents, by electric fields. Electrically-driven flow, known as electroosmosis, uses the spontaneous charge dissociation that is generated at the surface of a channel in contact with an aqueous solvent. Electroosmotic flow (EOF) typically occurs in glass capillaries when wall surfaces become negatively charged (at pH>4) as a consequence of de-protonated silanol (Si-OH) groups, $Si\text{-}O^-$. The positive ions contained in the solvent are attracted to the negatively-charged channel surface and form the so-called electric double layer (EDL), consisting of a fixed adsorbed cation layer and a diffuse layer of cations. When a voltage drop is applied externally between the two ends of the channel, it generates an electric field in the electrolyte. Consequently, a force towards the cathode is exerted on the diffuse charge layer, which imparts a uniaxial motion to the bulk of the fluid by viscous drag. In uniformly-charged channels, EOF thus gives rise to a plug-like flow, with shear located only in the close-surface region formed by the EDL (generally a few nm-thick). One of the reasons that EOF pumping is so often used in microfluidics is that much lower sample dispersions can be achieved compared to parabolic Poiseuille flows obtained with pressure-driven pumping. Electroosmosis plays an important role in electrophoresis, which is widely used to separate charged molecules (DNA, proteins, etc.) of different sizes according to their different migration velocities.[11]

Moreover, fluid flow in micron scale channels shows behaviors different from that observed in their macroscopic counterparts. For instance, due to the very low Reynolds' number (typically $Re \leq 1$) induced in part by the larger surface to volume ratios, fluid flow in microchannels is laminar rather than turbulent. Such properties can be exploited for specific applications but can become obstacles for other purposes. For example fluid mixing can no longer rely on turbulence, and therefore schemes enhancing molecular diffusion and/or off-axis fluid transport inside a microchannel are being developed for mixing purposes. In addition to continuous flow, droplet displacement is also a major issue in microfluidic systems.

For a number of biological processes, micrometer scale patterning is sufficient, in particular for cellular studies where the size of a typical cell is about 10 μm. However, the ability to fabricate structures with cross-sectional dimensions close to the size of biological macromolecules is of interest both for fundamental studies on single molecule behavior and for device applications. Improvements in performance are expected in many devices by a further scaling down of dimensions or else by including nanostructured elements in micron-scaled channels. Recently, size-dependent separation of long DNA molecules in a nanofluidic channel device consisting of a series of entropic traps was achieved.[12] This device was fabricated by standard photolithography and silicon etching to create alternating deep and shallow (75-100 nm) regions in a 30 μm-wide channel fixed over a flat glass substrate to allow fluorescence microscopy observation. Creating controlled artificial gels by patterning the surface of electrophoresis microchannels with an array of microposts has been investigated for some years[13,14]. These arrays of stationary obstacles force DNA molecules to sneak their way through at different velocities according to their size, and yield improved separation performances. Attempts at reducing the size of pores and posts down to the nanometer range for smaller biomolecule separation (single-stranded DNA, proteins, etc.) have been carried out in silica-based materials using electron-beam lithography.[15] More recently, optical fluorescence detection has also been achieved at a nanometer scale, which should be useful for single molecular studies.[16,17]

Most microfluidic devices until now have been fabricated in silica-based materials using standard cleanroom processes such as photolithography, electron beam lithography, wet etching, reactive ion etching, and anodic wafer bonding. Most of these techniques are expensive and not accessible to many researchers, especially biologists. To overcome these shortcomings, low-cost and flexible alternatives to conventional technologies are highly desirable. Moreover, for many applications cheap and disposable devices fabricated on bio-compatible substrates are greatly sought after. For instance, point-of-care clinical diagnostics, where cross-contamination is a critical issue, require disposable assays which can be used cost-effectively in hospitals and at the physician's clinic. Fabrication techniques based on cheaper and less-fragile polymeric materials offer attractive solutions to these requirements. Two main fabrication approaches have undergone considerable development these past few years.

G.W. Whitesides first proposed a set of fabrication techniques called soft lithography, based on elastomeric material molding and the use of elastomer stamps to achieve micropatterning of self-assembling molecules.[18] Soft lithography is now being used widely in research laboratories around the world. In particular, prototype device fabrication is fast and easy, accessible to most biologists, and can benefit from experience in surface biochemistry. A company, Fluidigm Inc. offering elastomer-based microfluidic devices has recently been launched.[19]

Moreover, plastic device fabrication also offers many advantages. A wide variety of biocompatible thermoplastic materials can be molded via replication technologies like hot embossing and injection molding[20–23] and thermoplastic imprinting-based techniques are undergoing rapid development. An increasing number of biotechnology companies are currently developing plastic microfluidic devices (ACLARA Biosciences, Micronics, Gyros AB, STEAG microParts, Mildendo, etc.).[9] Bonding procedures for sealing polymer devices are also much less demanding that time-and-cost-worthy anodic bonding used for glass and Si.

The development of related nanoimprint lithography[24] for the high resolution replication of patterns compatible with standard microelectronic processing offers a complementary approach and the possibility of integrating microfluidic devices on active semiconductor substrates including nano-electronic and nano-optical functionalities.

Other polymer-based technologies for the fabrication of microfluidic devices are also being investigated, among which laser ablation,[25–27] plasma etching,[28] use of stimuli-responsive hydrogels,[29,30] microstereolithography,[31] X-ray lithography[32,33] and direct fabrication in thick SU-8 photoresist.[34] More details on the diversity of techniques being developed for the fabrication of polymeric microfluidic chips can be found in recent review articles.[20–23,35,36]

We have chosen within the frame of this chapter to focus on soft lithography and imprint-based techniques, the two most advanced low-cost alternative lithography approaches, which furthermore have demonstrated nanofabrication capabilities.[24,37–39] Soft lithography approaches will be discussed first, including elastomeric replica molding, micro-contact printing and microfluidic network techniques. Thermoplastic polymer imprinting techniques will be presented next, including nanoimprint lithography, nanoembossing, hot-embossing, and injection molding. For each technique, a selection of examples of its application to micro- and nanofluidics and biological analysis will be presented. The advantages, disadvantages, and complementarity of both classes of approaches will be discussed.

17.2. SOFT LITHOGRAPHY

Soft lithography covers an ensemble of microfabrication techniques based on the printing and molding of elastomeric stamps with desired patterns.[18] The elastomer generally used is poly(dimethylsiloxane) (PDMS). This silicone polymer, consisting of repeated units of $-OSi(CH_3)_2O-$, is commercially available. PDMS possesses a Young's modulus of typically 1MPa, which makes it a moderately hard elastomer. PDMS has several advantages for microfluidics[40] and biological applications.[41] It is non-toxic, chemically and biologically inert. It is optically transparent between ~240 nm and 1000 nm, non-fluorescent, and therefore allows standard optical detection schemes. It also possesses a high bulk electrical resistivity which is important for electrokinetic applications. Moreover, its surface chemical properties can be modified. For instance, PDMS is naturally hydrophobic but can be converted into a hydrophilic surface by various treatments, including O_2 plasma oxidation[40,41] and boiling in diluted HCl.[42] Oxygen plasma is known to create ionizable silanol (Si-OH) surface bonds presumably by converting $-OSi(CH_3)_2O-$ surface groups into $-O_nSi(OH)_{4-n}$,[40,41] while HCl soaking is believed to create hydroxyl (-O-H) and silanol bonds by breaking siloxane (-O-Si-O-) bonds through hydrolysis. The higher surface energy of these silica-like PDMS surfaces is therefore more easily wet by polar liquids. However these effects are reported to be short-lived, especially if the treated PDMS surface is exposed to air and not kept in contact with an aqueous buffer.[40,41,43] This hydrophobicity recovery is not very well understood but current works suggest that it is due to the migration of lower molecular weight PDMS chains to the surface, thus covering the Si-OH bonds and/or to the reorientation of the surface hydroxyl bonds towards the PDMS bulk.[41,43] Finally, PDMS is also relatively permeable to gases. While this property can be an inconvenience for certain applications, permeability to O_2, N_2 and CO_2 in particular is essential when working with living cells, e.g. for mammalian cell culture. Another consequence of this permeability to gases is the possibility to eliminate air bubbles in PDMS microfluidic channels by applying sufficient pressure to force them into the PDMS bulk.

In this section, the fabrication of microfluidic devices using PDMS replica molding will be reviewed first. Microfluidic and bio-applications achieved via micro-contact printing will be presented next. Lastly, the development of microfluidic networks for biopatterning and cell treatment will be described.

17.2.1. Casting of elastomers

The step common to almost all soft lithography approaches is replica molding, i.e. casting and curing of an elastomer against a patterned mold. The master used for casting can be a silicon, metal or resist surface, bearing a pattern inverse of the one desired on the PDMS replica. For feature sizes larger than 1 μm, masters are fabricated by optical lithography using ultraviolet (UV) light, whereas molds with smaller feature sizes can be obtained by a higher resolution lithography technique such as electron beam, X-ray or nanoimprint lithography. In the case of silicon masters, lithography is followed by reactive ion etching (RIE), while metal molds are generally obtained by electroplating. For masters with patterns of feature sizes larger than 50 μm, the chromium-on-glass photomask can be replaced by a transparency sheet mask obtained by a commercial desktop printer, and exposure can be carried out by using directly a mercury rare gas UV lamp instead of an optical aligner/stepper,[6,36] thus allowing very low cost and rapid prototyping. The casting master is generally treated with an anti-adhesive coating prior to PDMS pouring in order to reduce its tendency to adhere to the elastomer. On silicon molds, a layer of trichlorosilanes is typically spin-coated or vapor-deposited.[36] Recently, an alternative prototyping technique where the liquid elastomer is poured directly on a photocopy has been proposed for microfluidic application.[44] PDMS casting over large 3D patterns (>250 μm) obtained by solid object printing was also recently reported.[45]

Commercially available Dow Corning Sylgard 184 or General Electrics RTV 615 are commonly used as the PDMS elastomer. Mixing of pre-polymer solution and cross-linking catalyzer is often followed by vacuum-enhanced degassing for air bubble removal. Pouring over the master is carried out immediately after elastomer mix preparation. After curing, at relatively low temperatures, typically an hour at 80°C or a few hours at 60°C, the solidified cross-linked PDMS stamp can be peeled off the master.

Microfluidic structures, containing channels and reservoirs, can therefore be cast in PDMS, and a closed device is obtained after sealing with a coverslip. Generally, access holes to the reservoirs are punched through the elastomer. Adhesion of PDMS to low roughness surfaces is excellent and a reversible water-tight seal can be obtained by conformal contact on glass, Si, plastic, as well as PDMS surfaces (for low-pressure flows). For irreversible bonding, surface treatments such as oxygen plasma exposure,[36] or corona discharge from a Tesla coil[43] are particularly efficient. It is believed that irreversible bonding occurs by the formation of bridging covalent siloxane Si-O-Si bonds through spontaneous dehydration of the silanol groups of two facing oxidized slabs (-SiOH + HOSi-).[40,43] Bonding protocols for sealing microfluidic devices based on PDMS are therefore much simpler and lower cost than standard anodic bonding used for glass and Si devices.

In the following, we give examples of both all-PDMS microfluidic devices, some involving multi-level processing, and of hybrid devices such as PDMS-over-glass and PDMS-over-Si configurations.

Elastomeric materials are deformable, so that with simple designs moving parts can be obtained. A successful example of mechanical actuation achieved in all-PDMS devices is the pneumatically-driven valves and peristaltic pump fabricated by S. Quake and coworkers.[46,47] They have developed a multilayer cross-channel architecture where a layer of fluid channels is separated from a top layer of air channels only by a thin membrane at crossing points. The two PDMS layers of different thicknesses and hardness are cast separately. Holes are punched through for fluid access, and the two layers are irreversibly bonded. By applying pressure in the top air channel, the interlayer membrane collapses and stops fluid flow in the bottom channel, thus acting as a valve. In particular, rotary flow in a closed loop, which cannot be easily achieved by electrokinetic flow, can be obtained with this multilayer elastomeric scheme, enabling such applications as mixing of reagents and on-chip PCR.[47,48]

Another approach to mixing was recently proposed by Stroock et al.[49,50] In contrast with the active mixer decribed above the proposed mechanism here is purely passive. It involves creating small-amplitude grooves on the floor (or a sidewall) of the channel, at 45° with respect to the long axis of the channel. Such a grooved floor naturally generates a helical flow of a pressure-driven fluid. The channels were obtained via PDMS replica molding over SU-8 resist patterned by photolithography, and sealed against a flat glass slide. A more complicated pattern than the simple parallel ridges, that uses a staggered herringbone shape, leads to a flow where the streamlines are successively stretched, separated and folded over. As could be successfully visualized by confocal microscopy, this results in chaotic mixing as the fluid is being transported. Clearly, this mixing effect, first demonstrated in PDMS devices, can easily be integrated in microfluidic devices of any kind of material.

Another interesting mixing scheme was recently demonstrated in cast PDMS microfluidic channels sealed on glass, based on electrokinetic flow instabilities generated by applying an AC field excitation to platinum electrodes inserted at the upstream and downstream reservoirs to create an oscillating electroosmotic flow.[51]

Like glass and silica, PDMS surfaces can exhibit a degree of EOF, but typically of a factor of 5 or less smaller, which is caused by either a low level of fixed negatively-charged surface functional groups or by ions adsorbed from the aqueous buffer. As mentioned earlier, surface treatments such as oxygen plasma can also activate PDMS surfaces by creating $Si\text{-}O^-$ surface groups. Specific surface depositions can also be carried out, [52,53] however stability of the electroosmotic properties of PDMS channels have been shown to vary with aqueous buffer ionic strength and pH.[41] Prolonged contact with air reduces EOF as it allows hydrophobicity recovery. However, EOF regeneration was reported by subsequent prolonged exposure of the channels to 1M NaOH.[41] Several research groups have used PDMS to make microfluidic devices capable of electrophoretic separations.[36] Ekström et al. were probably the first to propose elastomeric polymers to fabricate on-chip electrophoresis devices, in the early 1990s.[54] The first PDMS-based DNA capillary electrophoresis (CE) device was fabricated by Effenhauser and coworkers in 1997.[55] They demonstrated efficient separation of DNA fragments as well as peptides in gel-filled microchannels which were bonded reversibly and could therefore be dismantled and cleaned. Whitesides and coworkers also developed CE devices in PDMS for DNA and peptide charge ladders.[40,56] Most recently, a self-organizing nanomagnetic particle-based molecular sieve was demonstrated in a PDMS microfluidic device by P. Doyle and coworkers.[57] These nanoparticles line up when an external magnetic field is applied and form a regular array of nanocolumns, serving as an artificial gel for the separation of large DNA molecules.

A single-molecule DNA sizing and sorting microfluidic device,[58] as well as a flow cytometer or cell sorter, were developed by Quake and coworkers[47,60] using the multilayer PDMS fabrication technique described earlier. Both devices rely on pressure-driven flows and localized fluorescence intensity measurements in a small detection volume (less than a picoliter).

Concerning applications proteomics, prototype devices molded in PDMS have been successfully used for delivery of peptide mixtures to electrospray ionization mass

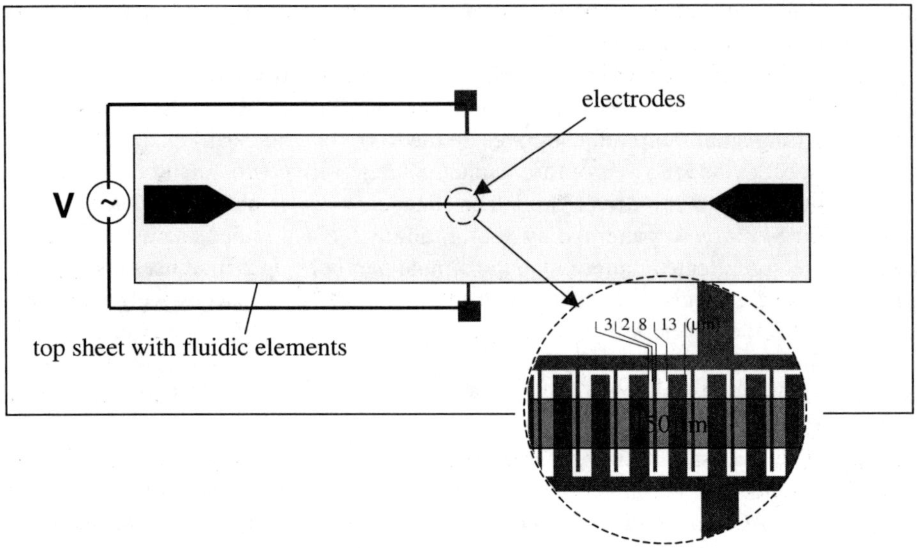

FIGURE 17.1. Schematic representation of a local fluid pumping device. The Ti/Au array of asymmetric electrodes is deposited on the bottom plate (glass). The fluidic elements (channel, reservoirs) are cast in the top plate (PDMS).

spectrometry (ESI-MS)[61] and more advanced ESI-MS emitters have been recently proposed.[62] Chromatographic separation of proteins was recently demonstrated by Seki et al. in an all-PDMS chip made by replica molding.[63] Pressure-driven flow was used in the bead-packed PDMS channels hydrolized by a concentrated HCl solution in order to prevent non-specific adsorption of proteins on the channel walls.

We developed recently a local fluid pump based on electroosmotic effects generated in a PDMS channel sealed over a glass substrate on which an array of interdigitated asymmetric microelectrodes is deposited.[64] Figure 1 gives a schematic top-view of the device.

We have used a mold produced by photolithography on a 20 μm-thick SU-8 negative photoresist to cast the PDMS microchannel, and the metal electrode array was obtained by standard photolithography and wet etching techniques, allowing a minimum feature size of 2 μm. Net fluid flow could be created in a microchannel by submitting the electrode array to a stationary AC voltage of low amplitude. Figure 2(a) shows the fluid pumping velocity dependence on frequency at fixed voltage amplitude, while Figure 2(b) displays the fluid velocity measurements carried out as a function of the applied voltage amplitude for a fixed frequency. As can be seen, for a 1V voltage amplitude, the pumping velocity reaches a maximum (around 18 μm/s) at a frequency of about 5 kHz, in good agreement with theoretical predictions.[65-67] Increasing voltage at a fixed frequency induces a non-linear rise of the pumping velocity.

It should be mentioned that the implementation of such a low-voltage local pumping system in microfluidic devices is a very interesting alternative to more standard pumping methods (DC EOF and pressure driven flows) which require creating a potential of high voltage amplitude or pressure differences between the two ends of the channel.

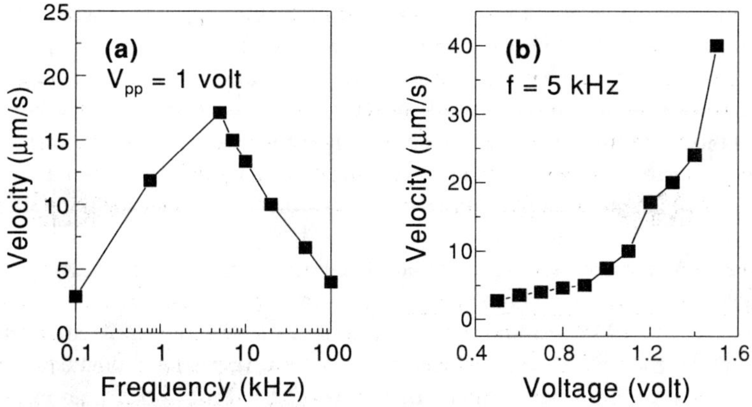

FIGURE 17.2. Measured fluid velocity as a function of (a) the applied frequency for a given 1V voltage amplitude, (b) the applied voltage amplitude, over an array of asymmetric interdigitated electrodes, in a PDMS-over-glass device

Flow velocity has been predicted to increase with decreasing array period and further downscaling of the electrode geometry is now under development.

17.2.2. Micro-contact printing

Micro-contact printing (μCP) is based on the use of an elastomeric stamp, obtained via replica molding and containing the patterns of interest in bas-relief, to ink a solid substrate with the help of molecular assembly.[18,68] As described previously, elastic rubber stamps are obtained by casting and curing PDMS against a patterned master. Molecular transfer occurs when the inked stamp is in contact with the substrate, resulting in the deposition of a self-assembled monolayer (SAM) on the substrate regions touched by the inked protrusions of the stamp. The resolution of μCP is basically limited by the master feature size, the mechanical deformation of the stamp, and the diffusion of the ink molecules around the contact areas. In practice, PDMS stamps provide a minimum feature size around 200 nm[18] but see recent development described in Chapter 8 and 10. Higher resolution can be achieved with stamps of higher stiffness[38] and a minimum critical dimension of 50 nm has been obtained by researchers at IBM Zurich.[37,38] μCP is particularly well-suited for patterning over large areas in a single printing step. A wide variety of SAM inks have been developed,[18,36] some of which had been discussed in Chapter 9. The commonly used inks are solutions of thiols in ethanol for noble metal substrates, and octadecyltrichlorosilanes (OTS) in toluene for oxidized silicon substrates.

The ability to generate surface-bound biomolecules is a key issue for driving applications such as DNA microarrays and diagnostic immunoassays. μCP can efficiently be used to place functional molecules as biological layers on a substrate with high precision. This can be achieved by using ligand-ended SAMs which adsorb specific biomolecules, including proteins. This ability to engineer surface properties via SAMs has indeed made it possible to study a variety of biological processes, such as protein interactions and cell adhesion.[36,69] Direct printing of proteins can also be achieved on several types of substrates such as glass, polystyrene or silicon, generally by using a PDMS stamp incubated with a solution of the desired proteins, since biomolecules spontaneously adhere to the hydrophobic surface of a PDMS stamp.[70-72] This technique was used for example to directly print antibody molecules, such as immunoglobulin.[38,70] The transfer efficiency is often measured quantitatively via radioactivity or the use of fluorescently-labeled antibodies. For most current biological use, patterning resolutions between 1 μm and 100 μm are typically required, however single protein deposition by μCP has been demonstrated by using a PDMS stamp of high Young's modulus which was patterned with 80-nm wide posts.[38]

A related patterning technique called affinity contact printing (αCP) involves extracting target proteins with high selectivity from crude biological sample solutions through the use of a PDMS stamp preriously coated with capture molecules, and transferring the target molecules on a surface by contact printing.[73] Due to an adequate force balance of the different binding interactions involved, affinity-bound target molecules can dissociate from their capture molecules and adsorb on the substrate. The strong chemical bond between capture molecules and elastomeric stamp allows the stamp to remain coated and the αCP process can be repeated several times. Typical applications include the transfer of antibodies and enzymes.

Proteins tend to adhere non-specifically to many surfaces and therefore accurate patterning often requires biospecific surface treatments rendering desired portions of the surfaces inert to protein adsorption. The most commonly-used procedure is to cover the desired protected surfaces with poly(ethylene glycol) (PEG). This is typically performed by dipping the sample in the PEG solution after having locally deposited proteins by μCP, but other SAM treatments have also been developed.[36] This step is crucial for cellular studies, since mammalian cell membrane proteins such as integrins also tend to adhere non-specifically. Cells can indeed be successfully patterned on substrates through biospecific adhesion to proteins such as fibronectin.[69]

As has been described above, the main advantage of contact printing over standard procedures for making bio-arrays is the capability of processing large bioactive surfaces with arbitrary patterns in a single run. But μCP can also have other applications in microfluidics. In particular, μCP was successfully used to create surfaces patterned with different charges, in order to create complicated flow profiles in electroosmotically pumped channels.[74] Indeed, since speed and direction of EOF depend directly on the charge of a channel's walls, flows can be tailored by controlling surface properties. Nonuniaxial fluid motion can thus be generated by creating specific charge patterns along the walls of microchannels and/or variable channel geometry. For instance, convection rolls can be generated inside a microchannel when an electric field is applied parallel to a surface patterned with successive regions of positive and negative charge. Most of these studies deal with patterns in the range of one to a hundred micrometers, and the promising submicrometer scale remains to be tackled.

17.2.3. Microfluidic networks

Instead of using an inked elastomeric stamp to deposit locally molecules at the contact area underneath the stamp protrusions as in μCP, the non-contact areas, i.e. cavities, between a stamp and a substrate can be used as guiding channels to transport solutions to desired locations and achieve patterning of molecules. This approach is known as micromolding in capillaries (MIMIC)[18] or microfluidic networks (μFN).[38]

In contrast with μCP, single-step deposition of multiple inks can be achieved via μFN by using different solutions in adjacent channels.[36,75,76] Accurate pattern replication requires good wetting of the channels with the fluids of interest. This is typically achieved with specific preliminary surface treatment of the PDMS, such as a brief oxygen plasma treatment in order to hydrophilize the channels and allow filling by capillary action. In the case of protein patterning, stable PEG coating is often preferentially chosen to guarantee hydrophilic protein-repellent surfaces. The μFN approach is indeed particularly suited for the patterning of biological molecules.

In particular, immunoassays relying on affinity binding between antibodies and antigens (or proteins) are commonly performed in macroscopic well plates for diagnostic purposes and can greatly benefit from the use of miniaturized arrays achieved by μFN. Micromosaic immunoassays have been achieved on flat PDMS substrates at IBM Zurich using PDMS μFNs or Au-coated microfabricated Si μFNs.[77,78] Two-step immunoassays using fluorescently-tagged analytes have been performed, as well as sandwich-type immunoassays where the recognition pattern is detected by binding a fluorescent-labeled or enzyme-conjugated antibody to the analyte, as in the case of Enzyme-Linked ImmunoSorbent Assays (ELISAs).

Whitesides and coworkers have exploited laminar flow , i.e. the fact that within a microchannel multiple fluid streams run side by side with no mixing except that due to molecular diffusion, to selectively pattern a substrate inside a single channel using μFNs.[36,79] Feature sizes of less than 5 μm and spatial localization to within 5 μm can be achieved by laminar flow patterning. Silver wires have been fabricated in this manner which can serve for in-channel electrical probing. Selective treatment of a single cell placed inside a microchannel was also achieved recently using laminar flow to deliver different membrane-permeable molecules at parallel locations on the cell.[36,80]

In contrast with these 2D μFNs, 3D architectures of PDMS μFNs have been developed by stacking stencil-like membranes containing through-holes with planar network layers.[81] More complex and discontinous patterns can be generated in this manner. Protein and cells could be deposited on substrates this way, and selectively treated with different substances.[81] The 3D μFN concept has also been adapted by Quake and coworkers to create DNA hybridization chips.[82]

17.3. IMPRINT-BASED TECHNIQUES

Fabrication of microfluidic devices and biological arrays based on plastic materials offers a number of advantages.[20–23] They are cheaper than other materials (Si, borofloat glass, etc.) and bonding protocols for sealing plastic microfluidic devices are simple, including gluing, lamination, ultrasonic or thermal bonding. Moreover, a large number of plastic materials have been approved for biomedical applications, and some of them for use in implantable devices for in vivo purposes. With further developments, plastic materials may be adapted to fit the optical, electrical, chemical and biological needs dictated by a given application.

Among the most commonly-used materials are thermoplastics such as polycarbonate (PC), poly(methyl methacrylate) (PMMA), and cyclo-olefin copolymers (COC). These plastics are transparent within a wide range of wavelengths (typically between 250nm and 900nm), emit low autofluorescence, and are thus well-suited for analyses involving optical detection. They also have good electrical insulating properties and high thermal conductivity, which are crucial for devices based on electrokinetic pumping.

Surface properties of plastics can be tailored chemically. As was the case for PDMS elastomer, thermoplastic polymers exhibit a degree of EOF, at a lower level than silica-based substrates. EOF in plastics is due to surface chemical groups, such as carboxylate ($-COO^-$) and hydroxyl groups which create negatively-charged surfaces when in contact with the polar solvent buffer carried in microchannels. Plastic substrates also present higher hydrolytic stability compared to PDMS substrates.[83] However, surface charge inhomogeneities depend upon material, processing technique, ion adsorption, etc., leading to non-uniformities in flow velocities already reported in plastics[84,85] and can affect performance in electrokinetic separations. Enhanced hydrophilicity and activation of ionizable surface groups is possible through different ways, including oxygen plasma treatment[86] and UV irradiation,[87] while different coatings can on the contrary help limit EOF. Furthermore, liquid and gas-phased methods using covalent attachment protocols have also been used to derivatize PMMA and PC surfaces in order to yield surfaces that are either hydrophobic, hydrophilic or contain a variety of surface moieties. These surface functional groups can be exploited to control the EOF or act as a basis for attachment of biomolecules or metal films.[88] Soane & Soane were probably among the first to propose using thermoplastics for chip capillary electrophoresis in the early 1990s.[89]

Plastic devices can be fabricated by various techniques, including direct processing by mechanical micromachining, laser ablation, plasma etching, or by replication techniques based on thermal molding, e.g. injection molding and hot embossing or compression molding. While plasma etching is carried out through a mask, laser ablation and micromaching are generally sequential techniques and thus have low throughput rates. In addition, these direct processing techniques have low resolution, and produce surfaces that are often too rough for many applications, in particular high resolution electrophoresis.[90] In contrast, injection molding and hot embossing are well-developed techniques offering lower cost capabilities for mass production. Moreover, as replication techniques, they benefit from the superior definition, high resolution and very low roughness that can be obtained on optimized molds. Injection molded or hot-embossed thermoplastic devices are therefore preferential choices for lower cost and disposable devices for single-use against cross-contamination, for higher throughput, etc. The nanoembossing technique is a nanoscale version of the hot embossing process which is suited for nanofluidic applications. This approach is inspired from nanoimprint lithography, which couples polymer thermomolding with conventional pattern transfer processes used in semiconductor microelectronics and can be exploited to produce cost-effective microfluidic devices in silica-based substrates, as will be described in the following section.

17.3.1. Nanoimprint lithography

In nanoimprint lithography (NIL) a rigid mold is used to physically deform a heat-softened thermoplastic polymer layer preriously spin-coated on a substrate.[24,91-96] The bas relief obtained in the thin polymer layer is then transferred to the substrate by RIE. The main assets of NIL are its high resolution, low cost and a good process compatibility to most of the conventional pattern transfer techniques.

The imprint step is carried out above the glass transition temperature (T_g) of the polymer. The most commonly used thermoplastic polymer is PMMA which has a T_g around 110°C and is typically imprinted at 180°C under a pressure of 50 bars for a few minutes. Structures below 10 nm have have been produced by NIL on PMMA.[24] We recently demonstrated room temperature and low pressure NIL using newly-developed hybrane resist. Sub-100nm resolution could be achieved with this semi-crystalline hyperbranched thermoplastic polymer.[92] Various alternative imprinting schemes have been proposed, such as UV-enhanced step-and-flash NIL, where UV-curable resist is imprinted at low temperature and pressure via an irradiated transparent quartz mold.[93]

In order to improve process latitude, we developed a tri-layer approach[94,95] where the top resist layer is imprinted while the thick bottom resist layer, separated from the imprinted film by a thin Ge layer, remains thermally stable. The pattern is then transferred to the bottom layers by sequential RIE giving a larger thickness contrast. Tri-layer NIL thus renders possible the fabrication of high aspect ratio structures as well as pattern transfer on non-planar substrates.

We exploited this tri-layer NIL process for the fabrication of high resolution patterns inside microfluidic channels for improved biomolecule separation.[96] The use of high density nanopillar arrays placed inside microchannels, serving as artificial gels in integrated capillary electrophoresis (CE) chips, has been proposed for single nucleotide polymorphism (SNP) DNA analysis and proteomics.[33] The required array area is typically several millimeters long and covers the total width of the channel, generally around 100 μm-wide. Although electron beam lithography can be used to produce such nanostructures areas,[15,97] a lower-cost replication technology is needed for mass production. A top view of the designed structure is schematically represented in Figure 3.

It contains a standard cross-shaped channel of 50 μm width with 500 μm diameter reservoirs at their ends. The nanostructured portion of the separation channel is 7 mm-long and contains a 320 nm period triangular array of 150 nm diameter pillars. The device

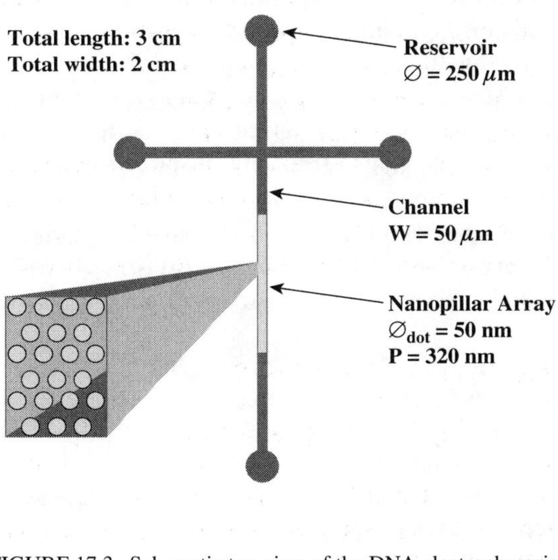

FIGURE 17.3. Schematic top view of the DNA electrophoresis microchip.

FIGURE 17.4. Since the fabrication process we developed is compatible with more conventional microfabrication techniques, other elements such as PCR reactors, mixing chambers, local electrokinetic fluid pumps (see section 2.2, as well as built-in optical fiber detection, could be integrated on the same chip.

is made of a top and bottom part later assembled together. The bottom part is an oxidized silicon wafer and holds the shallow etched structures, while the top part is in PDMS. The bottom part contains exactly the pattern described on Figure 3 with a homogeneous pattern depth of 500 nm and is obtained by tri-layer NIL, and subsequent metal lift-off and RIE of SiO_2 in a CHF_3/SF_6 gas mixture. The top part contains 10 μm-deep patterns corresponding to the design of Fig. 3 without the nanostructure array, therefore with a 7 mm-long interruption in the separation channel, and is obtained by elastomeric replica molding. This part also embeds the fluidic access to the reservoirs. Holes were punched through the PDMS layer for this purpose. The top PDMS part is fixed on the nanostructured SiO_2/Si substrate by adhesion, after treating the top and bottom parts with an oxygen plasma. Therefore, in the assembled device, the microchannel depth is 10 μm outside the array, and 500 nm inside the sieving array.

Imprinting molds of uniform pattern thickness 150 nm were fabricated in SiO_2/Si wafers by high resolution electron beam lithography (EBL) using a JEOL JBX 5DIIU vector scan generator working at 50 keV, followed by metal lift-off and CHF_3/SF_6 RIE in order to generate SiO_2 nanopillars. The molds have been used repeatedly to perform NIL without recording any damage. Figure 4 is a scanning electron micrograph (SEM) displaying a portion of a nanopillar array replicated in SiO_2/Si via tri-layer NIL. The resulting nanostructures show high aspect ratio and a very good homogeneity over the whole imprinted area.

Since the fabrication process we developed is compatible with more conventional microfabrication techniques, other elements such as PCR reactors, mixing chambers, local electrokinetic fluid pumps (see section 2.2, as well as built-in optical fiber detection, could be integrated on the same chip.

Recently, Austin et al. reported on the fabrication of sealed nanofluidic channels of width around 50 nm via NIL and observed stretching of large double-stranded megabase DNA molecules inside these channels.[98]

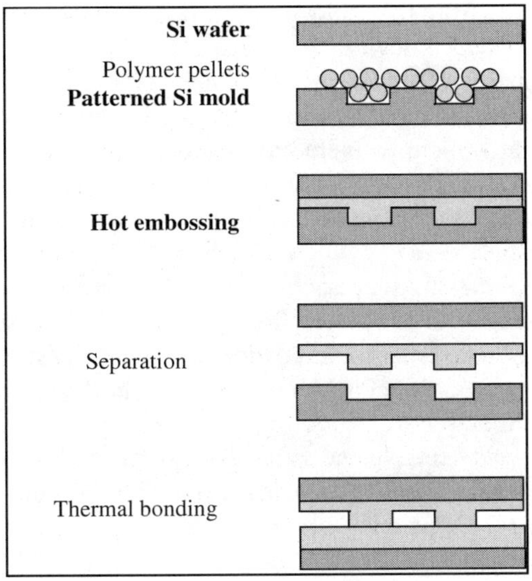

FIGURE 17.5. Schematic representation of the fabrication method involving hot-embossing of thermoplastic polymer pellets and thermal bonding.

Because of its high resolution and low-cost parallel processing capabilities, NIL holds strong promises for the fabrication of SiO_2/Si-based, or more generally, semiconductor-based microfluidic devices or biochips which could include advanced electronic, magnetic, microwave or optical components. NIL could for instance allow easy fabrication of cell-on-chip devices such as the patch-on-a-chip device developed by Fromherz and coworkers which utilize Si field-effect transistors.[99]

17.3.2. Nanoembossing

Nanoscale patterns can also be produced in bulk plastic by directly embossing polymer pellets and taking advantage of nanoimprint developments. This technique, which we named nanoembossing, offers an even lower-cost option for the fabrication of high resolution plastic microfluidic devices.[39,96]

Silicon molds fabricated by EBL and RIE are used to mold nanostructures, in a way similar to the NIL process. However here, pure thermoplastic polymer pellets are molded directly, instead of using a spin-coated thin film resist. Deep wide channels and high resolution shallow features can be molded in a single step with the proper mold. The fabricated microfluidic circuits can then be sealed by thermal bonding with a plastic cover sheet of the same material. Figure 5 shows a schematic representation of the fabrication procedure.

Polymer pellets are placed on the patterned silicon mold and covered with a flat silicon wafer. The sandwich is then inserted between the hot plates of a hydraulic press and temperature is raised to a value well above the T_g of the chosen polymer. After hot compression molding and cooling down below T_g, pressure is released and the patterned plastic sheet is removed from the mold. The fabricated sheet is then bonded to another plastic plate to form a sealed device. The same press we use routinely for NIL is here used for both hot embossing and thermal bonding.[39,96]

All our nanoembossing molds are made in Si using conventional microfabrication techniques. Deep microstructures including microchannels and reservoirs are defined by photolithography on thick SU-8 photoresist followed by standard CHF_3/SF_6 RIE. Channel depths of 10 to 20 μm can be easily obtained with a surface roughness below 5 nm. Nanostructures are defined by high resolution EBL. For the negative relief molds needed here, i.e. nanoholes in microstripes, the following trilayer process was used: a 200 nm-thick photoresist (AZ5214) was first spin-coated on the substrate, followed by sputtering of a 10 nm Ge layer and spin-coating of PMMA resist. After EBL, sequential RIE was performed to transfer the PMMA features into the Si substrate. The combination of EBL and photolithography on thick SU-8 resist allowed the fabrication of molds containing both wide and deep channels and shallower nanostructures.

Nanoembossing experiments were carried out with the same equipment routinely used for NIL. For PMMA pellets (Oroglas V825T, $T_g = 108°C$ from Atofina), the applied pressure was typically 100 bar after heating up to 180°C. The pressure was maintained for a few minutes and then released after cooling below T_g. The patterned PMMA sheet was then removed from the mold. For thick PMMA sheets, a metallic mold which includes slots for lateral fluidic connections was machined to provide lateral confinement. Typically, 4 to 5 pellets of roughly $3 \times 3 \times 3$ mm^3 are needed to produce a 3×4 cm^2 PMMA sheet of thickness around 200 μm, while roughly 40 pellets are needed for millimeter-thick plates.

The fabricated microfluidic structures show well-defined shapes and excellent surface quality, exhibiting a roughness less than 5 nm, as on mold. Figure 6 displays a SEM

FIGURE 17.6. Scanning electron micrograph of the intersection of two 10 mm-deep and 50 mm-wide microchannels embossed in PMMA, showing an excellent surface roughness.

FIGURE 17.7. SEM picture of nanoembossed PMMA nanostructures, showing a portion of a 320 nm-period triangular array of 150 nm diameter nanopillars of thickness 200 nm, extending over a total 50 mm × 7 mm surface.

of the intersection of two 10 μm-deep and 50 μm-wide microfluidic channels. These channels reproduce the high surface quality and vertical sides of the Si master.

Although PMMA is the less hydrophobic of common plastics, prior to thermal bonding, the two pieces are exposed to an oxygen plasma in order to ensure hydrophilic surfaces, and promote negatively-charged channel walls. The sealed devices are not clogged, do not leak and present good wetting properties. Preliminary EOF measurements carried in PMMA microchannels give a mobility in the $10^{-4}cm^2/V.s$ range which is in agreement with previous reports.[100]

Figure 7 shows a SEM of a 300 nm-period triangular array of 150 nm-diameter nanopillars of height 200 nm, created inside a 200 nm-deep and 50 μm-wide channel, and extending over a total length of 7 mm. This structure was designed as an artificial gel for CE in high performance DNA separation experiments, as described in the previous section. The embossed features show extremely good uniformity over the whole array. After thermal bonding with a plastic cover slip, observation of aqueous fluid transport inside the nanopillar array was performed with a fluorescein solution of 15 mM concentration.

Figure 8(a) shows the fluorescein penetration inside the nanopillar array. Although the penetration speed is low, no clogging was observed. A multiple array structure was also successfully fabricated by nanoembossing and thermal bonding, as shown on Figure 8(b), and faster dye penetration was observed. Further experiments are needed to demonstrate the applicability of the fabricated devices for DNA molecule separation. However penetration and partial stretching of YOYO-tagged λ-DNA inside 200 nm-wide and 200 nm-deep nanoembossed PMMA nanofluidic channel has been observed in epifluorescence.[101]

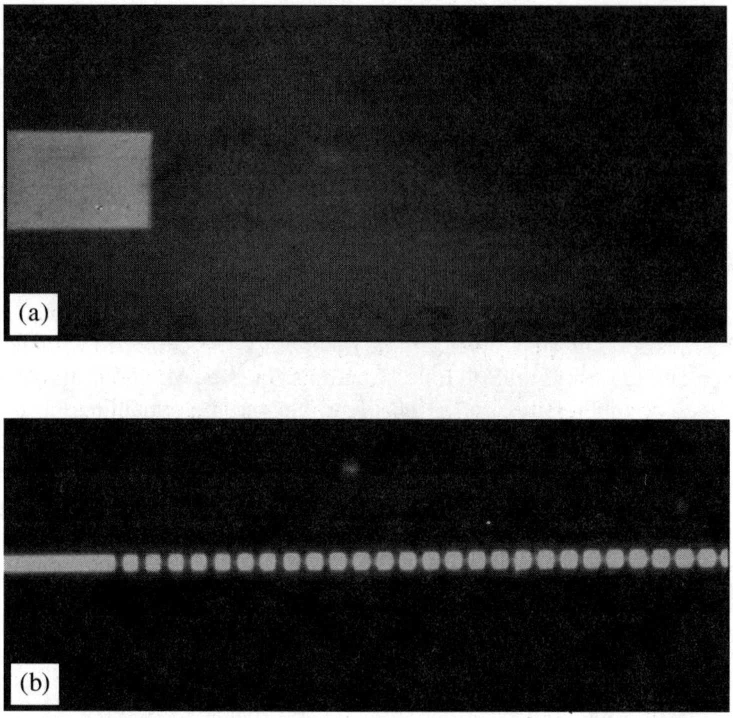

FIGURE 17.8. Fluorescence microscopy image of the penetration, from left to right, of a fluorescein solution (a) inside a 50 μm-wide channel containing a 200 nm-deep nanopillar array, (b) from a 50 μm-wide and 20 μm-deep channel to a series of 200 nm-deep nanopillar arrays.

Complex structures containing both deep micro-channels and shallower nanostructures can also be obtained by a single embossing step, as can be seen on the top portion of Figure 7. Homogeneous pattern replication over a wide range of scales is easily achieved with nanoembossing as compared to NIL, since viscous polymer flow takes place in bulk plastic and not in a thin polymer layer coated on a substrate. High resolution patterns can be easily obtained without carefully adjusting process parameters such as embossing temperature and pressure. Moreover, metal deposition can be achieved on plastic polymers if the T_g of the polymer is sufficiently high. This can allow the integration of electrodes in microchannels. One example is the integration of an array of interdigitated asymmetric electrodes for local AC electrokinetic fluid pumping, which we have described earlier in the soft lithography section and which could also be achieved on flat PMMA plates, thermally bonded to PMMA sheets containing microchannels.[63] Finally, thermoplastic polymers other than PMMA can be used, and nanopillars could be replicated in Topas COC.[102]

Nanoembossing can be viewed as the nano-regime extension of the hot embossing process. The latter technique, like injection molding, has been used widely these past few years for producing biological and biochemical analytical microdevices, and is presented in the following section.

17.3.3. Imprinting or hot embossing

Microfluidic elements such as microchannels and reservoirs can be imprinted directly in a thermoplastic plate.[20–23] Typically, a rigid mold is pressed into a heat-softened bulk polymer sheet at temperatures slighlty above T_g. Imprinting can be carried out slightly below T_g under elevated pressures, for soft plastic materials such as poly(etlylene terephthalate glycol (PETG).[21,85] Compression molding using pure polymer pellets deposited on a low resolution Si mold has recently been reported as an alternative approach.[87] As compared to nanoembossing, NIL and injection molding, hot embossing is easier to implement. Indeed, for the micron scale resolutions reported here, Si molds defined by photolithography -or alternatively metal molds obtained after an additional electroplating step- and a simple hydraulic press equipped with hot plates are sufficient to create plastic microfluidic structures. In general, photolithography masks can be offered commercially at a reasonable price, and the remaining fabrication steps can be performed at any laboratory without stringent cleanroom environment. Metal electroformed molds have also been fabricated via LIGA techniques for hot-embossing fabrication of plastic microfluidic devices.[103] An increasing number of academic and research laboratories are therefore getting equipped with either manual hydraulic presses or higher performance automated embossing tools now manufactured by several companies (EV Group, Obducat, etc).

L. Locascio and coworkers at NIST reported in 1997 the fabrication of imprinted plastic microchannels.[104] A chromium wire was pressed into the thermoplastic polymer surface in the early stages of fabrication and later replaced by machined Si embossing masters.[21,100,105] In particular, they investigated electroosmotic properties of native and chemically-treated channels embossed in PMMA or PETG, as well as hybrid channels made of embossed PMMA and PDMS cover lids,[84,85] through fluorescence mapping and EOF measurements. Electrochemical detection as well as contactless conductivity detection on hot-embossed PMMA CE devices were also recently demonstrated by another group.[106,107]

T.J. Johnson et al. recently proposed a passive microfluidic mixer fabricated in an imprinted PC T-microchannel subsequently etched by laser ablation in order to create a series of slanted wells on the floor of the microchannel after the T-junction.[108] This design is analogous to the grooved floor configuration proposed by A.D. Stroock et al. reported earlier,[49,50] however, no chaotic advection is induced here and the high level of lateral fluid transport and rapid mixing of the two confluent streams is obtained through EOF.

DNA amplification through PCR as well as CE separation could recently be achieved by researchers at Motorola in polycarbonate devices made by compression molding,[87] while a Cornell team has reported on the fabrication of a CE/ion spray mass spectrometry (MS) microfluidic device in Zeonor COC material using hot embossing.[109]

Greiner Bio-One and Mildendo are two German companies specializing in producing plastic microfluidic devices by hot embossing.[9] Mildendo in particular is fabricating a wide range of microdevices including multichannel CE for DNA analysis and sequencing, CE-MS chips, cell handling chips, continous-flow PCR.

Companies like ACLARA BioSciences and STEAG microParts develop both hot embossing and injection molding fabrication approaches for the production of plastic microfluidic devices. Among other applications, ACLARA has demonstrated DNA fragment sizing by CE as well as four-color DNA sequencing using hot-embossed acrylic microchannels filled with a non-cross-linked linear polymer as a sieving matrix.[20]

17.3.4. Injection molding

Injection molding is widely used for the mass production of various plastic pieces, with a throughput as high as several hundred pieces per hour. Melted thermoplastic polymer is forced inside a closed evacuated cavity of the desired shape by injecting it under high pressure. Injection temperatures are generally well above T_g for amorphous thermoplastic polymers such as PMMA and PC (and above melting point for semi-crystalline polymers). The cavity is typically heated at temperatures above T_g during filling, then cooled down, and the molded structure is ejected. While processing times are very short, the injection equipment is however expensive, needs large space, and fabrication of the metal molds (usually made of nickel) requires multilevel processing which include the following steps: patterning by lithography/etching or micromachining of a Si or glass master corresponding to the desired plastic structure, creation of a negative metal mold insert by electroforming, mounting of this patterned metal master against another metal support to form the closed injection cavity. For the latter reasons, injection molding is not a common laboratory equipment but rather an industrial processing tool. A well-known example of an injection-molded product is the compact disc (CD). Manufacturing of a CD made in PC or acrylic containing patterns below 1 μm nowadays costs less than €0.40. In addition to faster processing times, injection molding offers another advantage over hot embossing which is the possibility to create in a single molding step microchannels and integral through-holes acting as vertical access to the reservoirs. Injection molding also recently proved capable of replicating features below 100 nm.[110,111]

Several companies started a few years ago to fabricate plastic microfluidic devices using injection molding (Gyros AB, ACLARA, STEAG microParts, Mimotec, etc.).[9,20–23] ACLARA Biosciences Inc. were among the first to propose acrylic CE chips manufactured by injection molding.[112] Among others, separation of DNA sizing ladders was demonstrated in an injection molded acrylic CE device, as well as rapid parallel genotyping analyses.[20] Single-use electrophoretic-based enzyme assays for high throughput screening are also under development at ACLARA.[113,20]

The German company STEAG microParts uses a variety of techniques to produce the injection mold, including excimer laser machining, mechanical machining, e-beam, X-ray or UV lithographies. Typical applications include microtiter plates and capillarity-based microfluidic devices for point-of-care diagnostic.

Swedish Gyros AB is directly taking advantage of the injection-molded CD-like format to develop application-specific microfluidic platforms based on centrifugal force-induced pressure-driven flows and capillarity forces. These discs contain large numbers of identical microfluidic structures disposed radially for simultaneous processing. Samples are robot-dispensed to reservoirs at the center of the CD and as the disc spins at controlled speeds overcoming hybrophobic valves, the samples are pushed through different zones where they undergo any of a number of different treatments such as purification or analysis. The company is releasing a CD for the preparation of protein samples prior to identification with matrix-assisted laser desorption ionization (MALDI) mass spectrometry,[114,115] and is also working on discs for immunological and cell-based assays. Electrospray ionization mass spectrometry devices in injection-molded PMMA and PC have also been reported recently by another swedish group.[116]

Finally, Biosite Inc. recently commercialized a microfluidic protein chip fabricated by injection molding, in which microarray sandwich immunoassays using high affinity antibodies are performed directly from several drops of whole blood or plasma. Fluid flow in the microcapillaries is controlled passively by surface architecture and hydrophobicity. This protein chip was initially designed for rapid point-of-care diagnosis in hospital emergency rooms of drug overdose, myocardial infarction and congestive heart failure.[117]

17.4. CONCLUSIONS AND PERSPECTIVES

Alternative lithography techniques are now spurring progress in the microfluidics and on-chip bioanalysis arena. Compared to their applications in information technology, fully illustrated in the preceding chapters (microelectronics in Chapter 12, nanomagnetism in Chapter 13 and photonics in Chapter 15), soft lithography and imprint based approaches are perhaps even more interesting for microfluidics and microarrays, because of the huge demands on low-cost disposable devices and fast prototyping of a large variety of assays. We presented in this chapter some of the major bioapplications developed with these two classes of alternative fabrication techniques.

Soft lithography, based on casting of elastomers over patterned photoresist, is most convenient for the fabrication of prototype microfluidic devices and the rapid demonstration of a large variety of effects. More sophisticated schemes involving multilayer processing of cast PDMS layers are proving to be a powerful and reliable approach for the fabrication of integrated lab-on-a-chip systems. In addition, the patterned elastomer can be used as a stamp for surface bio-engineering with SAMs or proteins via microcontact printing, or as a guiding mold for surface micropatterning via microfluidic networks and laminar flow. These approaches are truly revolutionizing biological studies occurring on surfaces and allowing the production of high-throughput bioarrays and immunoassays.

Techniques based on imprinting of thermoplastic materials are now being used widely for microfluidic device fabrication. Injection molding offers the highest production rate for plastic components and should be the relevant choice for any large scale industrial implementation. However, for smaller production numbers, hot embossing is an attractive choice because of the lesser efforts required in system design and process validation. For cost-effective high resolution patterning, both nanoimprint lithography and nanoembossing are reliable approaches extending the possibility of microfluidic device fabrication to a true nano-regime. While nanoembossing allows the creation of bulk plastic devices containing nanostructures, NIL transfers the pattern imprinted into a thermoplastic polymer to a solid substrate, such as a semiconductor or silica, and therefore offers strong potentials for multicomponent integration.

We have selected for this chapter a number of application examples which illustrate how alternative micro and nanofabrication technologies based on polymer molding are beginning to profoundly impact Life Sciences. Table 1 summarizes the different

TABLE 17.1. Summary of the different approaches to nanofabrication in the life sciences described in this Chapter.

Category	Method	Resolution	Speed	Application	References
	Casting of elastomer	Medium	Medium	Rapid prototyping	6,18-20,35, 36-45,63-67,91
Soft lithography	Micro contact printing	Medium	High	Microarray printing	18,36,38,68-74,91
	Microfluidic networks	Low	Medium	2D or 3D fluidic network	18,36,38, 75-82
	Nanoimprint	High	High	Nanopatterning	24,91-96,98
	Nanoembossing	High	High	Plastic molding	39,64,96,101,102
Imprint-based	Imprinting/ embossing	Low	High	Molding into plates	9,20-23,85,87, 91, 100,103-109
	Injection molding	Medium	Very high	Mass production	20-23,91, 110-117
	LIGA process	Low	Medium	High aspect ratio	91,103
	Laser ablation	Low	Low	Local modification	20-23,25-27, 91,108
	Plasma etching	Medium	Medium	Prototyping	21-23,28
Others	Hydrogels	Medium	Low	Prototyping	29,30,35
	Stereolithography	Low	Low	Prototyping	21,23,31,91
	X-ray lithography	High	High	High aspect ratio	21-23,32,33,91
	SU-8 photolithography	Medium	High	Prototyping	21-23,34

approaches reported in this chapter. However, microfluidic lab-on-a-chip devices, as well as bioarrays, biosensors, and emerging cell-on-chip systems are still in their early development stage. Much effort has to be devoted to the multidisciplinary synergy between innovative fabrication techniques and biotechnologies.

For instance in the field of microfluidics, no fabrication standards have yet been set for integrated building blocks like pumps, valves, mixers and separators. The impact of lab-on-a-chip devices will also strongly depend on the development of innovative detection schemes, adapted data processing such as bio-informatics, and efficient external fluidic connections either through user-friendly fluid injection via syringes or by ink-jet delivery, already widely used for DNA microarrays and immunoassays, and which is itself a microfluidic technology.

In microfluidic systems, we have just reached the nanoscale. Although some promising examples have been presented in this chapter, this regime has not yet been exploited. Fluid behavior in nanochannels is not well understood and can differ strongly from that observed at the micron scale. Among expected phenomena, electrokinetically-driven flow profiles in channels approaching the thickness of the EDL can be significantly affected and thus reduce electrokinetic transport,[118,119] pressure drops and surface tensions will also be considerably higher, while molecular diffusion will be more efficient in nanochannels and should thus improve mixing. Practical tools for the analysis of biomolecules,[120] single molecule chemical reaction and detection, as well as biomimetism could emerge from nanofluidic developments.

Finally, at the molecular level, bottom-up engineering including self-assembling based microcontact printing and molecular imprinting[121] is an exciting area just starting to be explored, in conjunction with the top-down polymer-based techniques presented here.

Due to the rapid development of the microfluidics and bioanalytical field, it was not possible to cover the topic in an exhaustive way, however we hope that the various examples we chose to present here offer a comprehensive view of the subject and will encourage innovative exploration.

REFERENCES

1. S.P. Fodor, R.P. Rava, X.C. Huang, A.C. Pease, C.P. Holmes, C.L. Adams, *Nature,* (1993), 364, 555-556.
2. R.J. Lipshutz, S.P. Fodor, T.R. Gingeras, D.J. Lockhart, *Nature Genetics,* (1999), 21, 20-4.
3. http://www.affymetrix.com;www.nanogen.com; www.motorola.com/lifesciences/; http://www.biochipnet.de
4. M.F. Templin, D. Stoll, M. Schrenk, P.C. Traub, C.F. Vöhringer and T.O. Joos, *Trends Biotechnol.,* (2002), 20, 160-166.
5. see proceedings from the μTAS international conferences: *Micro Total Analysis Systems 2001*, J. M. Ramsey and A. Van den Berg (eds), Kluwer, Dortrecht, Netherlands (2001); *Micro Total Analysis Systems 2000*, A. Van den Berg, W. Olthuis and P. Bergveld (eds), Kluwer, Dortrecht, Netherlands (2000); *Micro Total Analysis Systems '98,* D.J. Harrison, A. Van den Berg (eds), Kluwer, Dortrecht, Netherlands (1998); *Micro Total Analysis Systems* , A. Van den Berg, P. Bergveld (eds), Kluwer, Dortrecht, Netherlands (1994).
6. *Microsystem Technology in Chemistry and Life Sciences*, A. Manz and H. Becker (eds), Berlin, Germany, Springer-Verlag (1998).
7. *Microfluidic Technology and Applications,* M. Koch, A. Evans and A. Brunnschweiler, Microtechnologies and Microsystems Series, R. Pethig (ed), Baldock England, Research Studies Press Ltd. (2000).
8. See *Lab-on-a-chip* journal, Royal Society of Chemistry (first issue October 2001).
9. http://www.calipertech.com;www.chem.agilent.com;www.aclara.com; www.gyros.com;www.cepheid.com;www.micralyne.com;www.micronics.net; www.nanostream.com;www.tecan.com;www.microparts.de; www.mildendo-fluidics.de;www.greiner-bio-one.com;etc.
10. D. J. Harrison, A.Manz, Z.Fan, H.Lüdi, H.M.Widmer, *Anal. Chem.*, (1992), 64, 1908-1919.
11. J. L. Viovy, *Rev. Mod. Phys.*, (2000), 72, 813-872 .
12. J. Han and H. G. Craighead, *Science,* (2000), 288, 1026-1029; J. Han, S. W. Turner, and H. G. Craighead, *Phys. Rev. Lett.*, (1999), 83, 1688-1691.
13. W. D. Volkmuth and R.H. Austin, *Nature,* (1992), 358, 600.
14. D. Ertas, *Phys. Rev. Lett.*, (1998), 80, 1548.
15. S. W. Turner, A. M. Perez, A. Lopez and H.G. Craighead, *J. Vac.Sci & Technol,,* (1998),*B* 16, 3835.
16. D. R. Larson, M. Levene, S. W. Turner, H. G. Craighead, and W. W. Webb, *Biophys. J.* 80 Part 2, (2001), 65509.
17. M.E. Foquet, S.W. Turner, J. Korlach, W.W. Webb and H.G. Craighead, *Micro Total Analysis Systems 2000,* A. Van den Berg, W. Olthuis and P. Bergveld (eds), Kluwer, Dortrecht, Netherlands, p. 549-552 (2000).
18. Y.N. Xia and G.M. Whitesides. *Angew. Chem. Int. Ed. Engl.*, (1998), 37, 550.
19. http://www.fluidigm.com
20. T.D. Boone, A. J. Ricco, Z.H. Fan, H. Tan, H.H. Hooper and S.J. Williams, *Anal. Chem.,* (2002),74, 78A-86A.
21. H. Becker and L.E. Locascio, *Talanta,* (2002), 56, 267-287.
22. A.de Mello, *Lab on a Chip* 2, 31N-36N (2002).
23. H. Becker and C.Gärtner, *Electrophoresis* 21, (2000), (1), 12-26.
24. S.Y. Chou, P.R. Krauss and P.J. Renstrom, *Appl. Phy. Lett.*,1995),76, 3114.
25. M.A. Roberts, J.S. Rossier, P. Bercier, H. Girault, *Anal. Chem.,*(1997), 69, 2035-2042.
26. J.S. Rossier, A. Schwarz, F. Reymond, R. Ferrigno, F. Bianchi and H.H. Girault, *Electrophoresis,* (1999), 20, 727-731.

27. P.M. Martin, D.W. Watson, W.D. Bennett, Y. Lin and D.J. Hammerstrom, *J. Vac. Sci. Technol.*, (1999),*A* 17, 2264.
28. J.S. Rossier, F. Reymond, and P.E. Michel, *Electrophoresis,* (2002), 23, 858-867.
29. D.J. Beebe, J.S. Moore, J.M. Bauer, Y. Qing, R.H. Liu, C. Devadoss and B-H. Jo, *Nature,* (2000), 404, 588-590.
30. D.J. Beebe, J.S. Moore, Q. Yu, R.H. Liu, M.L. Kraft, B-H. Jo, C. Devadoss,*Proc. Natl. Acad. Sci. USA* 97, (2000), (25), 13488-13493.
31. A. Bertsch, S. Heimgartner, P. Cousseau and P. Renaud, *Lab on a Chip,* 1,(2002), 56-60.
32. S. M. Ford, B. Kar, S. Mcwhorter, J. Davies, S.A. Soper, M. Klopf, J. Calderon and V. Saile, *J. Microcolumn Sep.,* (1998),10, 413-422.
33. M. Ueda, Y. Kiba, H. Abe, A. Arai, H. Nakanishi, and Y. Baba, *Electrophoresis,* (2000), 21, 176.
34. A. Bertsch, P. Bernhard, C. Vogt, P. Renaud, *Rapid Prototyping* 6, (2000), (4), 259.
35. G.M. Whitesides and A.D. Stroock, *Physics Today* 54, (2001), (6), 42.
36. G.M. Whitesides, E. Ostuni, S. Takayama, X. Jiang and D. E. Ingber, *Annu. Rev. Biomed. Eng.,*3, (2001), 335-373.
37. H. Schmid and B. Michel, *Macromolecules,* (2000), 33, 3042.
38. B. Michel et al. *IBM J. Res. & Dev.,* (2001), 45, 697-719.
39. V. Studer, A. Pépin, and Y. Chen, *Appl. Phys. Lett.,* 80, May 13 (2002), (19).
40. D.C. Duffy, J.C. McDonald, O.J.A. Schueller and G.M. Whitesides, *Anal. Chem.,*70, (1998), 4974-4984.
41. X. Ren, M. Bachman, C. Sims, G.P. Li, N. Allbritton, *J. Chromatogr. B.,* (2001), 762, 117-125.
42. H-P. Chou, C. Spence, A. Scherer and S.R. Quake, *Proc. Natl. Acad. Sci.,* (1999), *USA* 96, 11-13.
43. K.W. Ro, K. Lim, H. Kim and J.H. Hahn, *Electrophoresis,* (2002), 23, 1129-1137.
44. A.Tan, K.Rodgers, J.P. Murrihy, C.O'Mathuna and J.D. Glennon, *Lab on a Chip,*(2002), 1, 7-9.
45. J.C. McDonald, M.L. Chabinyc, S.J. Metallo, J.R. Anderson, A.D. Stroock and G.M. Whitesides, *Anal. Chem.,* (2002),74, 1537-1545.
46. M.A. Unger, H-P. Chou, T. Thorsen, A. Scherer and S.R. Quake, *Science,* (2000), 288, 113-116.
47. S. R. Quake and A. Scherer, *Science,* (2000), 290, 1536.
48. J. Liu, M. Enzelberger and S. Quake, *Electrophoresis,* (2002), 23, 1531-1536.
49. A.D. Stroock, S.K.W. Dertinger, A. Ajdari, I. Mezic, H.A. Stone, G.M. Whitesides, *Science,*(2002), 295, 647-651.
50. A.D. Stroock, S.K.W. Dertinger, G.M. Whitesides, A. Ajdari, to be published in *Anal. Chem.* (2002).
51. M. Oddy, J.G. Santiago and J.C. Mikkelsen, *Anal. Chem.,* (2001), 73, 5822-5832.
52. C. Donzel, M. Geissler, A. Bernard, H. Wolf, B. Michel, J. Hilborn and E. Delamarche, *Adv. Mater.,* (2001), 13 (15), 1164-1167.
53. Y. Lui, J.C. Fanguy, J.M. Bledsoe and C.S. Henry, *Anal. Chem.,* (2000), 72, 5939-5944.
54. B. Ekström, G. Jacobson, O. Ohman, H.Sjodin, *PCT 91/16966* (1991).
55. C.S. Effenhauser, G.J.M. Bruin, A. Paulus, M. Ehrat, *Anal. Chem.,* (1997), 69, 3451-3457.
56. J.C. McDonald, D.C. Duffy, J.R. Anderson, D.T. Chiu, H. Wu, O.J.A. Schueller and G.M. Whitesides, *Electrophoresis,* (2000), 21, 27-40.
57. P.S. Doyle, J. Bibette, A. Bancaud, and J. L. Viovy, *Science,*(2002), 295, 2237.
58. H-P. Chou, C. Spence, A. Scherer and S.R. Quake, *Proc. Natl. Acad. Sci. USA,* (1999),96, 11-13.
59. A.Y. Fu, C. Spence, A. Scherer, F.H. Arnold and S.R. Quake, *Nature Biotechnology,* (1999), 17, 1109-1111.
60. A.Y. Fu, H-P Chou, C. Spence, F.H. Arnold and S.R. Quake, *Anal. Chem.,*(2002), 74, (11), 2451-2457.
61. J.H. Chan, A.T. Timperman, D. Qin and R. Aebersold, *Anal. Chem.,* (1999),71, 4437.
62. J.S. Kim and D.R. Knapp, *Micro Total Analysis Systems 2001*, J. M. Ramsey and A. Van den Berg (eds), Kluwer, Dortrecht, Netherlands, p.333-334 (2001) and references therein
63. M. Seki, M. Yamada, R. Ezaki, R. Aoyama and J.W. Hong, *Micro Total Analysis Systems 2001*, J. M. Ramsey and A. Van den Berg (eds), Kluwer, Dortrecht, Netherlands, p.48 (2001).
64. V. Studer, A. Pépin, Y. Chen and A. Ajdari, *Microelectron. Eng.* (2002) *in press*
65. A. Ajdari, *Phys. Rev. E,* (2000), 61, R45.
66. A. Ajdari, *Appl. Phys. A,* 75, (2002), 271-274.
67. A.B.D. Brown, C.G. Smith and A.R. Rennie, *Phys. Rev. E,* 63, (2001), 016305.
68. G.M. Whitesides, J.P. Mathias, C.T. Seto,*Science,* (1991), 254, 1312-1319.

69. M. Mrksich, *Cell. Mold. Life. Sci.*, (1998), 54, 653-662.
70. A.Bernard, E.Delamarche, H.Schmid, B.Michel, H.R.Bosshard, H.A.Biebuyck, *Langmuir*, (1998), 14, 2225-2229.
71. A. Bernard, J.P. Renault, B. Michel, H.R. Bosshard and E. Delamarche, *Adv. Mater.* 12, (2000), 1067.
72. C.D. James, R.C. Davis, L. Kam, H.G. Craighead, M. Isaacson, J.N. Turner and W. Shain, *Langmuir,* (1998), 14, 741.
73. A. Bernard, D. Fitzli, P. Sonderegger, E. Delamarche, B. Michel, H.R. Bosshard and H. Biebuyck, *Nature Biotechnology,* (2001),19, 866-869.
74. A.D. Stroock, M. Weck, D.T. Chiu, W.T.S. Huck, P.J.A. Kenis, R.F. Ismagilov, and G.M. Whitesides, *Phys. Rev. Lett.*, (2000), 84, 3314-3317.
75. E. Delamarche, A. Bernard, H. Schmid, A. Bietsch, B. Michel, H.A. Biebuyck, *Science,* (1997), 276, 779-781.
76. E. Delamarche, A. Bernard, H. Schmid, A. Bietsch, B. Michel, H.A. Biebuyck, *J. Am. Chem. Soc.* (1998), 120, 500-508.
77. A. Bernard, B. Michel and E. Delamarche, *Anal. Chem.*, (2001),73, (1), 8-12.
78. A. Papra, A. Bernard, D. Juncker, N.B. Larsen, B. Michel and E. Delamarche, *Langmuir,* (2001), 17, 4090-4095.
79. P.J.A. Kenis, R.F. Ismagilov and G.M. Whitesides,*Science,* (1999), 285, 83-85.
80. S. Takayama, E. Ostuni, P. Leduc, K. Naruse, D.E. Ingber and G.M. Whitesides, *Nature,* (2001), 411, 1016.
81. D.T. Chiu, N.L. Jeon, S. Huang, R.S. Kane, C.J. Wargo, I.S. Choi, D. E. Ingber and G.M. Whitesides, *Proc. Natl. Acad. Sci. USA* , (2000), 97(6), 2408-2413.
82. H-P. Chou, M.A. Unger, A. Scherer and S.R. Quake, *Proceedings of the Solid State Actuator and Sensor Workshop, Hilton Head, S.C. USA* (2000).
83. P.J. Schneider and H. Engelhardt, *J. Chromatogr. A* 802, 17-22 (1998).
84. D. Ross, T.J. Johnson, and L. Locascio, *Anal. Chem.*, (2001), 73, 2509-2515.
85. A.C. Henry, E.A. Waddell, R. Shreiner and L. Locascio, *Electrophoresis,* (2002), 23, 791-798.
86. Z. Wu, N. Xianthopoulos, F. Reymond, J.S. Rossier and H. Girault, *Electrophoresis,* (2002), 23, 782-790.
87. Y. Liu, D. Ganser, A. Schneider, R. Liu, P. Grodzinski and N. Kroutchinina, *Anal. Chem.*, (2001), 73, 4196-4201.
88. R.L.McCarley, A.C. Henry, A.Smith, B.Vaidya, M.Galloway, Y.Wang and S.A.Soper, *Micro Total Analysis Systems 2001*, J. M. Ramsey and A. Van den Berg (eds), Kluwer, Dortrecht, Netherlands, p.607-609 (2001) and references therein
89. D.S. Soane, Z.M. Soane, *US Patent 5, 126, 022* (1992).
90. C. Backhouse, M. Caamano, F. Oaks, E. Nordman, A. Carrillo, B. Johnson, S. Bay *Electrophoresis,* (2000), 21, 150-156.
91. Y. Chen and A. Pépin, *Electrophoresis,* (2001), 22, 187-207.
92. A. Lebib, Y. Chen, E. Cambril, P. Youinou, V. Studer, M. Natali and A. Pépin, *Microelectron. Eng.* (2002) *in press*
93. M. Bender, M. Otto, B. Hadam, B. Vratzov, B. Spangenberg, H. Kurz, *Microelectron. Eng.*, (2000), 53, 233.
94. Y. Chen, A. Lebib, S. Li, A. Pépin, D. Peyrade, M. Natali and E. Cambril, *Eur. Phys. J. Appl. Phys.*, (2000), 12, 223-229.
95. A. Lebib, Y. Chen, F. Carcenac, E. Cambril, L. Manin, L. Couraud and H. Launois *Microelectron. Eng.*, (2000), 53, 175.
96. A. Pépin, P. Youinou, V. Studer, A. Lebib and Y. Chen, *Microelectron. Eng.* (2002) *in press*
97. S.W.P. Turner, M. Cabodi and H.G. Craighead, *Phys. Rev. Lett.*, (2002), 88 (12), 128103.
98. H. Cao, J. Tegenfeldt, L. Wanli, Z-N Yu, R.H. Austin and S.Y. Chou, *March Meeting 2002, Indianapolis, USA* (2002).
99. B. Straub, E. Meyer and P. Fromherz, *Nature Biotechnol.*, (2001), 19, 121-124.
100. J. Xu, L. Locascio, M. Gaitan and C.S. Lee, *Anal. Chem.*, (2000), 72, 1930-1933.
101. O. Rossier, V. Studer, A. Pépin and Y. Chen, *unpublished results*
102. A. Pépin, *unpublished results*
103. M.B. Wabuyele, S. M. Ford, W. Stryjewski, J. Barrow and S. Soper, *Electrophoresis,* (2001), 22, 3939-3948.

104. L. Martynova, L. Locascio, M. Gaitan, G.W. Kramer, R.G. Christensen and W.A. Maccrehan, *Anal. Chem.,* (1997), 69, 4783.
105. S.L. Barker, M.L. Tarlov, H. Canavan, J.J. Hickman, L.E. Locascio, *Anal. Chem.,* (2000), 72, 4899-4903.
106. M. Pumera, J. Wang, F. Opekar, I. Jelínek, J. Feldman, H. Löwe, and S. Hardt, *Anal. Chem.,* (2002), 74 (9), 1968-1971.
107. J. Wang, M. Pumera, M.P. Chatrathi, A. Escarpa, R. Konrad, A. Griebel, W. Dörner and H. Löwe, *Electrophoresis,* (2002), 23, 596-601.
108. T.J. Johnson, D. Ross, and L.E. Locascio, *Anal. Chem.,* (2002), 74, 45-51.
109. J. Kameoka, H.G. Craighead, H. Zhang, J. Henion, *Anal. Chem.,* (2001), 73, 1935-1941.
110. H. Schift, C. David, M. Gabriel, J. Gobrecht, L.J. Heydermann, W. Kaiser, S. Köppel and L. Scandella, *Microelectron. Eng,.* (2000), 53, 171-174.
111. D. MacIntyre and S. Thomas, *Microelectron. Eng.,* (1998), 41/42 211-214.
112. R.M. McCormick, R.J. Nelson, M.G. Alonso-Amigo, D.J. Benvegnu, and H.H. Hopper, Anal. Chem., (1997), 69, 2626-2630.
113. Q. Xue, A. Wainright, S. Gangakhedkar, I. Gibbons, *Electrophoresis,* (2001), 22, 4000-4007.
114. A. Palm, S.R. Wallenborg, M. Gustafsson, A. Hedström, E. Togan-Tekin and P. Andersson, *Micro Total Analysis Systems 2001*, J. M. Ramsey and A. Van den Berg (eds), Kluwer, Dortrecht, Netherlands, p. 216-218 (2001).
115. R. Ehrnström, *Lab on a Chip* 2, 26N-30N (2002).
116. M. Svedberg, A. Pettersson, F. Nikolajeff and K. Markides, *Micro Total Analysis Systems 2001*, J. M. Ramsey and A. Van den Berg (eds), Kluwer, Dortrecht, Netherlands, p.335-336 (2001).
117. K.F. Buechler, P. McPherson, J. Anderberg, K. Nakamura, S. Lesefko and B. Noar, *Micro Total Analysis Systems 2001*, J. M. Ramsey and A. Van den Berg (eds), Kluwer, Dortrecht, Netherlands, p.42-44 (2001).
118. S.C. Jacobson, J. P. Alarie and J.M. Ramsey, *Micro Total Analysis Systems 2001*, J. M. Ramsey and A. Van den Berg (eds), Kluwer, Dortrecht, Netherlands, p.57 (2001).
119. A. T. Conlisk, J. McFerran, Z. Zheng and D. Hansford,*Anal. Chem.*, (2002), 74(9), 2139-2150.
120. O. Saleh and L.L. Sohn, *Micro Total Analysis Systems 2001*, J. M. Ramsey and A. Van den Berg (eds), Kluwer, Dortrecht, Netherlands, p.54 (2001).
121. K. Haupt and K. Mosbach, *Trends Biotechnol.,* (1998), 16, 468.

Index